Encyclopedia of Bioenergetics

Encyclopedia of Bioenergetics

Edited by **Zoe Hooper**

New York

Published by Callisto Reference,
106 Park Avenue, Suite 200,
New York, NY 10016, USA
www.callistoreference.com

Encyclopedia of Bioenergetics
Edited by Zoe Hooper

International Standard Book Number: 978-1-63239-215-2 (Hardback)

Printed in the United States of America.

Contents

Preface

In order to function properly and to avoid death, a cell's life relies upon energy deposits, conversion, utilization and exchange. Over 200 years of study tells us how cells convert biological fuels into usable energy. This process is broadly known as bioenergetics which has generated traditions in describing origins of life, metabolism, ecological adaptation, homeostasis, aging and various other life processes. This book discusses some of these traditions via chapters contributed by international experts. Both beginners and professionals will find this book benefiting in linking scientific revolutions in the fields of organism biology, membrane physiology and molecular biology to enhance the subject of bioenergetics towards solving contemporary and future difficulties in metabolic diseases, life transitions and longevity, and performance optimization.

The information shared in this book is based on empirical researches made by veterans in this field of study. The elaborative information provided in this book will help the readers further their scope of knowledge leading to advancements in this field.

Finally, I would like to thank my fellow researchers who gave constructive feedback and my family members who supported me at every step of my research.

Editor

Part 1

Reviews of Bioenergetics Applied to Life Span and Disease

Mitochondrial Calcium Signalling: Role in Oxidative Phosphorylation Diseases

Oulès Bénédicte[1], Del Prete Dolores[2] and Chami Mounia[3*]
[1]INSERM U 807, Paris V University, Paris,
[2]Istituto Italiano di Tecnologia,
Department of Neuroscience and Brain Technologies, Genova,
[3]Institute of Cellular and Molecular Pharmacology,
Institute of Molecular Neuromedecine, UMR6097 CNRS/UNSA, Valbonne,
[1,3]France
[2]Italy

1. Introduction

Mitochondria are double membrane-bound organelles that not only constitute the "cellular power plants" but also are crucially involved in cell survival, apoptosis, redox control, Ca^{2+} homeostasis and many metabolic and biosynthetic pathways.

The mitochondria generate energy by oxidizing hydrogen derived from dietary carbohydrate (TCA: tricarboxylic acid cycle) and lipids (beta-oxidation) with oxygen to generate heat and energy in the form of ATP (Adenosine triphosphate). Energy generation in mitochondria occurs primarily through oxidative phosphorylation (OXPHOS), a process in which electrons are passed along a series of carrier molecules called the electron transport chain (ETC). This chain is composed of four multisubunit assemblies that are embedded in the mitochondrial inner membrane: complex I (NADH:ubiquinone oxidoreductase; EC 1.6.5.3), complex II (succinate:ubiquinone oxidoreductase; EC 1.3.5.1), complex III (ubiquinol:cytochrome-*c* oxidoreductase; EC 1.10.2.2) and complex IV (cytochrome-*c* oxidase; EC1.9.3.1). Complexes I, III and IV actively translocate protons from the matrix into the intermembrane space using energy extracted from electrons passing through the chain. These electrons are liberated from NADH and $FADH_2$, at complexes I and II, respectively, where they are donated to the lipophilic electron carrier coenzyme Q for further transport to complex III. From there, electrons are shuttled to complex IV by cytochrome-*c*. At this complex, electrons are finally used for the reduction of oxygen to water (Hatefi, 1985; Saraste, 1999) (Figure 1 A).

The energy released by the flow of electrons through the ETC and the flux of protons out of the mitochondrial inner membrane creates a capacitance across the mitochondrial inner membrane, the electrochemical gradient (ΔP) composed of an electrical potential ($\Delta\psi$) and a concentration ratio (ΔpH). The potential energy stored in ΔP is coupled to ATP synthesis by complex V (F_0/F_1-ATP-synthase; EC 3.6.1.34). As protons flow back into mitochondrial

*Corresponding Author

matrix through complex V, ADP and Pi are bound, condensed and released as ATP. With Complex V, the ETC complexes constitute the OXPHOS system. The OXPHOS system generates the vast majority of cellular ATP during oxidative metabolism. Some of the ATP is used for the mitochondrion's own needs, but most of it is transported outside the organelle by the adenine nucleotide translocator (ANT) and used for diverse cell functions (Hatefi, 1985; Saraste, 1999) (Figure 1 A).

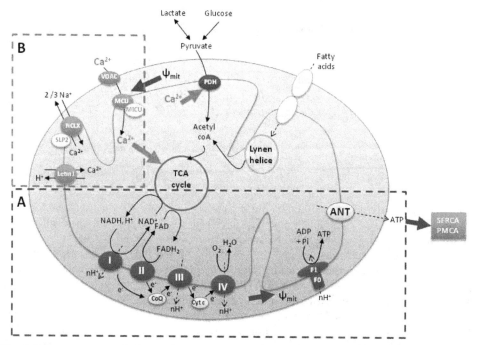

Fig. 1. Schematic view of of mitochondrial OXPHOS system (A), and. mitochondrial Ca²⁺ influx and efflux mechanisms (B). Large size arrows indicate the interplay between mitochondrial Ca²⁺ signaling and OXPHOS.

2. Genetics and pathology of OXPHOS

2.1 Genetics of OXPHOS system

The OXPHOS system is composed of more than 80 different proteins, 13 of which are encoded by the mitochondrial DNA (mtDNA) and the others by the nuclear genome (nDNA) (Chinnery & Turnbull, 2001; Wallace, 1992). There are seven mtDNA-encoded subunits in complex I, one in complex III, three in complex IV and two in complex V. Complex II consists solely of nDNA-encoded subunits.

The structure of human mitochondrial DNA (mtDNA) was reported ≈ 30 years ago (Clayton & Vinograd, 1967). The transcription products of mtDNA include 2 ribosomal RNA species (12S and 16S rRNA), 13 messenger RNAs and 22 transfer RNAs. Replication of mtDNA occurs independently from cell cycle phase and from replication of nuclear DNA. Mitochondrial DNA is present in 10^3–10^4 identical copies in each cell, with the exception of sperm and mature oocytes, in which mtDNA copy numbers are $\approx 10^2$ and $\approx 10^5$, respectively. In general, there are believed to be two to ten copies of DNA per mitochondrion. The sequences of mtDNAs from unrelated individuals in human populations typically differ by about 0.3 %. Most individuals, however, have a single mtDNA sequence variant in all their cells (homoplasmy). mtDNA transmission occurred exclusively through the maternal lineage. Almost all of the nDNA-encoded OXPHOS subunits have been characterized at the cDNA level and several at the genomic level in humans. In general, the chromosomal distribution of the genes seems to be random, and expression of most gene products is ubiquitous but predominates in tissues or organs with a high energy demand.

Richard Scarpulla and co-workers have provided important insight into the regulatory mechanisms that are involved in the transcriptional control of OXPHOS genes (Gugneja et al., 1996; Huo & Scarpulla, 1999; Wu et al., 1999). They identified the nuclear respiratory factors NRF1 and NRF2, which act on overlapping subsets of nuclear genes that are involved in the biogenesis of the respiratory chain. Recent mammalian studies have identified PGC1 as a crucial regulator of cardiac mitochondrial number and function in response to energy demand (Lehman et al., 2000). Analysis of the expression pattern of OXPHOS genes revealed that their regulation might also be exerted post-transcriptionally (Di Liegro et al., 2000).

2.2 OXPHOS diseases

Among the inborn errors of metabolism, mitochondrial disorders are the most frequent with an estimated incidence of at least 1 in 10,000 births (reviewed in (Smeitink et al., 2001)). Although the term mitochondrial disorder is very broad, it usually refers to diseases that are caused by disturbances in the OXPHOS system. After the first description, ≈ 40 years ago, of a patient with "loose coupling" – a defect in the coupling between mitochondrial respiration and phosphorylation – by Luft and collaborators (Luft et al., 1962), thousands of patients have been diagnosed by measurement of OXPHOS-system enzyme activities. The great complexity of the OXPHOS system, which consists of proteins, some encoded by the mitochondrial genome and others by the nuclear genome, may explain the wide variety of clinical phenotypes that are associated with genetic defects in oxidative phosphorylation. Disease-causing defects can occur in a single OXPHOS complex (isolated deficiency) or multiple complexes at the same time (combined deficiency). OXPHOS diseases give rise to a variety of clinical manifestations, particularly in organs and tissues with high-energy demand such as brain (encephalopathies), heart (cardiomyopathies), skeletal muscle (myopathies) and liver (hepatopathies) (reviewed in (Finsterer, 2006a, 2006b; Schaefer et al., 2004)).

We have also to consider the presence of fundamental differences between mitochondrial genetics and Mendelian genetics when studying human OXPHOS diseases. These differences are linked to maternal inheritance of mtDNA, polyplasmy, heteroplasmy and the threshold effect, whereby a critical number of mutated mtDNAs must be present for the OXPHOS system to malfunction (Wallace, 2005).

One of the frequent OXPHOS disorders is Leigh Syndrome (OMIM 256000), an early-onset progressive neurodegenerative disorder, leading to death mostly within a few years after the onset of the symptoms. This disorder is characterized by lesions of necrosis and capillary proliferation in variable regions of the central nervous system. Clinical signs and symptoms comprise muscular hypotonia, developmental delay, abnormal eye movements, seizures, respiratory irregularities and failure to thrive. Other mitochondrial disorders caused by

OXPHOS defects include Mitochondrial Encephalopathy, Lactic Acidosis and Stroke-like episodes (MELAS; OMIM 540000), Myoclonic Epilepsy with Ragged Red Fibers (MERRF; OMIM 545000), Neurogenic weakness, Ataxia, Retinitis Pigmentosa/Maternally Inherited Leigh Syndrome (NARP/MILS; OMIM 516060), Leber's Hereditary Optic Neuropathy (LHON; OMIM 535000), and Mohr–Tranebjaerg syndrome (a.k.a. Deafness Dystonia Syndrome; OMIM 304700).

Because of the genetic complexity of the energy-generating system, many other diseases have been shown to be associated with defect in mitochondrial function (DiMauro & Moraes, 1993; DiMauro & Schon, 2003). For example, there is increasing evidence that inherited OXPHOS dysfunction is also implicated in diabetes, age-related neurodegenerative diseases, such as Parkinson's, Alzheimer and Huntington's diseases, and various forms of cancers (Shoubridge, 2001; Zeviani & Carelli, 2007).

2.2.1 Mitochondrial DNA mutations linked to OXPHOS diseases

The complexity of mitochondrial DNA mutations linked to OXPHOS diseases is that one mutation can cause a broad spectrum of clinical manifestations. Conversely, different mutations can be associated with the same clinical phenotype. Specific phenotypes include forms of blindness, deafness, movement disorders, dementia, cardiovascular diseases, muscle weakness, renal dysfunction, and endocrine disorders including diabetes. In the past 20 years, more than 100 point mutations and innumerable rearrangements have been associated with human mitochondrial diseases. In this context, it is worth mentioning, however, that we still lack comprehensive and unbiased epidemiological data about the frequency of known mtDNA mutations. Although tRNA genes as a whole represent $\approx 10\%$ of the mtDNA, mutations in these genes account for $\approx 75\%$ of mtDNA-related diseases.

We can identify three categories of pathogenic mtDNA mutations: rearrangement mutations, polypeptide gene missense mutations, and protein synthesis (rRNA and tRNA) gene mutations (reviewed in (Wallace, 2005)).

- Rearrangement mutations of mtDNA can be either inherited or spontaneous. Inherited mtDNA rearrangements are primarily insertions. The first inherited insertion mutation to be identified caused maternally inherited diabetes and deafness (Ballinger et al., 1992, 1994). Spontaneous mtDNA deletions result in a related spectrum of symptoms, irrespective of the position of the deletion end points. This is because virtually all deletions remove at least one tRNA and thus inhibit protein synthesis (Moraes et al., 1989). Thus the nature and severity of the mtDNA deletion rearrangement is not a consequence of the nature of the rearrangement, but rather of the tissue distribution of the rearranged mtDNAs.

- Missense mutations in mtDNA polypeptide genes can also result in an array of clinical manifestations. Three relatively frequently observed point mutations are A3243G in the tRNA(Leu)(UUR) gene, A8344G in the tRNA(Lys) gene and T8993G in the ATPase 6 gene and are associated with NARP when present at lower percentage of mutants or with lethal Leigh syndrome when present at higher percentage of mutants (Holt et al., 1990; Tatuch et al., 1992). Mutations have also been identified in mtDNA genes that encode proteins of the OXPHOS system, such as the cytochrome b gene and the mitochondrial complex I genes. A prominent example of the latter group of mtDNA protein-coding gene mutations is LHON, which is a common cause of subacute bilateral optic neuropathy that usually presents in early adult life and that predominantly affects males. Most LHON patients harbor one of three point mutations that affect mtDNA complex I, or the NADH:ubiquinone oxidoreductase (ND) genes: G3460A in ND1, G11778A in ND4 and T14484C in ND6. Patrick Chinnery and colleagues showed that the mitochondrial ND6 gene is a hot spot for LHON mutations and suggested that the ND6 gene should be sequenced in all LHON patients who do not harbour one of the three common LHON mutations (Chinnery et al., 2001). Rare nonsense or frameshift mutants in Cytochrome oxydase subunit I (COI) have been associated with encephalomyopathies (Bruno et al., 1999; Comi et al., 1998).

- Pathogenic mtDNA proteins synthesis mutations can also result in multisystem disorders with wide range of symptoms. The most common mtDNA protein synthesis mutation is A3243G in the tRNA(Leu). This mutation is linked to a variety of clinical symptoms. When present at relatively low level (10%-30%) in the blood, the patient may manifest only type II diabetes. By contrast, when the mutation is present in > 70% of the mtDNA, it causes more severe symptoms including short stature, cardiomyopathy, Chronic Progressive External Ophthalmoplegia (CPEO; OMIM157640) and MELAS (Goto et al., 1990; van den Ouweland et al., 1994).

2.2.2 Nuclear DNA mutations linked to OXPHOS diseases

Nuclear DNA mutations linked to OXPHOS diseases includes defects in structural OXPHOS genes, faulty inter-genomic communication, and defects in OXPHOS assembly, homeostasis and import. Most nuclear gene mutations affect various protein subunits of complex I and complex II.

The first structural OXPHOS-gene mutation was reported in two sisters with Leigh syndrome and isolated complex II deficiency (Bourgeron et al., 1995). The pathogenic mutation was in the gene that encodes the flavoprotein: SDHA (succinate dehydrogenase subunit A). Subsequently, another family was found to have mutations in this subunit (Parfait et al., 2000). Very interestingly, two groups independently reported mutations of the complex II subunit D and C genes in hereditary paraganglioma — usually benign, vascularized tumours in the head and in the neck (Baysal et al., 2000; Niemann & Muller, 2000). This work has uncovered a new and surprising association between mitochondrial defects and carcinogenesis. Genetic characterization of Complex I deficiency in a patient with a Leigh-like presentation revealed a 5-base-pair (bp) duplication in NDUFS4 (NADH dehydrogenase (ubiquinone) Fe–S protein 4) that destroys the consensus phosphorylation site in the gene product and extends the length of the protein by 14 amino acids (van den Heuvel et al., 1998). Further studies have revealed that this duplication abolishes cyclic-AMP-dependent phosphorylation of NDUFS4, thereby impairing activation of the complex. Further complex I mutations have been identified and ≈ 40% of complex I deficiencies in children, in which the defect is detected in cultured skin fibroblasts, can now be explained by mutations in structural nuclear genes (Loeffen et al., 1998, 2000).

OXPHOS defects caused by defective interplay between the mitochondrial and nuclear genomes have also been described. The clinical features of the Mitochondrial Neuro-GastroIntestinal Encephalomyopathy syndrome (MNGIE) include ophthalmoparesis, peripheral neuropathy, leucoencephalopathy and gastrointestinal symptoms (chronic diarrhea and intestinal dysmotility). Muscle biopsy shows ragged red fibers (RRFs) and COX-negative fibers and either partial isolated complex IV deficiency or combined OXPHOS-complex deficiencies (Hirano et al., 1994). Mitochondrial DNA analysis in this autosomal recessive syndrome showed mtDNA deletions, depletion, or both. The MNGIE locus was mapped to chromosome 22q13.32-qter, a region that contains the thymidine phosphorylase (TP) gene (gene symbol ECGF1). Studies on patients showed that TP activity was markedly decreased. Ichizo Nishino and collaborators found various homozygous as well as compound heterozygous ECGF1 mutations in the genomic DNA of MNGIE patients (Nishino et al., 1999). The precise mechanism by which TP deficiency leads to mtDNA rearrangements have still to be explained, but imbalance of the mitochondrial nucleotide pool is likely to have a role. Autosomal dominant Progressive External Ophtalmogia (adPEO) is an adult-onset mitochondrial disorder that is characterized by progressive external ophthalmoplegia and variable additional features, including exercise intolerance, ataxia, depression, hypogonadism, hearing deficit, peripheral neuropathy and cataract (Zeviani et al., 1990). Some patients carry mtDNA deletions, although the disease is inherited in an autosomal fashion. Of the two autosomal loci for this disorder, the 4q-adPEO locus includes the gene for the heart and skeletal muscle isoform of the ANT1. Kaukonen and collaborators (Kaukonen et al., 2000) identified two heterozygous missense mutations in this gene in several families and in one sporadic patient with adPEO.

Enzyme complex I and IV deficiencies are by far the most frequently observed abnormalities of the OXPHOS system. In sharp contrast to isolated complex I deficiencies, no mutations have been found as yet in the ten nuclear genes that encode the structural proteins of complex IV (Adams et al., 1997). The discovery of mutations in a nuclear assembly gene that is associated with COX deficiency resulted from chromosomal transfer experiments. This approach identified mutations in the SURF1 gene in patients with COX-deficient Leigh syndrome (Tiranti et al., 1998; Zhu et al., 1998). SURF1 is part of a cluster of unrelated housekeeping genes and is the only gene of this cluster that is known or believed to be involved in COX assembly (Tiranti et al., 1999). Nuclear gene defects that are associated with isolated complex III or complex V deficiencies have not yet been discovered. In recent years, four inherited neurodegenerative diseases, Friedreich ataxia, hereditary spastic paraplegia, human DDP syndrom (deafness/dystonia peptide) and dominant optic atrophy (OPA1) have also been shown to be mitochondrial disorders that are caused by nuclear DNA mutations in the genes for frataxin, paraplegin, DDP and OPA1, respectively. Mitochondria obtained from heart biopsies of Friedreich ataxia patients disclosed specific defects in the citric-acid cycle enzyme aconitase, and complex I–III activities (Rotig et al., 1997). The causative Friedreich ataxia protein, frataxin, has an essential role in mitochondrial iron homeostasis, and Friedreich ataxia can therefore be considered as an OXPHOS homeostasis defect. Muscle biopsies from the autosomal recessive form of patients with hereditary spastic paraplegia revealed histochemical signs of a mitochondrial disorder, namely RRFs, COX-negative fibers and succinate dehydrogenase-positive hyperintense fibers (Casari et al., 1998). Linkage and subsequent mutation analysis revealed large deletions in a gene dubbed paraplegin (Casari et al., 1998). Owing to the homology with a yeast mitochondrial ATPase with both proteolytic and chaperone-like activities, it has been suggested that this form of hereditary spastic paraplegia could be a neurodegenerative disorder due to OXPHOS deficiency, attributing a putative function in the assembly or import of respiratory chain subunits or cofactors to paraplegin (Di Donato, 2000). The DDP syndrome, an X-linked recessive disorder also known as the Mohr–Tranebjaerg syndrome, is associated with a novel defect in mitochondrial protein import (Koehler et al., 1999). The defective gene is homologous to the yeast protein Tim8, which belongs to a family of proteins that are involved in intermembrane protein transport in mitochondria. Therefore, the DDP syndrome should be considered as the first example of a new group of mitochondrial import diseases (Koehler et al., 1999). Finally, OPA1 is caused by defects in a dynamin-related protein that is targeted to mitochondria and might exert its function in mitochondrial biogenesis and in stabilization of mitochondrial membrane complexes (Delettre et al., 2000).

3. Models to study OXPHOS diseases

3.1 Cybrids and Rho^0 cells

Cybrids, or "cytoplasmic hybrids," are cultured cells manipulated to contain introduced mitochondrial DNA (mtDNA). Cybrids have been a central tool to unravel effects of mtDNA mutations in OXPHOS diseases. In this way, the nuclear genetic complement is held constant so that observed effects on OXPHOS can be linked to the introduced mtDNA. The cybrids are produced by first treating mitochondrial donor cells with cytochalasin B to weaken the cytoskeleton, before subjecting the cells to a centrifugal force, either as attached cells or in suspension. The dense nuclei are extruded, leaving plasma membrane-bound "cytoplasts" containing cell cytoplasm and organelles, including mitochondria. These cytoplasts are then fused with a nuclear donor cell line. The first mammalian cultured cell phenotype identified to segregate with mtDNA was in human (HeLa) cells, where mtDNA imparted resistance to the antibiotic chloramphenicol (Spolsky & Eisenstadt, 1972). Several other mtDNA-linked drug-resistant phenotypes were identified in mammalian cells in the 1970s and 1980s, including resistance to the complex III inhibitors antimycin and myxothiazol (Howell & Gilbert, 1988) and to the complex I inhibitor rotenone (Bai & Attardi, 1998). The development of robust DNA-sequencing methods leads to the identification of single-base substitutions in the 16S rRNA gene of the mtDNA of independently derived yeast, mouse, and human chloramphenicol cell lines (Blanc & Dujon, 1980; Kearsey & Craig, 1981). These pioneering studies were in turn followed by identification of the first cytochrome b mutants (Howell & Gilbert, 1988) and more recently ND5, ND6, and COI mutants.

The second cellular OXPHOS model corresponds to the isolation of a human cell line without mtDNA (called Rho^0 cells). Employing an approach first used in yeast (Slonimski et al., 1968), cells were incubated with low levels of the drug ethidium bromide, which intercalates DNA. Low levels of the drug selectively inhibits the gamma-DNA polymerase responsible for mtDNA replication, and with ongoing cell division, the mtDNAs are "diluted" to the point where clones can be isolated without detectable organelle genomes. King and Attardi (King & Attardi, 1989) also discovered the absolute requirement for pyruvate gained by these cells and confirmed the previous observation from Paul Desjardins and collaborators (Desjardins et al., 1985) that mtDNA-less cells also required added uridine for growth. This allowed a selection regime to be used after cytoplast– Rho^0 cell fusion so that unfused Rho^0 cells could be eliminated and cybrids selected with the use of an appropriate nuclear drug-resistant marker (King & Attardi, 1989). Apart from their value in cybrid experiments, such Rho^0 cells represent a unique research tool by themselves. They are a surprising reminder that OXPHOS is dispensable, at least for some differentiated mammalian cell types. In this instance, ATP production is 100% from glycolysis, so the cells acidify culture media very rapidly by producing large quantities of lactate. They retain functional mitochondria (except lacking OXPHOS), which show a transmembrane potential (probably from the electrogenic exchange of ATP for ADP) and can import the hundreds of other proteins needed for non-OXPHOS functions. The pioneering cybrid work using the selectable markers was limited in the sense that endogenous mtDNAs were also present; that is, the cybrids were heteroplasmic. The Rho^0 cell approach allowed creation of homoplasmic or heteroplasmic cells, depending on the mtDNA donor cell(s) used.

3.2 Human fibroblasts

The use of individual patient's cells in tissue cultures enables the study of specific defects. With respect to cell type, myoblasts are most likely to express the phenotype observed in muscle, but it is generally not feasible to derive myoblasts for each diagnostic muscle biopsy, because most of the muscle tissue is used up for enzymatic, pathological and molecular workup. Moreover, myoblasts are not representative of some liver-specific phenotypes. An alternative to myoblasts, are fibroblasts, which are much easily obtained during a muscle biopsy or after (Robinson, 1996). Fibroblasts cultures are in general, the most obtainable and renewable source of cells for both diagnosis and research. The major drawback with fibroblasts in culture is that they sometimes fail to maintain the diseased phenotype. This is especially true for fibroblast cultures derived from tissue specific forms of mitochondrial diseases. Nevertheless, many patients do express mitochondrial dysfunction in primary fibroblasts albeit the defect is sometimes unmasked only under stressful growth conditions in culture media, devoid of glucose or serum (Iuso et al., 2006; Robinson, 1996; Taanman et al., 2003). Therefore, patient's fibroblast harboring nuclear encoded mutations can be a suitable tool to study OXPHOS diseases and a platform for the search for treatments by small molecules, using individual approaches tailored to a specific defect.

3.3 Mouse models

Despite some obvious limitations, our ability to mimic human disease in animal models is undoubtedly one of the most important technological breakthroughs in modern genetics.
Since the first knockout mice with impaired OXPHOS were generated in 1995 (reviewed in (Smeitink et al., 2001) and (Larsson & Rustin, 2001)), eight others have been described.
Classical knockout (KO) technology has been achieved for the manganese superoxide dismutase gene (SOD2) and the ANT1. These mice can be considered as secondary OXPHOS-deficient mice because the genes are only indirectly related to the OXPHOS system. SOD2 is an oxygen radical scavenger in the mitochondrial matrix, which acts as a first line of defense against the superoxide that is produced as a by-product of OXPHOS (Li et al., 1995). To gain further insight into the effects of the ANT1 mutation in particular, study the regulation of nuclear and mitochondrial genes in the skeletal muscle of mice KO of ANT1 (Murdock et al., 1999) revealed upregulation of 17 genes that fall into four categories: nuclear and mitochondrial genes that encode OXPHOS components; mitochondrial tRNA and ribosomal RNA genes; genes involved in intermediary metabolism; and an eclectic group of other genes, among which are genes previously unknown to be related to mitochondrial function.
Knockout mice for the mitochondrial transcription factor A (TFAM) can be considered primary OXPHOS mice, because TFAM has a direct role in the regulation of OXPHOS gene expression. Using a conditional knockout approach, three distinct TFAM knockout mice have been created (Larsson et al., 1998; Wang et al., 1999): one for skeletal and cardiac muscle; one for cardiac muscle alone; and one for pancreatic β-cells. TFAM is essential for

mitochondrial biogenesis and embryonic development, and the conditional knockouts have indicated that the OXPHOS system is crucial for normal heart function and insulin secretion. Five mouse models that were specifically designed to mimic isolated complex I deficiency in humans involve the NDUFS4 gene. This gene constitutes a mutational hotspot in humans. Four models are KO or conditioned KO for NDUFS4, the fifth one corresponds to a point mutation in NDUFS4. The whole-body and neuron-targeted NDUFS4 KO mice displayed small size and displayed weight loss. This was accompanied by ataxia, blindness, hearing loss, loss of motor skills and death from a fatal encephalomyopathy. The Purkinje cell specific KO mice only manifested mild behavioral and neuropathological abnormalities. Homozygote point mutation NDUFS4 mice were not viable, demonstrating that the presence of mutated NDUFS4 protein leads to a much more severe phenotype than complete absence of NDUFS4 (reviewed in (Roestenberg et al., 2011)).

Two mouse models for Friedreich ataxia have also been created (Puccio et al., 2001). Like the ANT1- and SOD2-deficient mice, these mice can also be considered as secondary OXPHOS-deficient mice. The frataxin-deficient mammals showed time-dependent iron accumulation and will allow the detailed study of the mechanism of frataxin involvement in iron metabolism and iron–sulphur biogenesis.

Finally, Jun-Ichi Hayashi's group, using a completely different approach, generated mice that carry large-scale mtDNA deletions (Inoue et al., 2000). Synaptosomes from mouse brains with naturally occurring somatic mtDNA mutations were fused with Rho^0 cells. Each fusion event introduced a variable number of mutant and wild-type mtDNAs, which then repopulate the Rho^0 cell, creating a cybrid cell line. Enucleated cybrid cells were fused to donor embryos and implanted in pseudo pregnant females. In this way, they generate heteroplasmic founder female animals in which mtDNA deletion transmission was obtained for three generations (Inoue et al., 2000).

4. Calcium signalling and mitochondrial OXPHOS physiology

Calcium (Ca^{2+}) is one of the most common second messengers in intracellular signalling networks. Periodic fluctuations in cytosolic calcium concentration ($[Ca^{2+}]_{cyt}$) is driven by electrical activation of voltage-gated Ca^{2+} channels (VGCC) or by agonist stimulation of plasma membrane receptors and the subsequent formation of Ca^{2+}-mobilizing second messengers, such as inositol 1,4,5-trisphosphate (IP_3). IP_3 binds to its receptor the IP_3R (inositol 1,4,5-trisphosphate) on the endoplasmic reticulum (ER) membrane leading to Ca^{2+} release from the ER to the cytosol. In excitable cells, Ca^{2+} release from the ER occurs also through ryanodine receptors (RyR) that function as Ca^{2+}-activated Ca^{2+} channels which further amplify Ca^{2+} signals originating from other sources.

The frequency, amplitude and/or duration of cytosolic $[Ca^{2+}]_{cyt}$ spikes can be detected and decoded by downstream Ca^{2+}-sensitive proteins providing a versatile pathway for extracellular stimuli to exert control over a wide range of metabolic pathways (Berridge et al., 2000).

Complex buffering systems that include multiple Ca^{2+}-buffering proteins, ATP-dependent Ca^{2+} pumps (SERCA (sarco-endoplasmic Reticulum Ca^{2+} ATPase) accumulating Ca^{2+} from the cytosol to the ER, and PMCA (Plasma membrane Ca^{2+} ATPase) extruding Ca^{2+} from cytosol to the extracellular space), and the sodium-Ca^{2+} exchanger (Na^+/Ca^{2+}), work together to restore $[Ca^{2+}]$ back to resting levels. Mitochondria also play an important role in shaping Ca^{2+} signals by utilizing potent mitochondrial Ca^{2+} uptake mechanisms. Ca^{2+}

uptake into mitochondria plays an important role in cellular physiology by stimulating mitochondrial metabolism and increasing mitochondrial energy production (Duchen, 1992). However, excessive Ca^{2+} uptake into mitochondria can lead to opening of a permeability transition pore (PTP) and apoptosis.

4.1 Interplay between Ca^{2+} and OXPHOS

Mitochondrial bioenergetics and Ca^{2+} shaping are mutually regulated. Indeed, on the one hand, mitochondria Ca^{2+} accumulation enables the activity of OXPHOS and ATP production; on the other hand, mitochondrial ATP favours the effective functioning of the two major Ca^{2+} pumps PCMA and SERCA and actively participates in shaping cytosolic Ca^{2+} signals (Figure 1 A and B).

One important target for Ca^{2+} signals is the activation of mitochondrial oxidative metabolism and the consequent increase in the formation of ATP. Studies performed in 1960-1970 led to the demonstration that four mitochondrial dehydrogenases are activated by Ca^{2+} ions. These are FAD-glycerol phosphate dehydrogenase, pyruvate dehydrogenase, NAD^+-isocitrate dehydrogenase and α-ketoglutarate dehydrogenase. FAD-glycerol phosphate dehydrogenase is located on the outer surface of the inner mitochondrial membrane and is influenced by changes in cytoplasmic Ca^{2+} ions concentrations. The other three enzymes are located within mitochondria and are regulated by matrix Ca^{2+} ions concentration. The effects of Ca^{2+} ions on FAD-isocitrate dehydrogenase involve binding to an EF-hand binding motif within this enzyme, leading to lowering of the K_m for glycerol phosphate very substantially (reviewed in (Denton, 2009)). Mitochondrial Ca^{2+} ions bind also directly to NAD^+-isocitrate dehydrogenase and α-ketoglutarate dehydrogenase to decrease the K_m for their respective substrates, whereas an increase in the dephosphorylated and active form of pyruvate dehydrogenase is regulated by a Ca^{2+}-sensitive phosphatase (Bulos et al., 1984; Denton & Hughes, 1978; Denton et al., 1972, 1978, 1996; McCormack et al., 1990; McCormack & Denton, 1979; Robb-Gaspers et al., 1998). Extramitochondrial Ca^{2+} regulates the glutamate-dependent state 3 respiration by the supply of glutamate to mitochondria via aralar, a mitochondrial glutamate/aspartate carrier (Gellerich et al., 2010).

A very recent finding suggests a novel paradigm in which the transcription of genes for mitochondrial enzymes that produce ATP and the genes that consume ATP is coordinately regulated by the same transcription factors (Watanabe et al., 2011). Thus, TFAM and TFB2M, recognized as mtDNA-specific transcription factors, were shown to regulate transcription of the SERCA2 gene (Watanabe et al., 2011).

It was also demonstrated that metabolites generated during energy production may influence IP_3R-mediated Ca^{2+} dynamics. Indeed, it was shown that reduced Nicotinamide adenine dinucleotide selectively stimulates the release of Ca^{2+} mediated by IP_3R (Kaplin et al., 1996). Another evidence of communication between cellular metabolism and Ca^{2+} signalling was reported recently by Bakowski and Parekh who showed that pyruvate, the precursor substrate for the Krebs cycle, directly increases the native I_{CRAC} (store operated Ca^{2+} influx channels at the plasma membrane) by reducing inactivation of the channel, thereby coupling oxidation of glucose and its own metabolism in the mitochondria to Ca^{2+} influx by the CRAC channel (Bakowski & Parekh, 2007).

In addition to serving as a target of Ca^{2+} signalling, the uptake of Ca^{2+} by mitochondria has important feedback effects to shape cytosolic Ca^{2+} signals. Rosario Rizzuto and collaborators (Rizzuto et al., 1993) were the first to make direct *in situ* measurements of mitochondrial

Ca^{2+}. They showed that receptor-activated Ca^{2+} signals caused rapid and large Ca^{2+} signals in the mitochondrial matrix (mechanisms of mitochondrial Ca2+ influx and efflux are detailed below).

4.2 Mechanisms of mitochondrial calcium influx and efflux
4.2.1 Mechanisms of mitochondrial calcium influx

Mitochondrial Ca^{2+} uptake is dependent on the strong driving force ensured by their membrane potential (-180 mV, negative inside) built by the respiratory chain (for review see (Bianchi et al., 2004)). It has been assumed that $[Ca^{2+}]_{cyt}$ far exceeding the micromolar range is required for net Ca^{2+} uptake, however, such $[Ca^{2+}]_{cyt}$ values have not been observed experimentally in the bulk cytoplasm. Ca^{2+} diffusion in the cytoplasm is also controlled by protein binding (Allbritton et al., 1992). Thus, local Ca^{2+} transients with amplitudes far exceeding those measured over the global cytoplasm are confined in cytosolic microdomains at the mouth of Ca^{2+} channels beneath the plasma membrane or ER internal store. This concept was consolidated by the demonstration that mitochondria, forming a complex cytoplasmic tubulovesicular system (Tinel et al., 1999), are frequently apposed to the smooth as well as the rough ER. These contact points, have been observed in several cell types by means of electron microscopy or tomography (Mannella et al., 1998). The experiments by Rosario Rizzuto and Tulio Pozzan definitively demonstrated that Ca^{2+} released through IP_3R in these microdomains, induce supramicromolar, or even submillimolar Ca^{2+} signals (Rizzuto et al., 1993).

Accordingly, the group of György Hajnoczky demonstrates that maximal activation of mitochondrial Ca^{2+} uptake is evoked by IP_3-induced perimitochondrial $[Ca^{2+}]$ elevations, which appear to reach values >20-fold higher than the global increases of $[Ca^{2+}]_{cyt}$. Incremental doses of IP_3 elicited $[Ca^{2+}]_{mit}$ elevations that followed the quantal pattern of Ca^{2+} mobilization, even at the level of individual mitochondria. These results and others by the same group allow concluding that each mitochondrial Ca^{2+} uptake site faces multiple IP_3R, a concurrent activation of which is required for optimal activation of mitochondrial Ca^{2+} uptake (Csordas et al., 1999; Hajnoczky et al., 1995) and reviewed in (Csordas et al., 2006). Targeting aequorin to the outer surface of the IMM in HeLa cells made the measurement of $[Ca^{2+}]$ in the mitochondrial intermembrane space possible. After stimulation with histamine $[Ca^{2+}]$ rose in the intermembrane space to significantly higher values than in the global cytosol (Rizzuto et al., 1998). This observation has given a strong support to the concept that net mitochondrial Ca^{2+} uptake occurs from high-Ca^{2+} peri-mitochondrial microdomains.

The existence of physical support for the ER–mitochondrial interface has been indicated by co-sedimentation of ER particles with mitochondria and electron microscopic observations of close associations between mitochondria and ER vesicles (Mannella et al., 1998; Meier et al., 1981; Shore & Tata, 1977). At these sites the shortest ER-OMM distance varies from 10 nm to 100 nm. In cells exposed to ER stress (serum starvation, tunicamycin) an increase in the ER–mitochondrial interface has been observed (Csordas et al., 2006). Also, coupling of the two organelles with a fusion protein increased the ER–mitochondria interface area, reduced the ER–mitochondrial distance to about 6 nm and greatly facilitated the transfer of cytosolic Ca^{2+} signal into the mitochondria of RBL-2H3 cells (Csordas et al., 2006). Accordingly, our team showed that the truncated variant of the sarco-endoplasmic reticulum Ca^{2+}-ATPase 1 (S1T) is induced under ER stress conditions. S1T is localized in the ER-mitochondria microdomains, increases number of ER-mitochondria contact sites, and inhibits mitochondria movements thus determining a privileged Ca^{2+} transfer from the ER to mitochondria leading to the activation of the mitochondrial apoptotic pathway (Chami et al., 2008).

Mitochondrial fission and fusion is another essential phenomenon for maintaining the metabolic function of these organelles as well as regulating their roles in cell signalling (Tatsuta & Langer, 2008; Yaffe, 1999; Chan, 2006). Changes in the relative rates of fusion and fission alter the overall morphology of the mitochondria affecting the function of the organelles both as regulators of survival/apoptosis and in Ca^{2+} handling. It has been shown that fusion is blocked (Karbowski & Youle, 2003) and mitochondria become fragmented during apoptosis (Frank et al., 2001). However, enhanced fission alone does not induce apoptosis and has even been shown to protect against Ca^{2+}-dependent apoptosis by preventing the propagation of harmful Ca^{2+} waves through the mitochondrial reticulum (Szabadkai et al., 2004).

The outer mitochondrial membrane is permeable to solutes and the inner mitochondrial membrane is impermeable to solutes that harbor the respiratory chain complexes. As described in chapter 1, the respiratory chain pumps protons against their concentration gradient from the matrix of the mitochondrion into the inter-membrane space, generating an electrochemical gradient in the form of a negative inner membrane potential and of a pH gradient, the matrix being more alkaline than the cytosol (Bernardi et al., 1999; Poburko et al., 2011).

Ca^{2+} import across the outer mitochondrial membrane (OMM) occurs through the voltage-dependent anion channels (VDAC) (Simamura et al., 2008). VDAC is as a large voltage-gated channel, fully opened with high-conductance and weak anion-selectivity at low transmembrane potentials (< 20–30 mV), but switching to cation selectivity and lower conductance at higher potentials (Colombini, 2009; Shoshan-Barmatz et al., 2010). The precise mechanisms of VDAC conductance are however still under debate.

Ca^{2+} import across the inner mitochondrial membrane (IMM) occurs through a Ca^{2+}-selective channel known as the mitochondrial Ca^{2+} uniporter (MCU) (Kirichok et al., 2004). Electrophysiological recordings of mitoplasts, small vesicles of inner mitochondrial membrane, revealed that the MCU is a highly Ca^{2+}-selective inward-rectifying ion channel (Kirichok et al., 2004). The MCU has a relatively low Ca^{2+} affinity (Kd $\sim$10 µM in permeabilized cells (Bernardi, 1999)). The activity of the MCU had been known for decades to be inhibited by ruthenium red and its derivative Ru360 (Vasington et al., 1972), but its molecular identity has only been unraveled very recently. It has been reported recently that the process of Ca^{2+} accumulation undergoes complex regulation by Ca^{2+} itself. Thus mitochondrial uptake of Ca^{2+} was significantly reduced by inhibitors of calmodulin, suggesting that a Ca^{2+}–calmodulin-mediated process is necessary for activation of the uniporter but Ca^{2+} also appeared to inhibit its own uptake. However, in contrast to the sensitization of mitochondrial Ca^{2+} uptake, the Ca^{2+}-dependent inactivation was not sensitive to calmodulin blockers (Moreau & Parekh, 2008).

In recent years, several molecules have been proposed to be either an essential or an accessory component of the MCU. In 2007, the uncoupling proteins (UCP) 2 and 3 (Trenker et al., 2007) were proposed to be essential for the MCU. Indeed, UCP2/3 overexpression increased mitochondrial Ca^{2+} elevations and the contrary is observed upon UCP2/3 depletion. In addition, mice lacking UCP2 exhibited a reduced sensitivity to the Ca^{2+} uptake inhibitor ruthenium red. However, these findings were disputed by another study that reported normal mitochondrial Ca^{2+} uptake in mice genetically ablated for UCP2 and UCP3 (Brookes et al., 2008). Furthermore, it was recently showed that UCP3 modulates the activity of sarco/endoplasmic reticulum Ca^{2+} ATPases by decreasing mitochondrial ATP production (De Marchi et al., 2011). The mitochondrial Ca^{2+} alterations associated with changes in UCP3 levels therefore reflect the exposure of mitochondria to abnormal cytosolic Ca^{2+} concentrations and do not reflect changes in MCU activity. These data indicate that UCP3 is not the mitochondrial Ca^{2+} uniporter. In 2009, Jiang and collaborators identified the leucine zipper EF hand containing transmembrane protein 1(Letm1) as a molecule that regulate both mitochondrial Ca^{2+} and H$^+$ concentrations (Jiang et al., 2009). Letm1 was reported to be a high-affinity mitochondrial Ca^{2+}/H$^+$ exchanger able to import Ca^{2+} at low (i.e. sub-micromolar) cytosolic concentrations into energized mitochondria. Earlier studies had however linked Letm1 to mitochondrial K$^+$/H$^+$ exchange and to the maintenance of ionic mitochondrial balance, the integrity of the mitochondrial network and cell viability (Dimmer et al., 2008; Nowikovsky et al., 2004). The high-affinity of Letm1 for Ca^{2+} and its postulated 1Ca^{2+}/1 H$^+$ stoichiometry are at odds with the known properties of the MCU. Thus, Letm1 is not the dominant mechanism of mitochondrial Ca^{2+} uptake. Instead, Letm1 might contribute to an alternate mode of mitochondrial Ca^{2+} uptake, known as rapid mode of uptake (RaM), that was first reported in isolated rat liver mitochondria by Gunter's group. It was reported that mitochondrial Ca^{2+} sequestration via a the RaM occurred at the beginning of each pulse and was followed by a slower Ca^{2+} uptake characteristic of the MCU (Sparagna et al., 1995; Szabadkai et al., 2001). The implications of the coexistence of low and high-affinity modes of Ca^{2+} uptake have been recently reviewed (Santo-Domingo & Demaurex, 2010).

In 2010, Palmer and Mootha reported that a new mitochondrial EF hand protein MICU1 (for mitochondrial Ca^{2+} uptake 1) was required for high capacity mitochondrial Ca^{2+} uptake, and proposed that MICU1 acts as a Ca^{2+} sensor that controls the entry of Ca^{2+} across the uniporter (Perocchi et al., 2010). Building up on this discovery, the same group and another simultaneously identified the mitochondrial Ca^{2+} uniporter (Baughman et al., 2011; De Stefani et al., 2011). Using *in silico* analysis combined with phylogenetic profiling and analysis of RNA and protein co-expressed with MICU1, the group of Vamsi Mootha isolated a novel protein that co-immunoprecipitated with MICU1 (Baughman et al., 2011). Using the same database, the group of Rosario Rizzuto independently identified the same protein. From the 14 proteins characterized by two or more transmembrane domains and known to exhibit or lack uniport activity domains, these authors identified a protein with a highly conserved domain encompassing two transmembrane regions separated by a loop bearing acidic residues. Functional analysis confirmed that this protein behaves as expected for the mitochondrial uniporter, and it was therefore assigned the defining name of MCU. Mitochondrial Ca^{2+} uptake was strongly reduced by MCU silencing in cultured cells and in purified mouse liver mitochondria, whereas MCU overexpression enhanced ruthenium red-sensitive mitochondrial Ca^{2+} uptake in intact and permeabilized cells (De Stefani et al., 2011). The MCU is a 45 kDa protein that can forms oligomers (Baughman et al., 2011). Both studies mapped the MCU to the inner mitochondrial membrane, but disagreed on whether the N and C termini face the matrix of the inter-membrane space (Baughman et al., 2011; De

Stefani et al., 2011). Mutations of conserved acidic residues within the short sequence linking the two transmembrane domains abrogated the ability of MCU to reconstitute mitochondrial Ca^{2+} uptake, whereas mutation of a nearby serine residue (S_{259}) conferred resistance to Ru360, indicating that the acidic residues are required for Ca^{2+} uptake and that S_{259} is critical for MCU sensitivity to ruthenium red (Baughman et al., 2011). Finally, and most convincingly, expression of the purified protein in planar lipid bilayers was sufficient to reconstitute ion channel activity in solutions containing only Ca^{2+} (De Stefani et al., 2011). The currents were carried by a channel of small conductance (6–7 pS), fast opening/closing kinetics, and low opening probability, and were inhibited by ruthenium red, as expected for the MCU. Proteins mutated at two of the conserved acidic residues failed to generate Ca^{2+} currents when inserted into bilayers and acted as dominant negative when expressed in HeLa cells. These data clearly identified MCU as mitochondrial Ca^{2+} uniporter. In accordance to the notion that mitochondrial Ca^{2+} overload enhances the sensitivity to apoptosis, it was also demonstrated that cells overexpressing MCU were more sensitive to apoptosis after treatment with ceramide and H_2O_2 (De Stefani et al., 2011) (Figure 1B).

4.2.2 Mechanisms of mitochondrial calcium efflux

Compared to the MCU, the proteins that catalyze the efflux of Ca^{2+} from mitochondria have received much less attention. The extrusion of Ca^{2+} from mitochondria is coupled to the entry of Na^+ across an electrogenic $1Ca^+:3Na^+$ exchanger (Dash & Beard, 2008) that is inhibited by the benzothiazepine derivative CGP37157 ((Cox et al., 1993), and reviewed in (Bernardi, 1999)). The subsequent efflux of sodium ions by the mitochondrial $1Na^+:1H^+$ exchanger (mNHE) eventually results in the entry of three protons into the matrix for each Ca^{2+} ion that leaves mitochondria. Ca^{2+} extrusion thus has a high energetic cost, as it dissipates the proton gradient generated by the respiratory chain (reviewed in (Bernardi, 1999)). The molecule catalyzing mitochondrial Na^+/Ca^{2+} exchange has been recently identified as NCLX/NCKX6, a protein localized in mitochondrial cristae (Palty et al., 2010), whereas stomatin-like protein 2 (SLP-2), an inner membrane protein, was shown to negatively modulate the activity of the mitochondrial Na^+/Ca^{2+} exchanger (Da Cruz et al., 2010). Functional evidence from knock-down and overexpression studies indicate that NCLX is an essential part of the mitochondrial sodium Ca^{2+} exchanger whereas SLP-2 is an accessory protein that negatively regulates mitochondrial Ca^{2+} extrusion (Figure 1B).

4.2.3 Mitochondrial calcium overload: Activation of the permeability transition pore

When mitochondrial Ca^{2+} loads exceed the buffering capacity of inner membrane exchangers, an additional pathway for Ca^{2+} efflux from mitochondria may exist through opening of the permeability transition pore (PTP). The PTP is a voltage-dependent, cyclosporin A (CsA)-sensitive, high-conductance channel of the inner mitochondrial membrane (for reviews, see (Bernardi et al., 2006; Rasola & Bernardi, 2007)). Indeed, the interplay between the rate of mitochondrial Ca^{2+} influx and efflux modulates mitochondrial matrix Ca^{2+}, which in turn is widely considered to be a key factor for the regulation of the PTP open–closed transitions (Bernardi et al., 1999). Although opening of the PTP in response to Ca^{2+} has been documented in isolated mitochondria and permeabilized cells (Bernardi et al., 2006; Rasola & Bernardi, 2007), assessing opening of the PTP in intact neurons and other primary cells in response to physiological activators that dictate cytosolic Ca^{2+} has remained a major challenge. Yet, opening of the PTP is often thought to be associated with pathophysiological processes (for reviews see (Hajnoczky et al., 2006; Rizzuto et al., 2003)). In these scenarios, activation of the PTP leads to respiratory inhibition, and thus ATP depletion, and the release of mitochondrial Ca^{2+} stores and apoptotic activators, ultimately resulting in cell death (Bernardi et al., 1999; Di Lisa & Bernardi, 2009). These have led to the idea that opening of the PTP by elevated mitochondrial Ca^{2+} is a terminal, pathologic event. However, it has been reported recently that CyPD-dependent PTP may participate in non-lethal Ca^{2+} homeostasis in cells and neurons (Barsukova et al., 2011).

5. Calcium deregulation in OXPHOS diseases

The direct consequences of OXPHOS defects include alteration of mitochondrial membrane potential, ATP/ADP ratio, ROS production and mitochondrial Ca^{2+} homeostasis. The varied biochemical changes that occur in cases of OXPHOS deficiencies have a direct effect on cellular functions. Yet, they are also key underlying mediators of the (retrograde) communication between the mitochondrion and the nucleus, which results in specific gene expression of both nuclear and mitochondrial genomes (see review (Reinecke et al., 2009)).

We will review in this chapter only Ca^{2+} deregulation in OXPHOS. We will discuss the consequences of such deregulation on mitochondrial function and the cross regulation between Ca^{2+} and bioenergetics in the development of cellular pathology. We summarized in Table 1 the alterations of subcellular Ca^{2+} signals in OXPHOS related diseases (Table 1).

Decreased proton pumping due to respiratory chain defects can result in reduced mitochondrial membrane potential and proton gradient, which are used to generate ATP. Deregulation of the membrane potential secondary to a deficiency in the respiratory chain may modify the kinetics and/or accumulation capacity of Ca^{2+} in the mitochondria, with possible consequences not only at the level of respiratory chain function (loop effect) and of the mitochondria in general, but also at the level of the ER function, which is largely dependent on Ca^{2+} concentrations, and at the level of cytosolic Ca^{2+} signalling, which plays a major role in regulating cell functions. Deficiencies of OXPHOS also result in other immediate and downstream metabolic, structural, and functional effects. These effects are closely associated with mitochondrial dysfunction. The nicotinamide dinucleotide (NAD) redox balance, which is converted to the reduced state in OXPHOS deficiencies, is a fundamental mediator of several biological processes, such as energy metabolism, Ca^{2+} homeostasis, cellular redox balance, immunological function, and gene expression (Munnich & Rustin, 2001; Ying, 2008).

It is important to mention that analyses of Ca^{2+} signalling targeting OXPHOS diseases are sporadic, partial and incomplete. This situation can be explained by : 1) the recent development of new techniques permitting detailed and specific subcellular Ca^{2+} analyses such as recombinant "aequorin" probes developed by the group headed by Professors Rizzuto and Pozzan, and the latest generation of GFP-based Ca^{2+} probes (camgaroos, cameleons and pericams) characterized by a great potential to analyse Ca^{2+} dynamics in mitochondria at the single cell level; 2) Absence of suitable "easy" study models (see chapter 3); and 3) the difficulty in the characterization of OXPHOS deficiencies (see chapter 2-2).

Disease	Gene	Mutation/Deficiency	Study model	Mitochondrial/Cellular pathology	Ca²⁺ deregulation	Ca²⁺ probe	References
MERRF	tRNA_Lys	nt 8356 T/C	Cybrids	↓[ATP]_mit	↓[Ca²⁺]_mit; unchanged [Ca²⁺]_cyt	Aequorin	(Brini, 1999)
NARP	ATPase6	nt 8993 T/G	Cybrids	↓[ATP]_mit	Unchanged [Ca²⁺]_mit & [Ca²⁺]_cyt	Aequorin	(Brini, 1999)
NARP	ATPase6	nt 8993 T/G Rho⁰	Cybrids	Disturbed mitochondrial network and Actin cytoskeleton organization; ↓Δ_mit	↓ Ca²⁺ influx in NARP & Rho⁰	Fura 2, AM	(Szczepanowska, 2004)
MELAS	tRNA_Leu	nt 3243 A/G, 3271 T/C, 8344 A/G	Fibroblasts		↑ baseline level of [Ca²⁺]_cyt; ↓sequestration of [Ca²⁺]_mit	Fura 2, AM	(Moudy, 1995)
MELAS	tRNA_Leu	nt 3243 A/G, 3202 A/G Rho⁰	Cybrids	Complex I, III, IV and V deficiencies; ↓Δ_mit	↑ time to clear up [Ca²⁺]_cyt	Indo 1, AM ; Aequorin	(von Kleist-Retzow, 2007)
Leigh	NDUFS7	nt C364A	Fibroblasts	↓[ATP]_mit	↓[Ca²⁺]_ER ; ↓ [Ca²⁺]_cyt; ↓ [Ca²⁺]_ER	Fura 2, AM; Aequorin	(Visch, 2004)
NC	NDUFS1	nt G1854A nt C1668T	Fibroblasts	↓[ATP]_mit	↓[Ca²⁺]_mit ; ↓[[Ca²⁺]_cyt ; ↓ [Ca²⁺]_ER	Fura 2, AM; Aequorin	(Visch, 2004)
NC	NDUFS2	nt T1237C	Fibroblasts	↓[ATP]_mit	↓[Ca²⁺]_cyt	Fura 2, AM; Aequorin	(Visch, 2004)
NC	NDUFS2	nt C686A	Fibroblasts	↑[ATP]_mit	↓[Ca²⁺]_mit; unchanged [Ca²⁺]_cyt and [Ca²⁺]_ER	Fura 2, AM; Aequorin	(Visch, 2006)
NC	NDUFS2	nt G683A	Fibroblasts	Unchanged [ATP]_mit	unchanged [Ca²⁺]_mit, [Ca²⁺]_cyt and [Ca²⁺]_ER	Fura 2, AM; Aequorin	(Visch, 2006)
NC	NDUFS2	NC	Fibroblasts	↓[ATP]_mit	↓[Ca²⁺]_mit ; ↓ [Ca²⁺]_mb; ↓ [Ca²⁺]_ER	Fura 2, AM; Aequorin	(Visch, 2006)
NC	NDUFS2	NC	Fibroblasts	Unchanged [ATP]_mit	unchanged [Ca²⁺]_mit, [Ca²⁺]_cyt & [Ca²⁺]_ER	Fura 2, AM; Aequorin	(Visch, 2006)
NC	NDUFS4	AAGTC471(1)	Fibroblasts	Unchanged [ATP]_mit	unchanged [Ca²⁺]_mit, [Ca²⁺]_cyt & [Ca²⁺]_ER	Fura 2, AM; Aequorin	(Visch, 2006)
NC	NDUFS4	nt C316T	Fibroblasts	↓[ATP]_mit	↓[Ca²⁺]_cyt	Fura 2, AM; Aequorin	(Visch, 2006)
NC	NDUFS4	nt C316T	Fibroblasts	↓[ATP]_mit	↓[Ca²⁺]_mit ; ↓[Ca²⁺]_cyt ; ↓[Ca²⁺]_ER	Fura 2, AM; Aequorin	(Visch, 2006)
NC	NDUFS4	C202G/C203(2)	Fibroblasts	↓[ATP]_mit	↓[Ca²⁺]_mit; ↓[Ca²⁺]_cyt ; ↓[Ca²⁺]_ER	Fura 2, AM; Aequorin	(Visch, 2006)
NC	NDUFS7	nt C364A	Fibroblasts	↓[ATP]_mit	↓[Ca²⁺]_mit; ↓[Ca²⁺]_cyt; ↓[Ca²⁺]_ER	Fura 2, AM; Aequorin	(Visch, 2006)
NC	NDUFS8	nt C280T	Fibroblasts	↓[ATP]_mit	↓[Ca²⁺]_mit; ↓[Ca²⁺]_cyt; ↓[Ca²⁺]_ER	Fura 2, AM; Aequorin	(Visch, 2006)
NC	NDUFV1	nt C175T/ nt C1268T	Fibroblasts	↓[ATP]_mit	↓[Ca²⁺]_mit; ↓[Ca²⁺]_cyt; ↓[Ca²⁺]_ER	Fura 2, AM; Aequorin	(Visch, 2006)
NC	NDUFV1	nt C175T/ C1268T	Fibroblasts	Unchanged [ATP]_mit	unchanged [Ca²⁺]_mit, [Ca²⁺]_cyt & [Ca²⁺]_ER	Fura 2, AM; Aequorin	(Visch, 2006)
Leigh	SDHA	nt C1684T	Fibroblasts	↓[ATP]_mit ; ↓Δ_mit ; ↑ROS ; ↓ mitochondrial movement; ↑ ER-mitochondria contact sites	↑[Ca²⁺]_mit; ↑[Ca²⁺]_cyt; ↓[Ca²⁺]_ER	Fluo 4, AM; X-Rhod-1, AM; Aequorin	(M'Baya, 2010)
Leigh	SURF1	KO	SURF-/- mouse	COX deficiency; increased lifespan; protection from Ca²⁺-dependent neurotoxicity;	↓[Ca²⁺]_mit ; ↓[Ca²⁺]_cyt	Fura, FF ; Aequorin	(Dell'agnello, 2007)
Leigh	SURF1		Fibroblasts	COX deficiency	↓SOC	Fura 2, AM	(Wasniewska, 2001)
Freidreich ataxia	Frataxin	GAA(3)	Fibroblasts	ND	↑[Ca²⁺]	BAPTA-AM	(Wong & Cortopassi, 1997)
Huntington	Htt	CAG(3)	Lymphoblasts Brain from Tg mice	↓Δ_mit	↓[Ca²⁺]_mit	Green-5N	(Panov, 2002)
Huntington	Htt	CAG(3)	Immortalized striatal cells from Tg mouse	Normal mitochondrial function ; increased ROS	↓ basal [Ca²⁺] ; ↓ [Ca²⁺]mit ; ↓ IP₃, ↑ P2Y1/2 expression ; ↑ BK 1/2 expression	Fura 2, AM; Aequorin	(Lim, 2008)
NC	NC	COX deficiency	Fibroblasts	COX deficiency	Unchanged resting [Ca²⁺]_mit	Fura 2, AM	(Handran, 1997)
NC	NC	PDH deficiency	Fibroblasts	↓Δ_mit	↓[Ca²⁺]_mit	Aequorin	(Padua, 1998)

BK: bradykinin; COX: cytchrome oxidase Htt: Huntingtin; NC: non communicated; ND: not determined; ROS: reactive oxygen species; SOC: store operated Ca²⁺ entry; PDH: Pyruvate dehydrogenase; KO: knock out; [Ca²⁺]cyt, cytosolic calcium-concentration; [Ca²⁺]er, endoplasmic reticulum calcium-concentration; [Ca²⁺]mt, mitochondrial calcium-concentration; Ca²⁺, calcium. (1) Insertion; (2) Deletion; (3) repeat.

Table 1. Calcium deregulation in OXPHOS diseases

5.1 Calcium deregulation in MELAS, MERRF, NARP and LHON

Calcium deregulation was first reported in OXPHOS diseases linked to mitochondrial mutation. Brini and collaborators monitored subcellular Ca^{2+} signalling in cybrid cells with 0% and 100% of the MERRF (nt 8356 T/C) and NARP (nt 8993 T/G) mutations using cytosolic aequorin and aequorin probe targeted to the mitochondria. They showed a reduced mitochondrial $[Ca^{2+}]$ ($[Ca^{2+}]_{mit}$) transient in MERRF cells but not in NARP cells upon stimulation with IP3-generating agonist, whereas cytosolic Ca^{2+} responses ($[Ca^{2+}]_{cyt}$) were normal in both cell types (Brini et al., 1999).

In another study, cybrid cells with 98 % of NARP mutation (nt 8993 T/G) and Rho^0 cells show a disturbed mitochondrial network and actin cytoskeleton. These cells show also a slower Ca^{2+} influx rates in comparison to parental cells. Authors postulate that proper actin cytoskeletal organization is important for CCE (capacitative Ca^{2+} entry) in these cells (Szczepanowska et al., 2004).

Abnormal Ca^{2+} homeostasis and mitochondrial polarization was also reported in fibroblasts from patients with MELAS syndrome. These cells showed an increased Ca^{2+} influx associated to a decreased mitochondrial potential (Moudy, 1995).

A comparative study was performed to establish sensitivity to oxidant in cybrid cells bearing the LHON, MELAS, or MERRF. The order of sensitivity to H_2O_2 exposure was MELAS>LHON>MERRF>controls. Consistent with the hypothesis that death induced by oxidative stress is Ca^{2+} dependent, depletion of Ca^{2+} from the medium protected all cells from cell death. This study reveals indirectly that LHON as well as MELAS and MERRF show an increased basal Ca^{2+} load (Wong & Cortopassi, 1997).

In 2007, another study performed on cybrid cells incorporating two pathogenic mitochondrial mutations (nt 3243 A/G, nt 3202 A/G) reveal that the decreased ATP production by oxidative phosphorylation was compensated by a rise in anaerobic glycolysis. Regarding Ca^{2+} homeostasis, these cells did not show any alteration of Ca^{2+} signals in the cytosol but take longer to clear up the histamine induced Ca^{2+} signal in the mitochondria (von Kleist-Retzow et al., 2007).

All over, these studies revealed a deranged Ca^{2+} homeostasis in OXPHOS diseases linked to mitochondrial mutations. These alteration are not solely at the level of mitochondria but were also observed in the cytosol. Depending on the study model and/or mutation, increased cytosolic Ca^{2+} levels are linked to increased Ca^{2+} influx through the plasma membrane or reduced Ca^{2+} uptake capacity by the mitochondria.

5.2 Calcium deregulation in Complex I deficiency

The consequences of mitochondrial complex I deficiency on Ca^{2+} homeostasis was first studied in a genetically characterized human complex I deficient fibroblast cell lines harbouring nuclear NDUFS7 (nt 364G/A) mutation linked to Leigh's syndrome. These cells show a reduced mitochondrial Ca^{2+} accumulation and consequent ATP synthesis (Visch et al., 2004). In 2006, the same group investigated the mechanism(s) underlying this impaired response. The study was conducted in fibroblasts from 6 healthy subjects and 14 genetically characterized patients expressing mitochondria targeted luciferase. The results revealed that the agonist-induced increase in mitochondrial ATP ($[ATP]_{mit}$) was significantly, but to a variable degree, decreased in 10 patients. They also reported a reduced agonist-evoked mitochondrial $[Ca^{2+}]$ signal, measured with mitochondria targeted aequorin, and cytosolic $[Ca^{2+}]$ signal, measured with Fura-2, AM. Measurement of Ca^{2+} content of the ER, calculated from the increase in $[Ca^{2+}]_{Cyt}$ evoked by thapsigargin, an inhibitor of the ER Ca^{2+} ATPase revealed also a decrease in mutated cells as compared to controls. Regression analysis revealed that the increase in $[ATP]_{mit}$ was directly proportional to the increases in $[Ca^{2+}]_{cyt}$ and $[Ca^{2+}]_{mit}$ and to the ER Ca^{2+} content. This was the first report showing a pathological ER Ca^{2+} homeostasis in OXPHOS disease models. The authors postulated that the reduced ER Ca^{2+} content could be the direct cause of the impaired agonist-induced increase in $[ATP]_{mit}$ in human complex I deficiency (Visch et al., 2006). However, the molecular mechanisms underlying ER Ca^{2+} deregulation were not revealed.

Another key cellular feature that was extensively investigated in patient fibroblasts harboring complex I deficiency is mitochondrial morphology. The quantification of mitochondrial morphology in a cohort of 14 patients fibroblast cell lines revealed two distinct classes of patient fibroblasts, one in which the cells mainly contained short circular fragmented mitochondria, and one in which the cells displayed a normal filamentous mitochondrial morphology (Koopman et al., 2007). Authors postulated that these differences are linked to ROS levels (Koopman et al., 2007). In a second report, the authors analyzed the relationship between mitochondrial dynamics and structure and Ca^{2+}/ATP handling in the same cohort. Regression analysis of the agonist-induced Ca^{2+}/ATP handling and mitochondrial morphology shows that increased mitochondrial number is associated to reduced Ca^{2+}-stimulated mitochondrial ATP and reduced stimulation of cytosolic Ca^{2+} removal rate (Willems et al., 2009).

5.3 Calcium deregulation in Complex II deficiency

The investigation of Ca^{2+} deregulation linked to complex II deficiency were largely performed upon complex II inhibition by 3-nitropropionic acid (3NP) . The inhibition of complex II by 3NP is related to neuronal death, anatomic and neurochemical changes similar to those occurring in Huntington's disease (HD).

In primary cultures of rodent central nervous system, 3NP elicits an early increase in neuronal $[Ca^{2+}]_{cyt}$, and both apoptotic and necrotic neuronal death (Greene et al., 1998). 3NP treatment produces a long term potentiation of the NMDA-mediated synaptic excitation in striatal spiny neurons. This also involves increased intracellular Ca^{2+} (Calabresi et al., 2001).

To the mechanisms underlying increased $[Ca^{2+}]_{cyt}$ upon 3NP treatment, it was shown that short treatment-induced $[Ca^{2+}]_{cyt}$ increase occurs through NMDA-GLUR (Glutamate receptor) and VGCC and implicates also internal stores (Lee et al., 2002). In astrocyte cultures, Tatiani, R. Rosenstock and collaborators showed that 3NP is also able to release mitochondrial Ca^{2+} independently from internal stores and from Ca^{2+} entry through the plasma membrane (Rosenstock et al., 2004). Another group showed that 3NP-induced necrosis in primary hippocampal neurons is associated with an increase in both cytosolic and mitochondrial $[Ca^{2+}]$, decreased ATP and rapid mitochondrial potential depolarization. In this context, the increased $[Ca^{2+}]$ was shown to result from Ca^{2+} influx through NMDA receptors (Nasr et al., 2003).

The occurrence of mitochondrial permeability transition (PT) was shown to be the cause of the loss of neuronal viability induced by complex II inhibition (Maciel et al., 2004). This is in line with studies showing increased susceptibility of striatal mitochondria to Ca^{2+}- induced PT (Brustovetsky et al., 2003) and that cyclosporine A (inhibitor of PT) protects against 3NP toxicity in striatal neurons (Leventhal et al., 2000) and astrocytes (Rosenstock et al., 2004). Accordingly, inhibition of mitochondrial Ca^{2+} influx by ruthenium red significantly reduces 3NP-induced cell death (Ruan et al., 2004).

The data obtained upon complex II inhibition by 3NP are in accordance with those obtained from Huntington's patients and transgenic mice. Mitochondria isolated from lymphoblasts of individuals with HD showed reduced mitochondrial potential and increased sensitivity to depolarization upon Ca^{2+} addition. Similar results were obtained in transgenic HD mice expressing mutated huntingtin (Panov et al., 2002). In addition, mitochondria from HD mice showed lower Ca^{2+} retention capacity. These mitochondrial abnormalities preceded the onset of pathological or behavioural tract by months, suggesting that mitochondrial Ca^{2+} deregulation occurs early in HD (Panov et al., 2002). In a recent study, Lim and collaborators explore Ca^{2+} homeostasis and mitochondrial dysfunction in clonal striatal cell lines established from a transgenic HD mouse model and showed transcriptional changes in the components of the phosphatidylinositol cycle and in receptors for myo-inositol triphosphate-linked agonist. The overall result of such changes is to decrease basal Ca^{2+} in mutant cells. Mitochondria from mutant cells failed to handle large Ca^{2+} loads and this seems to be due to increased Ca^{2+} sensitivity of the permeability transition. This study reveals a compensatory attempt to prevent the Ca^{2+} stress that would exacerbate mitochondrial damage in HD (Lim et al., 2008).

Our group was the first to investigate Ca^{2+} homeostasis in human fibroblasts isolated from a patient with Leigh's syndrome harbouring a homozygous R554W substitution in the flavoprotein subunit of the complex II (SDHA). Our study was conducted in parallel in control fibroblasts and in neuroblastoma SH-SY5Y cells upon inhibition of complex II with 3NP or Atpenin A5 at doses which did not induce cell death, thus affording to study complex II deficiency independently from cell death. We showed that mutation or chronic inhibition of complex II determined a large increase in basal and agonist-evoked Ca^{2+} signals in the cytosol and mitochondria, in parallel with mitochondrial dysfunction (membrane potential loss, ATP reduction and increased ROS). Cytosolic and mitochondrial Ca^{2+} overload are linked to increased ER Ca^{2+} leakage, and to PMCA and SERCA2b proteasome-dependent degradation. Increased mitochondrial Ca^{2+} load is also contributed by decreased mitochondrial motility and increased ER-mitochondrial contacts. These findings are interesting since they link for the first time OXPHOS-related mitochondrial pathology to the regulation of the stability of two major actors in Ca^{2+} signalling regulation, namely PMCA and SERCA. We postulate that SERCA2b and PMCA degradation is predictably related to a decrease of mitochondrial ATP production, since SERCA2b and PMCA degradation was also observed upon ATP synthase inhibition by rotenone. This phenomenon could be interpreted as an adaptation response to ATP demise in OXPHOS diseases. Our study revealed also the activation of a compensatory attempt to restore total ATP level through the activation of anaerobic glycolysis in a Ca^{2+}-dependent manner (M'Baya et al., 2010). This study revealed a double hint of Ca^{2+} signalling deregulation in complex II deficiency. On the one hand Ca^{2+} overload may favour the activation of glycolytic ATP production and on the other hand favoured Ca^{2+}-mediated mitochondrial pathology (M'Baya et al., 2010).

5.4 Calcium deregulation in OXPHOS diseases linked to defects in OXPHOS assembly and iron homeostasis: COX and frataxin deficiencies

Leigh's syndrome associated with COX deficiency is usually caused by mutations of SURF1, a gene coding a putative COX assembly factor. Fibroblasts isolated from patients harboring SURF1 mutation displayed a low Ca^{2+} influx through SOC (store operated Ca^{2+} channels) as compared to control fibroblast (Wasniewska et al., 2001). The energy state of the mitochondrial membrane in mutated cells is naturally decreased. Accordingly, it was demonstrated that mitochondria can control SOC in a numerous cell types and that the collapse of mitochondrial membrane potential, either by an uncoupler or an inhibitor of the respiratory chain, greatly reduces the SOC (Makowska et al., 2000). In an earlier study, Handran and collaborators failed to document either mitochondrial morphology alteration or intracellular Ca^{2+} deregulation in COX-deficient human fibroblasts (Handran et al., 1997). This discrepancy between these results may be accounted on the partial recovery of COX enzyme activity in COX deficient fibroblasts. Fibroblasts are not a robust system for the study of mitochondrial dysfunction and cultured cells relays less on mitochondria for ATP production. It was thus concluded that this deficiency is not detrimental to fibroblast or that anaerobic respiration rescues the phenotype. In a strange manner, SURF1-/- KO mouse displayed mild reduction of COX activity in all tissues and did not show encephalopathy. These mice show a complete protection from in vivo neurodegeneration induced by exposure to high doses of kainic acid (a glutamatergic epiloptogenic agonist). Thus the ablation of SURF1 drastically reduces the glutamate-induced increase of Ca^{2+} both in the cytosol and the mitochondria. Authors postulate that reduced buffering capacity by SURF1-/- mitochondria in the contact sites between mitochondria and plasma membrane or the ER may promote the feedback closure of the Ca^{2+} channels thus inhibiting the cytosolic Ca^{2+} transient rise (Dell'agnello et al., 2007).

As introduced in chapter 2-2-2, Friedreich's ataxia (FA) is an autosomal recessive disease caused by decreased expression of the mitochondrial protein frataxin. The biological function of frataxin is unclear. The homologue of frataxin in yeast, YFH1, is required for cellular respiration and was suggested to regulate mitochondrial iron homeostasis. Patients suffering from FA exhibit decreased ATP production in skeletal muscle. Accordingly, overexpression of frataxin in mammalian cells causes a Ca^{2+}-induced up-regulation of tricarboxylic acid cycle flux and respiration, which, in turn, leads to an increased mitochondrial membrane potential and results in an elevated cellular ATP content. Thus, frataxin appears to be a key activator of mitochondrial energy conversion and oxidative phosphorylation (Ristow et al., 2000).

It was reported that mean mitochondrial iron content was increased in FA fibroblasts harboring expansion of intronic GAA repeat in frataxin leading to its reduced expression, and that staurosporine-induced caspase 3 activity was higher in FA fibroblasts than controls. Treatment of cells with BAPTA, AM rescued FA from oxidant-induced death. These data indirectly demonstrate that FA fibroblasts displayed an increased cytosolic Ca^{2+} content leading to increased sensitivity to oxidative stress (Wong & Cortopassi, 1997).

5.5 Calcium deregulation linked to mitochondrial DNA polymorphism

mtDNA is highly polymorphic and its variation in humans may contribute to individual differences in function as well as susceptibility to various diseases such as neurodegenerative diseases. Kazuno and collaborators searched for mtDNA polymorphisms that have mitochondrial functional significance using cybrid cells. Increased mitochondrial basal Ca^{2+} levels and increased agonist evoked cytosolic Ca^{2+} signals were observed in two closely linked nonsynonymous polymorphisms. Interestingly, these data highlight the role of mitochondrial polymorphisms in the pathology of neurodegenerative diseases by affecting Ca^{2+} dynamics (Kazuno et al., 2006).

5.6 Calcium deregulation in Pyruvate Dehydrogenase deficiency

Aerobic metabolism may also affect mitochondrial Ca^{2+} homeostasis. Thus, deregulation of Ca^{2+} handling was also reported in human fibroblasts from a patient with an inherited defect in pyruvate dehydrogenase (PDH). Indeed, these cells show a decrease ability to sequester cytosolic Ca^{2+} into mitochondria without affecting basal cytosolic and mitochondrial Ca^{2+} levels. It was postulated that reduced mitochondrial uptake is linked to decreased mitochondrial potential (Padua et al., 1998).

6. OXPHOS therapies: The place for Ca^{2+} modulating drugs

OXPHOS disorders are complex and heterogeneous group of multisystem diseases. The fact that they can result from mutations in hundreds of genes distributed across all of the chromosomes as well as the mtDNA, render the understanding of causative factors and the identification of common disease-related factors difficult. Accordingly effective therapeutic interventions are still not readily available. There are two main approaches to mitochondrial disease therapy: genetic and metabolic pharmacological (for recent review see (Roestenberg et al., 2011) and (Wallace et al., 2010)).

New approaches for genetic therapies for nDNA-encoded mitochondrial diseases as well as for mtDNA diseases are beginning to offer alternatives for individuals suffering from these devastating disorders. For mtDNA, these approaches include: (*a*) import of normal mtDNA polypeptides into the mitochondrion to complement the mtDNA defect, (*b*) reduction of the proportion of mutant mtDNAs (heteroplasmy shifting), and (*c*) direct medication of the mtDNA. Researchers are focusing also on the possible use of stem cell as a medication of OXPHOS disorders. However, these approaches are not as likely to relieve the devastating symptoms suffered by individuals with bioenergetic diseases.

The pharmacological approach includes the use of: (a) cofactors that increase the production of ATP (coQ, Idebenone, and succinate), (b) vitamins and metabolic supplements (thiamine, riboflavine, carnitine and L-arginine), (c) reactive oxygen species scavengers and mitochondrial antioxidants (CoQ/Idebenone, Vitamin E and Vitamin C), (d) modulators of PTP (cyclosporin A), and (e) regulators of mitochondrial biogenesis (bezafibrate and sirtuin analogs).

Current interventions based on metabolic correction include the use of mitochondrial-targeted drugs (compounds and peptides targeted to the mitochondrial matrix) such as mitoquinone "MitoQ", a derivative of coenzyme Q10, and SS-peptides, Szesto-Schiller peptides, a novel class of small cell permeable peptide antioxidants.

Another alternative to rescue mitochondrial bioenergetics defects is the use the mitochondrial Na^+/Ca^{2+} exchanger inhibitor benzothiazepine CGP37157 (Cox & Matlib, 1993). CGP37157 normalized aberrant mitochondrial Ca^{2+} handling during hormone stimulation of cybrid cells carrying the tRNALys mutation associated with MERRF syndrom (Brini et al., 1999). Short-term pre-treatment with CGP37157 (1 µM, 2 min) fully normalized the amplitude of the hormone-induced mitochondrial Ca^{2+} signal in fibroblasts from patients with isolated complex I deficiency (Visch et al., 2004), without altering this parameter in healthy fibroblasts. Similar result was obtained recently in a study including a large number of patient fibroblasts with complex I deficiency (Willems et al., 2009). Also the reduced maximal [ATP] in the mitochondrial matrix and cytosol were fully normalized by CGP37157 treatment. The effect of CGP37157 was independent of the presence of extracellular Ca^{2+}, excluding a stimulatory effect on Ca^{2+} entry across the plasma membrane (Willems et al., 2009).

It is worth to mention that CGP37157 may also stimulate the IP_3-induced release of Ca^{2+} from intracellular stores. In addition to these effects, CGP37157 was demonstrated to inhibit capacitative store refilling (Malli et al., 2005; Poburko et al., 2007). As far as its specificity is concerned, recent studies suggest that CGP37157 can also directly act on L-type Ca^{2+} channels (Thu le et al., 2006). Thus the use of this drug will hamper Ca^{2+}-stimulated processes that depend on Ca^{2+} entry across the plasma membrane (Luciani et al., 2007).

All over, these findings suggest that the mitochondrial Na^+/Ca^{2+} exchanger is a potential target for drugs aiming to restore or improve Ca^{2+}-stimulated mitochondrial ATP synthesis in OXPHOS deficiencies and highlight the role of Ca^{2+} deregulation in the development of mitochondrial and cellular pathology in OXPHOS diseases.

7. Conclusion

This literature analysis highlights the broad Ca^{2+} deregulation in different models of OXPHOS diseases and demonstrates the cross regulation between Ca^{2+} and bioenergetics in the development of mitochondrial and cellular pathologies. Some studies revealed also the potential use of Ca^{2+} modulating drugs to reveres mitochondrial pathology. These studies may encourage researcher to investigate systematically Ca^{2+} deregulation in OXPHOS and help to reveal new targets for the development of new or combined therapies to rescue mitochondrial pathology in these diseases.

8. Acknowledgements

Work presented in this review has been supported by INSERM, CNRS, (AFM (11456 and 13291) and FRM (DEQ20071210550). We gratefully acknowledge INSERM for supporting the MD-PhD curriculum (Ecole de l'INSERM) of B. Oulès and the Italian Institute of Technology, Genova, Italy for supporting the PhD curriculum of Dolores Del Prete.

9. Abbreviations

ANT, adenine nucleotide translocator; ATP, adenosine triphosphate; $[Ca^{2+}]_{cyt}$, cytosolic calcium-concentration; $[Ca^{2+}]_{er}$, endoplasmic reticulum calcium-concentration; $[Ca^{2+}]_{mt}$, mitochondrial calcium-concentration; Ca^{2+}, calcium; DNA, Deoxyribonucleic acid, ETC, electron transport chain; ER, endoplasmic reticulum; $\Delta\psi$, electrical potential; IMM, inner mitochondrial membrane; IP_3, inositol 1,4,5-triphosphate; IP_3R, inositol triphosphate receptor; MCU, mitochondrial Ca^{2+} uniporter; NCX/HCX, Na^+/Ca^{2+} exchanger and H^+/Ca^{2+} exchanger; OMM, outer mitochondrial membrane; OXPHOS, oxidative phosphorylation; PMCA, plasma membrane Ca^{2+}-ATPase; SERCA, sarco-Endoplasmic reticulum Ca^{2+}-ATPase; RYR, ryanodine receptor; SERCA, sarco-Endoplasmic reticulum Ca^{2+}-ATPase; SOC, store operated channel.

10. References

Adams, P. L., Lightowlers, R. N. &Turnbull, D. M. (1997). Molecular analysis of cytochrome c oxidase deficiency in Leigh's syndrome. *Ann Neurol*, 41, 2, pp. (268-70).

Allbritton, N. L., Meyer, T. &Stryer, L. (1992). Range of messenger action of calcium ion and inositol 1,4,5-trisphosphate. *Science*, 258, 5089, pp. (1812-5).

Bai, Y. &Attardi, G. (1998). The mtDNA-encoded ND6 subunit of mitochondrial NADH dehydrogenase is essential for the assembly of the membrane arm and the respiratory function of the enzyme. *EMBO J*, 17, 16, pp. (4848-58).

Bakowski, D. &Parekh, A. B. (2007). Regulation of store-operated calcium channels by the intermediary metabolite pyruvic acid. *Curr Biol*, 17, 12, pp. (1076-81).

Ballinger, S. W., Shoffner, J. M., Hedaya, E. V., Trounce, I., Polak, M. A., Koontz, D. A. &Wallace, D. C. (1992). Maternally transmitted diabetes and deafness associated with a 10.4 kb mitochondrial DNA deletion. *Nat Genet*, 1, 1, pp. (11-5).

Ballinger, S. W., Shoffner, J. M., Gebhart, S., Koontz, D. A. &Wallace, D. C. (1994). Mitochondrial diabetes revisited. *Nat Genet*, 7, 4, pp. (458-9).

Barsukova, A., Komarov, A., Hajnoczky, G., Bernardi, P., Bourdette, D. &Forte, M. (2011). Activation of the mitochondrial permeability transition pore modulates Ca2+ responses to physiological stimuli in adult neurons. *Eur J Neurosci*, 33, 5, pp. (831-42).

Baughman, J. M., Perocchi, F., Girgis, H. S., Plovanich, M., Belcher-Timme, C. A., Sancak, Y., Bao, X. R., Strittmatter, L., Goldberger, O., Bogorad, R. L., Koteliansky, V. &Mootha, V. K. (2011). Integrative genomics identifies MCU as an essential component of the mitochondrial calcium uniporter. *Nature*, 476, 7360, pp. (341-5).

Baysal, B. E., Ferrell, R. E., Willett-Brozick, J. E., Lawrence, E. C., Myssiorek, D., Bosch, A., van der Mey, A., Taschner, P. E., Rubinstein, W. S., Myers, E. N., Richard, C. W., 3rd, Cornelisse, C. J., Devilee, P. &Devlin, B. (2000). Mutations in SDHD, a mitochondrial complex II gene, in hereditary paraganglioma. *Science*, 287, 5454, pp. (848-51).

Bernardi, P. (1999). Mitochondrial transport of cations: channels, exchangers, and permeability transition. *Physiol Rev*, 79, 4, pp. (1127-55).

Bernardi, P., Scorrano, L., Colonna, R., Petronilli, V. &Di Lisa, F. (1999). Mitochondria and cell death. Mechanistic aspects and methodological issues. *Eur J Biochem*, 264, 3, pp. (687-701).

Bernardi, P., Krauskopf, A., Basso, E., Petronilli, V., Blachly-Dyson, E., Di Lisa, F. &Forte, M. A. (2006). The mitochondrial permeability transition from in vitro artifact to disease target. *FEBS J*, 273, 10, pp. (2077-99).

Berridge, M. J., Lipp, P. &Bootman, M. D. (2000). The versatility and universality of calcium signalling. *Nat Rev Mol Cell Biol*, 1, 1, pp. (11-21).

Bianchi, K., Rimessi, A., Prandini, A., Szabadkai, G. &Rizzuto, R. (2004). Calcium and mitochondria: mechanisms and functions of a troubled relationship. *Biochim Biophys Acta*, 1742, 1-3, pp. (119-31).

Blanc, H. &Dujon, B. (1980). Replicator regions of the yeast mitochondrial DNA responsible for suppressiveness. *Proc Natl Acad Sci U S A*, 77, 7, pp. (3942-6).

Bourgeron, T., Rustin, P., Chretien, D., Birch-Machin, M., Bourgeois, M., Viegas-Pequignot, E., Munnich, A. &Rotig, A. (1995). Mutation of a nuclear succinate dehydrogenase gene results in mitochondrial respiratory chain deficiency. *Nat Genet*, 11, 2, pp. (144-9).

Brini, M., Pinton, P., King, M. P., Davidson, M., Schon, E. A. &Rizzuto, R. (1999). A calcium signaling defect in the pathogenesis of a mitochondrial DNA inherited oxidative phosphorylation deficiency. *Nat Med*, 5, 8, pp. (951-4).

Brookes, P. S., Parker, N., Buckingham, J. A., Vidal-Puig, A., Halestrap, A. P., Gunter, T. E., Nicholls, D. G., Bernardi, P., Lemasters, J. J. &Brand, M. D. (2008). UCPs--unlikely calcium porters. *Nat Cell Biol*, 10, 11, pp. (1235-7).

Bruno, C., Martinuzzi, A., Tang, Y., Andreu, A. L., Pallotti, F., Bonilla, E., Shanske, S., Fu, J., Sue, C. M., Angelini, C., DiMauro, S. &Manfredi, G. (1999). A stop-codon mutation in the human mtDNA cytochrome c oxidase I gene disrupts the functional structure of complex IV. *Am J Hum Genet*, 65, 3, pp. (611-20).

Brustovetsky, N., Brustovetsky, T., Purl, K. J., Capano, M., Crompton, M. &Dubinsky, J. M. (2003). Increased susceptibility of striatal mitochondria to calcium-induced permeability transition. *J Neurosci*, 23, 12, pp. (4858-67).

Bulos, B. A., Thomas, B. J. &Sacktor, B. (1984). Calcium inhibition of the NAD+-linked isocitrate dehydrogenase from blowfly flight muscle mitochondria. *J Biol Chem*, 259, 16, pp. (10232-7).

Calabresi, P., Gubellini, P., Picconi, B., Centonze, D., Pisani, A., Bonsi, P., Greengard, P., Hipskind, R. A., Borrelli, E. &Bernardi, G. (2001). Inhibition of mitochondrial complex II induces a long-term potentiation of NMDA-mediated synaptic excitation in the striatum requiring endogenous dopamine. *J Neurosci*, 21, 14, pp. (5110-20).

Casari, G., De Fusco, M., Ciarmatori, S., Zeviani, M., Mora, M., Fernandez, P., De Michele, G., Filla, A., Cocozza, S., Marconi, R., Durr, A., Fontaine, B. &Ballabio, A. (1998). Spastic paraplegia and OXPHOS impairment caused by mutations in paraplegin, a nuclear-encoded mitochondrial metalloprotease. *Cell*, 93, 6, pp. (973-83).

Chami, M., Oules, B., Szabadkai, G., Tacine, R., Rizzuto, R. &Paterlini-Brechot, P. (2008). Role of SERCA1 truncated isoform in the proapoptotic calcium transfer from ER to mitochondria during ER stress. *Mol Cell*, 32, 5, pp. (641-51).

Chan, D. C. (2006). Mitochondria: dynamic organelles in disease, aging, and development. *Cell*, 125, 7, pp. (1241-52)

Chinnery, P. F., Brown, D. T., Andrews, R. M., Singh-Kler, R., Riordan-Eva, P., Lindley, J., Applegarth, D. A., Turnbull, D. M. &Howell, N. (2001). The mitochondrial ND6 gene is a hot spot for mutations that cause Leber's hereditary optic neuropathy. *Brain*, 124, Pt 1, pp. (209-18).

Chinnery, P. F. &Turnbull, D. M. (2001). Epidemiology and treatment of mitochondrial disorders. *Am J Med Genet*, 106, 1, pp. (94-101)

Clayton, D. A. &Vinograd, J. (1967). Circular dimer and catenate forms of mitochondrial DNA in human leukaemic leucocytes. *Nature*, 216, 5116, pp. (652-7).

Colombini, M. (2009). The published 3D structure of the VDAC channel: native or not? *Trends Biochem Sci*, 34, 8, pp. (382-9).

Comi, G. P., Bordoni, A., Salani, S., Franceschina, L., Sciacco, M., Prelle, A., Fortunato, F., Zeviani, M., Napoli, L., Bresolin, N., Moggio, M., Ausenda, C. D., Taanman, J. W. &Scarlato, G. (1998). Cytochrome c oxidase subunit I microdeletion in a patient with motor neuron disease. *Ann Neurol*, 43, 1, pp. (110-6).

Cox, D. A., Conforti, L., Sperelakis, N. &Matlib, M. A. (1993). Selectivity of inhibition of Na(+)-Ca2+ exchange of heart mitochondria by benzothiazepine CGP-37157. *J Cardiovasc Pharmacol*, 21, 4, pp. (595-9).

Cox, D. A. &Matlib, M. A. (1993). Modulation of intramitochondrial free Ca2+ concentration by antagonists of Na(+)-Ca2+ exchange. *Trends Pharmacol Sci*, 14, 11, pp. (408-13).

Csordas, G., Thomas, A. P. &Hajnoczky, G. (1999). Quasi-synaptic calcium signal transmission between endoplasmic reticulum and mitochondria. *EMBO J*, 18, 1, pp. (96-108).

Csordas, G., Renken, C., Varnai, P., Walter, L., Weaver, D., Buttle, K. F., Balla, T., Mannella, C. A. &Hajnoczky, G. (2006). Structural and functional features and significance of the physical linkage between ER and mitochondria. *J Cell Biol*, 174, 7, pp. (915-21).

Da Cruz, S., De Marchi, U., Frieden, M., Parone, P. A., Martinou, J. C. &Demaurex, N. (2010). SLP-2 negatively modulates mitochondrial sodium-calcium exchange. *Cell Calcium*, 47, 1, pp. (11-8).

Dash, R. K. &Beard, D. A. (2008). Analysis of cardiac mitochondrial Na+-Ca2+ exchanger kinetics with a biophysical model of mitochondrial Ca2+ handling suggests a 3:1 stoichiometry. *J Physiol*, 586, 13, pp. (3267-85).

De Marchi, U., Castelbou, C. &Demaurex, N. (2011). Uncoupling protein 3 (UCP3) modulates the activity of sarco/endoplasmic reticulum Ca2+ ATPase (SERCA) by decreasing mitochondrial ATP production. *J Biol Chem*, pp.).

De Stefani, D., Raffaello, A., Teardo, E., Szabo, I. &Rizzuto, R. (2011). A forty-kilodalton protein of the inner membrane is the mitochondrial calcium uniporter. *Nature*, 476, 7360, pp. (336-40).

Delettre, C., Lenaers, G., Griffoin, J. M., Gigarel, N., Lorenzo, C., Belenguer, P., Pelloquin, L., Grosgeorge, J., Turc-Carel, C., Perret, E., Astarie-Dequeker, C., Lasquellec, L., Arnaud, B., Ducommun, B., Kaplan, J. &Hamel, C. P. (2000). Nuclear gene OPA1, encoding a mitochondrial dynamin-related protein, is mutated in dominant optic atrophy. *Nat Genet*, 26, 2, pp. (207-10).

Dell'agnello, C., Leo, S., Agostino, A., Szabadkai, G., Tiveron, C., Zulian, A., Prelle, A., Roubertoux, P., Rizzuto, R. &Zeviani, M. (2007). Increased longevity and refractoriness to Ca(2+)-dependent neurodegeneration in Surf1 knockout mice. *Hum Mol Genet*, 16, 4, pp. (431-44).

Denton, R. M., Randle, P. J. &Martin, B. R. (1972). Stimulation by calcium ions of pyruvate dehydrogenase phosphate phosphatase. *Biochem J*, 128, 1, pp. (161-3).

Denton, R. M. &Hughes, W. A. (1978). Pyruvate dehydrogenase and the hormonal regulation of fat synthesis in mammalian tissues. *Int J Biochem*, 9, 8, pp. (545-52).

Denton, R. M., Richards, D. A. &Chin, J. G. (1978). Calcium ions and the regulation of NAD+-linked isocitrate dehydrogenase from the mitochondria of rat heart and other tissues. *Biochem J*, 176, 3, pp. (899-906).

Denton, R. M., McCormack, J. G., Rutter, G. A., Burnett, P., Edgell, N. J., Moule, S. K. &Diggle, T. A. (1996). The hormonal regulation of pyruvate dehydrogenase complex. *Adv Enzyme Regul*, 36, pp. (183-98).

Denton, R. M. (2009). Regulation of mitochondrial dehydrogenases by calcium ions. *Biochim Biophys Acta*, 1787, 11, pp. (1309-16).

Desjardins, P., Frost, E. &Morais, R. (1985). Ethidium bromide-induced loss of mitochondrial DNA from primary chicken embryo fibroblasts. *Mol Cell Biol*, 5, 5, pp. (1163-9).

Di Donato, S. (2000). Disorders related to mitochondrial membranes: pathology of the respiratory chain and neurodegeneration. *J Inherit Metab Dis*, 23, 3, pp. (247-63).

Di Liegro, C. M., Bellafiore, M., Izquierdo, J. M., Rantanen, A. &Cuezva, J. M. (2000). 3'-untranslated regions of oxidative phosphorylation mRNAs function in vivo as enhancers of translation. *Biochem J*, 352 Pt 1, pp. (109-15).

Di Lisa, F. &Bernardi, P. (2009). A CaPful of mechanisms regulating the mitochondrial permeability transition. *J Mol Cell Cardiol*, 46, 6, pp. (775-80).

DiMauro, S. & Moraes, C. T. (1993). Mitochondrial encephalomyopathies. *Arch Neurol*, 50, 11, pp. (1197-208).

DiMauro, S. & Schon, E. A. (2003). Mitochondrial respiratory-chain diseases. *N Engl J Med*, 348, 26, pp. (2656-68).

Dimmer, K. S., Navoni, F., Casarin, A., Trevisson, E., Endele, S., Winterpacht, A., Salviati, L. &Scorrano, L. (2008). LETM1, deleted in Wolf-Hirschhorn syndrome is required for normal mitochondrial morphology and cellular viability. *Hum Mol Genet*, 17, 2, pp. (201-14).

Duchen, M. R. (1992). Ca(2+)-dependent changes in the mitochondrial energetics in single dissociated mouse sensory neurons. *Biochem J*, 283 (Pt 1), pp. (41-50).

Finsterer, J. (2006). Central nervous system manifestations of mitochondrial disorders. *Acta Neurol Scand*, 114, 4, pp. (217-38).

Finsterer, J. (2006). Overview on visceral manifestations of mitochondrial disorders. *Neth J Med*, 64, 3, pp. (61-71).

Frank, S., Gaume, B., Bergmann-Leitner, E. S., Leitner, W. W., Robert, E. G., Catez, F., Smith, C. L. &Youle, R. J. (2001). The role of dynamin-related protein 1, a mediator of mitochondrial fission, in apoptosis. *Dev Cell*, 1, 4, pp. (515-25).

Gellerich, F. N., Gizatullina, Z., Trumbeckaite, S., Nguyen, H. P., Pallas, T., Arandarcikaite, O., Vielhaber, S., Seppet, E. &Striggow, F. (2010). The regulation of OXPHOS by extramitochondrial calcium. *Biochim Biophys Acta*, 1797, 6-7, pp. (1018-27).

Goto, Y., Nonaka, I. &Horai, S. (1990). A mutation in the tRNA(Leu)(UUR) gene associated with the MELAS subgroup of mitochondrial encephalomyopathies. *Nature*, 348, 6302, pp. (651-3).

Greene, J. G., Sheu, S. S., Gross, R. A. &Greenamyre, J. T. (1998). 3-Nitropropionic acid exacerbates N-methyl-D-aspartate toxicity in striatal culture by multiple mechanisms. *Neuroscience*, 84, 2, pp. (503-10).

Gugneja, S., Virbasius, C. M. &Scarpulla, R. C. (1996). Nuclear respiratory factors 1 and 2 utilize similar glutamine-containing clusters of hydrophobic residues to activate transcription. *Mol Cell Biol*, 16, 10, pp. (5708-16).

Hajnoczky, G., Robb-Gaspers, L. D., Seitz, M. B. &Thomas, A. P. (1995). Decoding of cytosolic calcium oscillations in the mitochondria. *Cell*, 82, 3, pp. (415-24).

Hajnoczky, G., Csordas, G., Das, S., Garcia-Perez, C., Saotome, M., Sinha Roy, S. &Yi, M. (2006). Mitochondrial calcium signalling and cell death: approaches for assessing the role of mitochondrial Ca2+ uptake in apoptosis. *Cell Calcium*, 40, 5-6, pp. (553-60).

Handran, S. D., Werth, J. L., DeVivo, D. C. &Rothman, S. M. (1997). Mitochondrial morphology and intracellular calcium homeostasis in cytochrome oxidase-deficient human fibroblasts. *Neurobiol Dis*, 3, 4, pp. (287-98).

Hatefi, Y. (1985). The mitochondrial electron transport and oxidative phosphorylation system. *Annu Rev Biochem*, 54, pp. (1015-69).

Hirano, M., Silvestri, G., Blake, D. M., Lombes, A., Minetti, C., Bonilla, E., Hays, A. P., Lovelace, R. E., Butler, I., Bertorini, T. E. & et al. (1994). Mitochondrial neurogastrointestinal encephalomyopathy (MNGIE): clinical, biochemical, and genetic features of an autosomal recessive mitochondrial disorder. *Neurology*, 44, 4, pp. (721-7).

Holt, I. J., Harding, A. E., Petty, R. K. &Morgan-Hughes, J. A. (1990). A new mitochondrial disease associated with mitochondrial DNA heteroplasmy. *Am J Hum Genet*, 46, 3, pp. (428-33).

Howell, N. &Gilbert, K. (1988). Mutational analysis of the mouse mitochondrial cytochrome b gene. *J Mol Biol*, 203, 3, pp. (607-18).

Huo, L. &Scarpulla, R. C. (1999). Multiple 5'-untranslated exons in the nuclear respiratory factor 1 gene span 47 kb and contribute to transcript heterogeneity and translational efficiency. *Gene*, 233, 1-2, pp. (213-24).

Inoue, K., Nakada, K., Ogura, A., Isobe, K., Goto, Y., Nonaka, I. &Hayashi, J. I. (2000). Generation of mice with mitochondrial dysfunction by introducing mouse mtDNA carrying a deletion into zygotes. *Nat Genet*, 26, 2, pp. (176-81).

Iuso, A., Scacco, S., Piccoli, C., Bellomo, F., Petruzzella, V., Trentadue, R., Minuto, M., Ripoli, M., Capitanio, N., Zeviani, M. &Papa, S. (2006). Dysfunctions of cellular oxidative metabolism in patients with mutations in the NDUFS1 and NDUFS4 genes of complex I. *J Biol Chem*, 281, 15, pp. (10374-80).

Jiang, D., Zhao, L. &Clapham, D. E. (2009). Genome-wide RNAi screen identifies Letm1 as a mitochondrial Ca2+/H+ antiporter. *Science*, 326, 5949, pp. (144-7).

Kaplin, A. I., Snyder, S. H. &Linden, D. J. (1996). Reduced nicotinamide adenine dinucleotide-selective stimulation of inositol 1,4,5-trisphosphate receptors mediates hypoxic mobilization of calcium. *J Neurosci*, 16, 6, pp. (2002-11).

Karbowski, M. &Youle, R. J. (2003). Dynamics of mitochondrial morphology in healthy cells and during apoptosis. *Cell Death Differ*, 10, 8, pp. (870-80).

Kaukonen, J., Juselius, J. K., Tiranti, V., Kyttala, A., Zeviani, M., Comi, G. P., Keranen, S., Peltonen, L. &Suomalainen, A. (2000). Role of adenine nucleotide translocator 1 in mtDNA maintenance. *Science*, 289, 5480, pp. (782-5).

Kazuno, A. A., Munakata, K., Nagai, T., Shimozono, S., Tanaka, M., Yoneda, M., Kato, N., Miyawaki, A. &Kato, T. (2006). Identification of mitochondrial DNA polymorphisms that alter mitochondrial matrix pH and intracellular calcium dynamics. *PLoS Genet*, 2, 8, pp. (e128).

Kearsey, S. E. &Craig, I. W. (1981). Altered ribosomal RNA genes in mitochondria from mammalian cells with chloramphenicol resistance. *Nature*, 290, 5807, pp. (607-8).

King, M. P. &Attardi, G. (1989). Human cells lacking mtDNA: repopulation with exogenous mitochondria by complementation. *Science*, 246, 4929, pp. (500-3).

Kirichok, Y., Krapivinsky, G. &Clapham, D. E. (2004). The mitochondrial calcium uniporter is a highly selective ion channel. *Nature*, 427, 6972, pp. (360-4).

Koehler, C. M., Leuenberger, D., Merchant, S., Renold, A., Junne, T. &Schatz, G. (1999). Human deafness dystonia syndrome is a mitochondrial disease. *Proc Natl Acad Sci U S A*, 96, 5, pp. (2141-6).

Koopman, W. J., Hink, M. A., Verkaart, S., Visch, H. J., Smeitink, J. A. &Willems, P. H. (2007). Partial complex I inhibition decreases mitochondrial motility and increases matrix protein diffusion as revealed by fluorescence correlation spectroscopy. *Biochim Biophys Acta*, 1767, 7, pp. (940-7).

Larsson, N. G., Wang, J., Wilhelmsson, H., Oldfors, A., Rustin, P., Lewandoski, M., Barsh, G. S. &Clayton, D. A. (1998). Mitochondrial transcription factor A is necessary for mtDNA maintenance and embryogenesis in mice. *Nat Genet*, 18, 3, pp. (231-6).

Larsson, N. G. &Rustin, P. (2001). Animal models for respiratory chain disease. *Trends Mol Med*, 7, 12, pp. (578-81).

Lee, W. T., Itoh, T. &Pleasure, D. (2002). Acute and chronic alterations in calcium homeostasis in 3-nitropropionic acid-treated human NT2-N neurons. *Neuroscience*, 113, 3, pp. (699-708).

Lehman, J. J., Barger, P. M., Kovacs, A., Saffitz, J. E., Medeiros, D. M. &Kelly, D. P. (2000). Peroxisome proliferator-activated receptor gamma coactivator-1 promotes cardiac mitochondrial biogenesis. *J Clin Invest*, 106, 7, pp. (847-56).

Leventhal, L., Sortwell, C. E., Hanbury, R., Collier, T. J., Kordower, J. H. &Palfi, S. (2000). Cyclosporin A protects striatal neurons in vitro and in vivo from 3-nitropropionic acid toxicity. *J Comp Neurol*, 425, 4, pp. (471-8).

Li, Y., Huang, T. T., Carlson, E. J., Melov, S., Ursell, P. C., Olson, J. L., Noble, L. J., Yoshimura, M. P., Berger, C., Chan, P. H., Wallace, D. C. &Epstein, C. J. (1995). Dilated cardiomyopathy and neonatal lethality in mutant mice lacking manganese superoxide dismutase. *Nat Genet*, 11, 4, pp. (376-81).

Lim, D., Fedrizzi, L., Tartari, M., Zuccato, C., Cattaneo, E., Brini, M. &Carafoli, E. (2008). Calcium homeostasis and mitochondrial dysfunction in striatal neurons of Huntington disease. *J Biol Chem*, 283, 9, pp. (5780-9).

Loeffen, J. L., Triepels, R. H., van den Heuvel, L. P., Schuelke, M., Buskens, C. A., Smeets, R. J., Trijbels, J. M. &Smeitink, J. A. (1998). cDNA of eight nuclear encoded subunits of NADH:ubiquinone oxidoreductase: human complex I cDNA characterization completed. *Biochem Biophys Res Commun*, 253, 2, pp. (415-22)

Loeffen, J. L., Smeitink, J. A., Trijbels, J. M., Janssen, A. J., Triepels, R. H., Sengers, R. C. &van den Heuvel, L. P. (2000). Isolated complex I deficiency in children: clinical, biochemical and genetic aspects. *Hum Mutat*, 15, 2, pp. (123-34).

Luciani, D. S., Ao, P., Hu, X., Warnock, G. L. &Johnson, J. D. (2007). Voltage-gated Ca(2+) influx and insulin secretion in human and mouse beta-cells are impaired by the mitochondrial Na(+)/Ca(2+) exchange inhibitor CGP-37157. *Eur J Pharmacol*, 576, 1-3, pp. (18-25).

Luft, R., Ikkos, D., Palmieri, G., Ernster, L. &Afzelius, B. (1962). A case of severe hypermetabolism of nonthyroid origin with a defect in the maintenance of mitochondrial respiratory control: a correlated clinical, biochemical, and morphological study. *J Clin Invest*, 41, pp. (1776-804).

M'Baya, E., Oulès, B., Caspersen, C., Tacine, R., Massinet, H., Chrétien, D., Munnich, A., Rötig, A., Rizzuto, R., Rutter, G.A., Paterlini-Bréchot, P. &Chami, M. (2010). Calcium signalling-dependent mitochondrial dysfunction and bioenergetics regulation in respiratory chain Complex II deficiency *Cell Death Differ*, 17 12, pp. (1855-66).

Maciel, E. N., Kowaltowski, A. J., Schwalm, F. D., Rodrigues, J. M., Souza, D. O., Vercesi, A. E., Wajner, M. &Castilho, R. F. (2004). Mitochondrial permeability transition in neuronal damage promoted by Ca2+ and respiratory chain complex II inhibition. *J Neurochem*, 90, 5, pp. (1025-35).

Makowska, A., Zablocki, K. &Duszynski, J. (2000). The role of mitochondria in the regulation of calcium influx into Jurkat cells. *Eur J Biochem*, 267, 3, pp. (877-84).

Malli, R., Frieden, M., Trenker, M. &Graier, W. F. (2005). The role of mitochondria for Ca2+ refilling of the endoplasmic reticulum. *J Biol Chem*, 280, 13, pp. (12114-22).

Mannella, C. A., Buttle, K., Rath, B. K. &Marko, M. (1998). Electron microscopic tomography of rat-liver mitochondria and their interaction with the endoplasmic reticulum. *Biofactors*, 8, 3-4, pp. (225-8).

McCormack, J. G. &Denton, R. M. (1979). The effects of calcium ions and adenine nucleotides on the activity of pig heart 2-oxoglutarate dehydrogenase complex. *Biochem J*, 180, 3, pp. (533-44).

McCormack, J. G., Halestrap, A. P. &Denton, R. M. (1990). Role of calcium ions in regulation of mammalian intramitochondrial metabolism. *Physiol Rev*, 70, 2, pp. (391-425)

Meier, P. J., Spycher, M. A. &Meyer, U. A. (1981). Isolation and characterization of rough endoplasmic reticulum associated with mitochondria from normal rat liver. *Biochim Biophys Acta*, 646, 2, pp. (283-97).

Moraes, C. T., DiMauro, S., Zeviani, M., Lombes, A., Shanske, S., Miranda, A. F., Nakase, H., Bonilla, E., Werneck, L. C., Servidei, S. &et al. (1989). Mitochondrial DNA deletions in progressive external ophthalmoplegia and Kearns-Sayre syndrome. *N Engl J Med*, 320, 20, pp. (1293-9).

Moreau, B. &Parekh, A. B. (2008). Ca2+ -dependent inactivation of the mitochondrial Ca2+ uniporter involves proton flux through the ATP synthase. *Curr Biol*, 18, 11, pp. (855-9).

Moudy, A. M., Handran, S. D., Goldberg, M. P., Ruffin, N., Karl, I., Kranz-Eble, P., DeVivo, D. C. &Rothman, S. M. (1995). Abnormal calcium homeostasis and mitochondrial polarization in a human encephalomyopathy. *Proc Natl Acad Sci U S A*, 92, 3, pp. (729-33).

Munnich, A. &Rustin, P. (2001). Clinical spectrum and diagnosis of mitochondrial disorders. *Am J Med Genet*, 106, 1, pp. (4-17).

Murdock, D. G., Boone, B. E., Esposito, L. A. &Wallace, D. C. (1999). Up-regulation of nuclear and mitochondrial genes in the skeletal muscle of mice lacking the heart/muscle isoform of the adenine nucleotide translocator. *J Biol Chem*, 274, 20, pp. (14429-33).

Nasr, P., Gursahani, H. I., Pang, Z., Bondada, V., Lee, J., Hadley, R. W. &Geddes, J. W. (2003). Influence of cytosolic and mitochondrial Ca2+, ATP, mitochondrial membrane potential, and calpain activity on the mechanism of neuron death induced by 3-nitropropionic acid. *Neurochem Int*, 43, 2, pp. (89-99).

Niemann, S. &Muller, U. (2000). Mutations in SDHC cause autosomal dominant paraganglioma, type 3. *Nat Genet*, 26, 3, pp. (268-70).

Nishino, I., Spinazzola, A. &Hirano, M. (1999). Thymidine phosphorylase gene mutations in MNGIE, a human mitochondrial disorder. *Science*, 283, 5402, pp. (689-92).

Nowikovsky, K., Froschauer, E. M., Zsurka, G., Samaj, J., Reipert, S., Kolisek, M., Wiesenberger, G. &Schweyen, R. J. (2004). The LETM1/YOL027 gene family encodes a factor of the mitochondrial K+ homeostasis with a potential role in the Wolf-Hirschhorn syndrome. *J Biol Chem*, 279, 29, pp. (30307-15).

Padua, R. A., Baron, K. T., Thyagarajan, B., Campbell, C. &Thayer, S. A. (1998). Reduced Ca2+ uptake by mitochondria in pyruvate dehydrogenase-deficient human diploid fibroblasts. *Am J Physiol*, 274, 3 Pt 1, pp. (C615-22).

Palty, R., Silverman, W. F., Hershfinkel, M., Caporale, T., Sensi, S. L., Parnis, J., Nolte, C., Fishman, D., Shoshan-Barmatz, V., Herrmann, S., Khananshvili, D. &Sekler, I. (2010). NCLX is an essential component of mitochondrial Na+/Ca2+ exchange. *Proc Natl Acad Sci U S A*, 107, 1, pp. (436-41).

Panov, A. V., Gutekunst, C. A., Leavitt, B. R., Hayden, M. R., Burke, J. R., Strittmatter, W. J. &Greenamyre, J. T. (2002). Early mitochondrial calcium defects in Huntington's disease are a direct effect of polyglutamines. *Nat Neurosci*, 5, 8, pp. (731-6).

Parfait, B., Chretien, D., Rotig, A., Marsac, C., Munnich, A. &Rustin, P. (2000). Compound heterozygous mutations in the flavoprotein gene of the respiratory chain complex II in a patient with Leigh syndrome. *Hum Genet*, 106, 2, pp. (236-43

Perocchi, F., Gohil, V. M., Girgis, H. S., Bao, X. R., McCombs, J. E., Palmer, A. E. &Mootha, V. K. (2010). MICU1 encodes a mitochondrial EF hand protein required for Ca(2+) uptake. *Nature*, 467, 7313, pp. (291-6).

Poburko, D., Liao, C. H., Lemos, V. S., Lin, E., Maruyama, Y., Cole, W. C. &van Breemen, C. (2007). Transient receptor potential channel 6-mediated, localized cytosolic [Na+] transients drive Na+/Ca2+ exchanger-mediated Ca2+ entry in purinergically stimulated aorta smooth muscle cells. *Circ Res*, 101, 10, pp. (1030-8)

Poburko, D., Santo-Domingo, J. &Demaurex, N. (2011). Dynamic regulation of the mitochondrial proton gradient during cytosolic calcium elevations. *J Biol Chem*, 286, 13, pp. (11672-84).

Puccio, H., Simon, D., Cossee, M., Criqui-Filipe, P., Tiziano, F., Melki, J., Hindelang, C., Matyas, R., Rustin, P. &Koenig, M. (2001). Mouse models for Friedreich ataxia exhibit cardiomyopathy, sensory nerve defect and Fe-S enzyme deficiency followed by intramitochondrial iron deposits. *Nat Genet*, 27, 2, pp. (181-6).

Rasola, A. &Bernardi, P. (2007). The mitochondrial permeability transition pore and its involvement in cell death and in disease pathogenesis. *Apoptosis*, 12, 5, pp. (815-33).

Reinecke, F., Smeitink, J. A. &van der Westhuizen, F. H. (2009). OXPHOS gene expression and control in mitochondrial disorders. *Biochim Biophys Acta*, 1792, 12, pp. (1113-21).

Ristow, M., Pfister, M. F., Yee, A. J., Schubert, M., Michael, L., Zhang, C. Y., Ueki, K., Michael, M. D., 2nd, Lowell, B. B. &Kahn, C. R. (2000). Frataxin activates mitochondrial energy conversion and oxidative phosphorylation. *Proc Natl Acad Sci U S A*, 97, 22, pp. (12239-43).

Rizzuto, R., Brini, M., Murgia, M. &Pozzan, T. (1993). Microdomains with high Ca2+ close to IP3-sensitive channels that are sensed by neighboring mitochondria. *Science*, 262, 5134, pp. (744-7).

Rizzuto, R., Pinton, P., Carrington, W., Fay, F. S., Fogarty, K. E., Lifshitz, L. M., Tuft, R. A. &Pozzan, T. (1998). Close contacts with the endoplasmic reticulum as determinants of mitochondrial Ca2+ responses. *Science*, 280, 5370, pp. (1763-6).

Rizzuto, R., Pinton, P., Ferrari, D., Chami, M., Szabadkai, G., Magalhaes, P. J., Di Virgilio, F. &Pozzan, T. (2003). Calcium and apoptosis: facts and hypotheses. *Oncogene*, 22, 53, pp. (8619-27).

Robb-Gaspers, L. D., Burnett, P., Rutter, G. A., Denton, R. M., Rizzuto, R. &Thomas, A. P. (1998). Integrating cytosolic calcium signals into mitochondrial metabolic responses. *EMBO J, 17, 17, pp.* (4987-5000).

Robinson, B. H. (1996). Use of fibroblast and lymphoblast cultures for detection of respiratory chain defects. *Methods Enzymol, 264, pp.* (454-64).

Roestenberg, P., Manjeri, G. R., Valsecchi, F., Smeitink, J. A., Willems, P. H. &Koopman, W. J. (2011). Pharmacological targeting of mitochondrial complex I deficiency: The cellular level and beyond. *Mitochondrion, Jul 2.* [Epub ahead of print].

Rosenstock, T. R., Carvalho, A. C., Jurkiewicz, A., Frussa-Filho, R. &Smaili, S. S. (2004). Mitochondrial calcium, oxidative stress and apoptosis in a neurodegenerative disease model induced by 3-nitropropionic acid. *J Neurochem, 88, 5, pp.* (1220-8).

Rotig, A., de Lonlay, P., Chretien, D., Foury, F., Koenig, M., Sidi, D., Munnich, A. &Rustin, P. (1997). Aconitase and mitochondrial iron-sulphur protein deficiency in Friedreich ataxia. *Nat Genet, 17, 2, pp.* (215-7).

Ruan, Q., Lesort, M., MacDonald, M. E. &Johnson, G. V. (2004). Striatal cells from mutant huntingtin knock-in mice are selectively vulnerable to mitochondrial complex II inhibitor-induced cell death through a non-apoptotic pathway. *Hum Mol Genet, 13, 7, pp.* (669-81).

Santo-Domingo, J. &Demaurex, N. (2010). Calcium uptake mechanisms of mitochondria. *Biochim Biophys Acta, 1797, 6-7, pp.* (907-12)

Saraste, M. (1999). Oxidative phosphorylation at the fin de siecle. *Science, 283, 5407, pp.* (1488-93).

Schaefer, A. M., Taylor, R. W., Turnbull, D. M. &Chinnery, P. F. (2004). The epidemiology of mitochondrial disorders--past, present and future. *Biochim Biophys Acta, 1659, 2-3, pp.* (115-20).

Shore, G. C. &Tata, J. R. (1977). Two fractions of rough endoplasmic reticulum from rat liver. I. Recovery of rapidly sedimenting endoplasmic reticulum in association with mitochondria. *J Cell Biol, 72, 3, pp.* (714-25).

Shoshan-Barmatz, V., Keinan, N., Abu-Hamad, S., Tyomkin, D. &Aram, L. (2010). Apoptosis is regulated by the VDAC1 N-terminal region and by VDAC oligomerization: release of cytochrome c, AIF and Smac/Diablo. *Biochim Biophys Acta, 1797, 6-7, pp.* (1281-91).

Shoubridge, E. A. (2001). Nuclear genetic defects of oxidative phosphorylation. *Hum Mol Genet, 10, 20, pp.* (2277-84).

Simamura, E., Shimada, H., Hatta, T. &Hirai, K. (2008). Mitochondrial voltage-dependent anion channels (VDACs) as novel pharmacological targets for anti-cancer agents. *J Bioenerg Biomembr, 40, 3, pp.* (213-7).

Slonimski, P. P., Perrodin, G. &Croft, J. H. (1968). Ethidium bromide induced mutation of yeast mitochondria: complete transformation of cells into respiratory deficient non-chromosomal "petites". *Biochem Biophys Res Commun, 30, 3, pp.* (232-9).

Smeitink, J., van den Heuvel, L. &DiMauro, S. (2001). The genetics and pathology of oxidative phosphorylation. *Nat Rev Genet, 2, 5, pp.* (342-52).

Sparagna, G. C., Gunter, K. K., Sheu, S. S. &Gunter, T. E. (1995). Mitochondrial calcium uptake from physiological-type pulses of calcium. A description of the rapid uptake mode. *J Biol Chem, 270, 46, pp.* (27510-5).

Spolsky, C. M. &Eisenstadt, J. M. (1972). Chloramphenicol-resistant mutants of human HeLa cells. *FEBS Lett,* 25, 2, pp. (319-324).

Szabadkai, G., Pitter, J. G. &Spat, A. (2001). Cytoplasmic Ca2+ at low submicromolar concentration stimulates mitochondrial metabolism in rat luteal cells. *Pflugers Arch,* 441, 5, pp. (678-85).

Szabadkai, G., Simoni, A. M., Chami, M., Wieckowski, M. R., Youle, R. J. &Rizzuto, R. (2004). Drp-1-dependent division of the mitochondrial network blocks intraorganellar Ca2+ waves and protects against Ca2+-mediated apoptosis. *Mol Cell,* 16, 1, pp. (59-68).

Szczepanowska, J., Zablocki, K. &Duszynski, J. (2004). Influence of a mitochondrial genetic defect on capacitative calcium entry and mitochondrial organization in the osteosarcoma cells. *FEBS Lett,* 578, 3, pp. (316-22).

Taanman, J. W., Muddle, J. R. &Muntau, A. C. (2003). Mitochondrial DNA depletion can be prevented by dGMP and dAMP supplementation in a resting culture of deoxyguanosine kinase-deficient fibroblasts. *Hum Mol Genet,* 12, 15, pp. (1839-45).

Tatsuta, T. &Langer, T. (2008). Quality control of mitochondria: protection against neurodegeneration and ageing. *EMBO J,* 27, 2, pp. (306-14).

Tatuch, Y., Christodoulou, J., Feigenbaum, A., Clarke, J. T., Wherret, J., Smith, C., Rudd, N., Petrova-Benedict, R. &Robinson, B. H. (1992). Heteroplasmic mtDNA mutation (T---G) at 8993 can cause Leigh disease when the percentage of abnormal mtDNA is high. *Am J Hum Genet,* 50, 4, pp. (852-8).

Thu le, T., Ahn, J. R. &Woo, S. H. (2006). Inhibition of L-type Ca2+ channel by mitochondrial Na+-Ca2+ exchange inhibitor CGP-37157 in rat atrial myocytes. *Eur J Pharmacol,* 552, 1-3, pp. (15-9).

Tinel, H., Cancela, J. M., Mogami, H., Gerasimenko, J. V., Gerasimenko, O. V., Tepikin, A. V. &Petersen, O. H. (1999). Active mitochondria surrounding the pancreatic acinar granule region prevent spreading of inositol trisphosphate-evoked local cytosolic Ca(2+) signals. *EMBO J,* 18, 18, pp. (4999-5008).

Tiranti, V., Hoertnagel, K., Carrozzo, R., Galimberti, C., Munaro, M., Granatiero, M., Zelante, L., Gasparini, P., Marzella, R., Rocchi, M., Bayona-Bafaluy, M. P., Enriquez, J. A., Uziel, G., Bertini, E., Dionisi-Vici, C., Franco, B., Meitinger, T. &Zeviani, M. (1998). Mutations of SURF-1 in Leigh disease associated with cytochrome c oxidase deficiency. *Am J Hum Genet,* 63, 6, pp. (1609-21).

Tiranti, V., Galimberti, C., Nijtmans, L., Bovolenta, S., Perini, M. P. &Zeviani, M. (1999). Characterization of SURF-1 expression and Surf-1p function in normal and disease conditions. *Hum Mol Genet,* 8, 13, pp. (2533-40).

Trenker, M., Malli, R., Fertschai, I., Levak-Frank, S. &Graier, W. F. (2007). Uncoupling proteins 2 and 3 are fundamental for mitochondrial Ca2+ uniport. *Nat Cell Biol,* 9, 4, pp. (445-52).

van den Heuvel, L., Ruitenbeek, W., Smeets, R., Gelman-Kohan, Z., Elpeleg, O., Loeffen, J., Trijbels, F., Mariman, E., de Bruijn, D. &Smeitink, J. (1998). Demonstration of a new pathogenic mutation in human complex I deficiency: a 5-bp duplication in the nuclear gene encoding the 18-kD (AQDQ) subunit. *Am J Hum Genet,* 62, 2, pp. (262-8).

van den Ouweland, J. M., Lemkes, H. H., Trembath, R. C., Ross, R., Velho, G., Cohen, D., Froguel, P. &Maassen, J. A. (1994). Maternally inherited diabetes and deafness is a distinct subtype of diabetes and associates with a single point mutation in the mitochondrial tRNA(Leu(UUR)) gene. *Diabetes,* 43, 6, pp. (746-51).

Vasington, F. D., Gazzotti, P., Tiozzo, R. &Carafoli, E. (1972). The effect of ruthenium red on Ca 2+ transport and respiration in rat liver mitochondria. *Biochim Biophys Acta,* 256, 1, pp. (43-54).

Visch, H. J., Rutter, G. A., Koopman, W. J., Koenderink, J. B., Verkaart, S., de Groot, T., Varadi, A., Mitchell, K. J., van den Heuvel, L. P., Smeitink, J. A. &Willems, P. H. (2004). Inhibition of mitochondrial Na+-Ca2+ exchange restores agonist-induced ATP production and Ca2+ handling in human complex I deficiency. *J Biol Chem,* 279, 39, pp. (40328-36).

Visch, H. J., Koopman, W. J., Leusink, A., van Emst-de Vries, S. E., van den Heuvel, L. W., Willems, P. H. &Smeitink, J. A. (2006). Decreased agonist-stimulated mitochondrial ATP production caused by a pathological reduction in endoplasmic reticulum calcium content in human complex I deficiency. *Biochim Biophys Acta,* 1762, 1, pp. (115-23).

von Kleist-Retzow, J. C., Hornig-Do, H. T., Schauen, M., Eckertz, S., Dinh, T. A., Stassen, F., Lottmann, N., Bust, M., Galunska, B., Wielckens, K., Hein, W., Beuth, J., Braun, J. M., Fischer, J. H., Ganitkevich, V. Y., Maniura-Weber, K. &Wiesner, R. J. (2007). Impaired mitochondrial Ca2+ homeostasis in respiratory chain-deficient cells but efficient compensation of energetic disadvantage by enhanced anaerobic glycolysis due to low ATP steady state levels. *Exp Cell Res,* 313, 14, pp. (3076-89).

Wallace, D. C. (1992). Mitochondrial genetics: a paradigm for aging and degenerative diseases? *Science,* 256, 5057, pp. (628-32).

Wallace, D.C. (2005). A mitochondrial paradigm of metabolic and degenerative diseases, aging, and cancer: a dawn for evolutionary medicine. *Annu Rev Genet,* 39, pp. (359-407).

Wallace, D. C., Fan, W. &Procaccio, V. (2010). Mitochondrial energetics and therapeutics. *Annu Rev Pathol,* 5, pp. (297-348).

Wang, J., Wilhelmsson, H., Graff, C., Li, H., Oldfors, A., Rustin, P., Bruning, J. C., Kahn, C. R., Clayton, D. A., Barsh, G. S., Thoren, P. &Larsson, N. G. (1999). Dilated cardiomyopathy and atrioventricular conduction blocks induced by heart-specific inactivation of mitochondrial DNA gene expression. *Nat Genet,* 21, 1, pp. (133-7).

Wasniewska, M., Karczmarewicz, E., Pronicki, M., Piekutowska-Abramczuk, D., Zablocki, K., Popowska, E., Pronicka, E. &Duszynski, J. (2001). Abnormal calcium homeostasis in fibroblasts from patients with Leigh disease. *Biochem Biophys Res Commun,* 283, 3, pp. (687-93).

Watanabe, A., Arai, M., Koitabashi, N., Niwano, K., Ohyama, Y., Yamada, Y., Kato, N. &Kurabayashi, M. (2011). Mitochondrial transcription factors TFAM and TFB2M regulate Serca2 gene transcription. *Cardiovasc Res,* 90, 1, pp. (57-67).

Willems, P. H., Smeitink, J. A. &Koopman, W. J. (2009). Mitochondrial dynamics in human NADH:ubiquinone oxidoreductase deficiency. *Int J Biochem Cell Biol,* 41, 10, pp. (1773-82).

Wong, A. &Cortopassi, G. (1997). mtDNA mutations confer cellular sensitivity to oxidant stress that is partially rescued by calcium depletion and cyclosporin A. *Biochem Biophys Res Commun,* 239, 1, pp. (139-45)

Wu, Z., Puigserver, P., Andersson, U., Zhang, C., Adelmant, G., Mootha, V., Troy, A., Cinti, S., Lowell, B., Scarpulla, R. C. &Spiegelman, B. M. (1999). Mechanisms controlling mitochondrial biogenesis and respiration through the thermogenic coactivator PGC-1. *Cell,* 98, 1, pp. (115-24).

Yaffe, M. P. (1999). Dynamic mitochondria. *Nat Cell Biol,* 1, 6, pp. (E149-50).

Ying, W. (2008). NAD+/NADH and NADP+/NADPH in cellular functions and cell death: regulation and biological consequences. *Antioxid Redox Signal,* 10, 2, pp. (179-206).

Zeviani, M., Bresolin, N., Gellera, C., Bordoni, A., Pannacci, M., Amati, P., Moggio, M., Servidei, S., Scarlato, G. &DiDonato, S. (1990). Nucleus-driven multiple large-scale deletions of the human mitochondrial genome: a new autosomal dominant disease. *Am J Hum Genet,* 47, 6, pp. (904-14).

Zeviani, M. &Carelli, V. (2007). Mitochondrial disorders. *Curr Opin Neurol,* 20, 5, pp. (564-71).

Zhu, Z., Yao, J., Johns, T., Fu, K., De Bie, I., Macmillan, C., Cuthbert, A. P., Newbold, R. F., Wang, J., Chevrette, M., Brown, G. K., Brown, R. M. &Shoubridge, E. A. (1998). SURF1, encoding a factor involved in the biogenesis of cytochrome c oxidase, is mutated in Leigh syndrome. *Nat Genet,* 20, 4, pp. (337-43).

Antioxidant Action of Mobile Electron Carriers of the Respiratory Chain

Iseli L. Nantes[1], Tiago Rodrigues[1], César H. Yokomizo[2],
Juliana C. Araújo-Chaves[3], Felipe S. Pessoto[2],
Mayara K. Kisaki[1] and Vivian W. R. Moraes[1]

[1]Universidade Federal do ABC
[2]Universidade Federal de São Paulo
[3]Universidade de Mogi das Cruzes
Brazil

1. Introduction

1.1 Evolutionary aspects

Both oxidative photophosphorylation and oxidative phosphorylation are dependent on electron transport chains sharing similarities that are suggestive of evolution of a chemolithotrophy-based common ancestor (conversion hypothesis). Therefore, an early form of electron transport chain with oxidative phosphorylation that is known as prerespiration was able of donating electrons to terminal acceptors available in the primitive reducing biosphere. In the evolutionary pathway this apparatus was supplemented by a photocatalyst capable of a redox reaction. Therefore, oxygenic photosynthesis was a late event during evolution that was preceded by anoxygenic photosynthesis. The development of the manganese complex able to promote water oxidation was a key event in developing oxygenic photosynthesis (Xiong & Bauer, 2002; Bennnown, 1982; Castresanal et al., 1994).

The development of oxygenic photosynthesis was one of the most important events in the biological evolution because it changed the redox balance on Earth and created conditions for the biological evolution to more complex life forms. Molecular data showing cytochrome oxidase in the common ancestor of Archaea and Bacteria and an existing cytochrome oxidase in nitrogen-fixing bacteria living in an environment where the level of oxygen was very low are indicia that aerobic metabolism could be present in an ancient organism, prior to the appearance of eubacterial oxygenic photosynthetic organisms. Although the hypothesis that aerobic metabolism arose several times in evolution after oxygenic photosynthesis is not sustained by the above mentioned data, the widespread use of molecular oxygen as final acceptor of electrons resulting from the oxidation of biological fuels was an evolutionary acquisition subsequent to the oxygen photosynthesis. The use of molecular oxygen as final acceptor of electrons removed from biological fuels resulted in a significant improvement of energy yield, a crucial event for the rise of complex heterotrophic organisms. According to the endosymbiotic theory, the respiratory chain present in prokaryotes was transferred to eukaryotes and resulted in cells bearing mitochondria. At the present step of the biological evolution, the aerobic oxidation of biological fuels occurs in the respiratory chain apparatus of the cell membrane of prokaryotes and in the inner mitochondrial membrane of eukaryotes (Xiong & Bauer, 2002; Bennnown, 1982; Castresanal et al., 1994).

Figure 1 illustrates the more recent view of the evolution pathway of electron chain transport correlated to the arising of more complex living organisms.

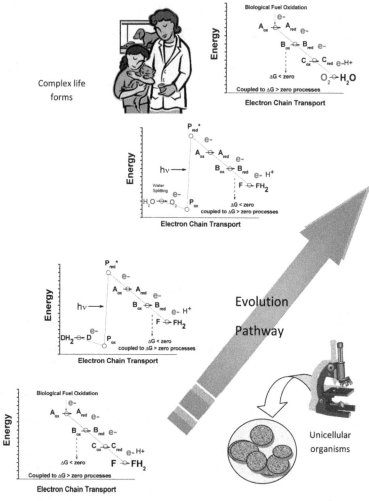

Fig. 1. Evolution pathway of electron transport chain that created conditions for the arising of more complex life forms. A, B and C are representative intermediates in a generic electron transport chain. F represents an electron final acceptor that made feasible the electron chain transport in the primitive reductive atmosphere. P represents a photocatalyst pigment responsible for light harvesting in non-oxygenic and oxygenic photophosphylation. D represents the electron donor in the non-oxygenic photophosphorylation. The oxygenic photophosphorylation was not represented as Z scheme for clarity. The energy scale is arbitrary. ATP synthesis (phosphorylation) are among the $\Delta G > $ zero processes coupled to electron transport chains.

1.2 Respiratory chain

The oxidation of biological fuels such as glucose, lipids and amino acids proceeds by the electron transfer to coenzymes NAD$^+$ and FAD. These metabolic pathways such as glucolysis, citric acid cycle and β-oxidation of fatty acids are totally dependent on the continuous recycling of NADH and FADH$_2$ coenzymes to the oxidized forms. In aerobic organisms, the recycling of NADH and FADH$_2$ was done by the electron transfer to respiratory protein complexes I and II, respectively. In the following, electrons are transported through a sequence of redox centers, most of them composed by hemeproteins that are known as respiratory cytochromes (Hatefi, 1985; Nantes & Mugnol, 2008). Similarly to the mechanism operating in the photosynthetic apparatus, spontaneous electron transfer through respiratory chain complexes is coupled to proton ejection from the matrix to the intermembrane space resulting in the protomotive force (Eq. 1).

$$\Delta p = \Delta \psi - 60\Delta pH \tag{1}$$

Δp supports ATP synthesis and other energy requiring processes in mitochondria such as ion transport and transhydrogenation. ATP synthesis is done by the enzyme ATP synthase that encloses a membrane extrinsic F$_1$ and a transmembrane F$_0$ subunits (Solaini et al, 2002; Zanzami et al, 2007).

As described above, the respiratory chain comprises proteins assembled as supramolecular complexes; most of them are composed by integral proteins inserted in the inner mitochondrial membrane. This redox system encompasses four complexes: NADH:ubiquinone oxidoreductase (complex I), succinate:ubiquinone oxidoreductase (complex II), ubiquinol:ferricytochrome c oxidoreductase (complex III), ferrocytochrome c:oxygen oxidoreductase (complex IV) that assembled with ATP synthase constitute the so-called respirasome (Hatefi, 1985; Nantes & Mugnol, 2008; Duchen, 1999; O'Reilly, 2003; Wittig et al, 2006; Fernandez-Vizarra et al, 2009 and Dudkina et al, 2008), (Fig. 2). However, the electron transport among the respiratory complexes is mediated by two mobile electrons carriers: coenzyme Q (CoQ) and cytochrome c.

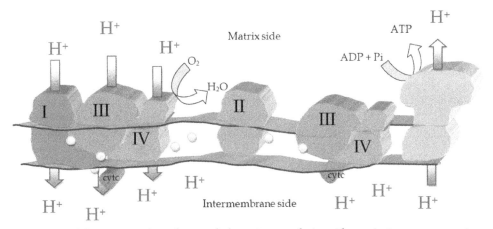

Fig. 2. Pictorial representation of a coupled respiratory chain with respiratory components assembled as respirasomes. CoQ is represented as yellow shadowed spheres.

1.2.1 Fundamental concepts about CoQ

The structure of CoQ was determined by Wolf et al. in 1958. The compound is a 2,3-dimethoxy-5-methylbenzoquinone with the redox active benzoquinone ring connected to a long isoprenoid side chain. According to the isoprenoid chain, five quinones are designated as members of a coenzyme Q group, i.e., CoQ6, CoQ7, CoQ8, CoQ9 and CoQ10 (Fig. 3). Ubiquinol is the product of two-electron reduction of ubiquinone with an ubisemiquinone intermediary form (Fig. 3). The predominant form of ubiquinone in humans presents 10 isoprenoid units in the side chain and it is referred as coenzyme Q_{10} (CoQ$_{10}$) or ubiquinone-10. The first studies about coenzyme Q were published in the end 50's with the isolation of a beef heart quinone (Crane et al., 1957) and sequential studies on its redox properties (Moore, 1959; Gale et al, 1963).

The hydrophobicity of this coenzyme results in its partition into the lipid bilayer (Littarru & Tiano, 2007).

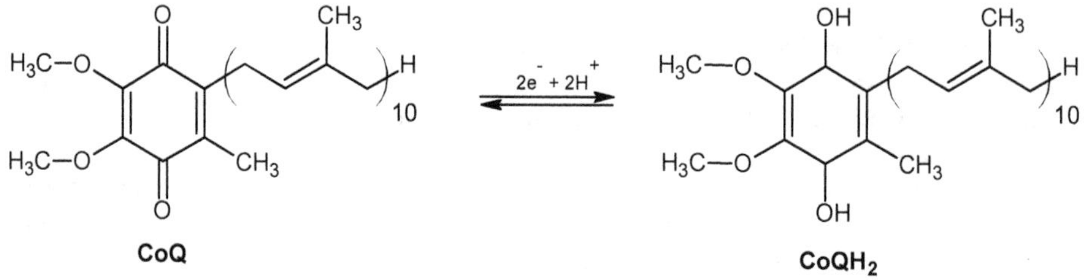

Fig. 3. Chemical structure coenzyme Q in its oxidized (CoQ) and reduced (CoQH$_2$) forms.

CoQ10 is found in almost all cellular membranes as those of Golgi apparatus and lysosomes. In the inner mitochondrial membrane, CoQ carries electrons from complexes I and II to $bc1$ complex but its participation in the respiratory chain involves a redox cycle that also contributes to the generation of the proton motive force. The CoQ redox cycle involves the interaction of the coenzyme with of the $bc1$ complex. Several studies are concerned about the mechanism of proton translocation through the cytochrome $bc1$ complex related to CoQ cycle and the function of individual subunits of the enzyme in the energy transduction process. Unlike the electron transfer pathway through the $bc1$ complex, there is not a consensus on the mechanism that couples the electron transfer to a transmembrane proton electrochemical potential. Two mechanisms of proton translocation by respiratory complexes have been described: the redox loop and the proton pump mechanism. The redox loop mechanism was the mechanism proposed by Mitchell, (1966). This mechanism requires concomitant acceptance of protons from the matrix side followed by proton release at the intermembrane space associated to the redox changes of some respiratory redox centers. The proton pump mechanism requires that the reduction and re-oxidation of protein redox centers would be accompanied by changes in the conformation of proteins with consequent alterations of the pK_a of amino acid side chains and leading to the exposure of these residues alternately at the internal and external side of the membrane (Erecinska, 1982; Trumpower, 1990; Boyer, 1993). Considering exclusively a redox loop mechanism, the CoQ molecules solved inside the membrane lipid fraction are converted to the completely reduced form (CoQH$_2$) by Complex I or II and the high potential $b562$ of Complex III. This process is accompanied by the uptake of two protons from the mitochondrial matrix. The reduction occurs in two steps and consequently semiquinone is generated as intermediate. The reoxidation of $CoQH_2$ results from one electron transfer to cytochrome $c1$ via the iron sulfur protein (ISP) and one electron transfer to heme $b566$ that recycles it to heme $b562$ that reinitiates the cycle by transferring one electron to oxidized CoQ. The oxidation of $CoQH_2$ releases two protons in the intermembrane space. The ratio H^+/electron transferred to cytochrome $c1$ and consequently to molecular oxygen is 2/1 (Figure 4).

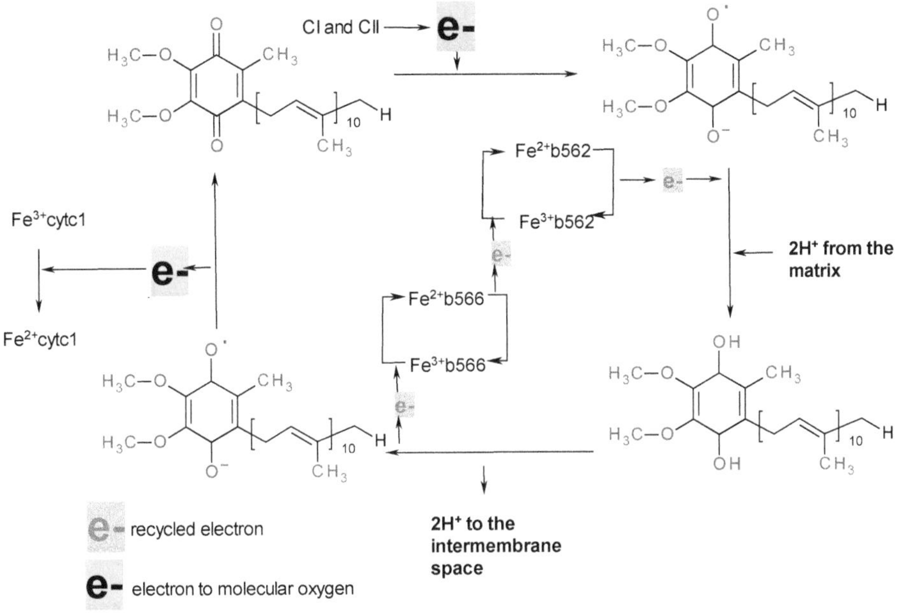

Fig. 4. Coenzyme Q cycle.

Besides the participation in the respiratory chain, literature data has reported, several other important functions of CoQ. The functions include participation in the uncoupling of oxidative phosphorylation and production of heat (Echtay et al., 2001), signaling for gene expression (Doring et al., 2007, Chew et al., 2007) and antioxidant activity (Gomez-Diaz et al., 1997, Papucci et al., 2003). This latter role is the focus of the present chapter and it will be discussed herein.

1.2.2 Fundamental concepts about cytochrome c

Respiratory cytochrome c is a nuclear-encoded protein located at the external side of the inner mitochondrial membrane. In mammals this protein contains 104 amino acids and a single heme group covalently bound to the protein and with a reduction potential of +260 mV. Unlike the other respiratory cytochromes that assembled in large and membrane bound complexes, cytochrome c is a small peripheral protein located at the external side of the inner mitochondrial membrane. Thus, cytochrome c is considered a diffusible carrier with pool function in the aqueous phase.

Cytochrome c is a basic protein bearing 19 lysine and 5 arginine residues giving a highly positively charged with pI = 9.6 and conferring to this protein a high affinity to acidic phospholipids such as cardiolipin, a lipid component of the inner mitochondrial membrane.

The electrostatic interaction is an important factor for the association of cytochrome c with phospholipid membranes and has been focus of several studies. (Rytömaa, 1995, Tuominen, 2002 Zucchi, 2003). The interaction of cytochrome c with acidic phospholipids involves both electrostatic and lipid extended interactions, the latter resulting from the insertion of one phospholipid chain in a hydrophobic channel, present in the cytochrome c structure, in the region of the heme crevice. Other important aspect about the interaction of cytochrome c with the inner mitochondrial membrane is the existence of two membrane binding sites in the cytochrome c structure (Kawai et al, 2005, Kawai, 2009).

The biological role of cytochrome is beyond the cell respiration and involves also apoptosis and redox cell balance (Yong-Ling et al, 2008, Huttemann et al, 2011). Cytochrome c participates in the mitochrondrial electron-transport chain as a mobile electron carrier that shuttles electrons between cytochrome c1 of Complex III and cytochrome c oxidase. The reduction of molecular oxygen to water catalyzed by cytochrome c via complex IV has a $\Delta G^{0\prime} = -100$ kJ/mol that is around twice higher as compared to the redox reactions catalyzed by complexes I and III (Hinkle et al., 1991)

Besides the participation in the respiratory chain, cytochrome c is a key protein for the intrinsic pathway of apoptosis triggered by some stimulus such as DNA damage, metabolic stress or the presence of unfolded proteins (Yong-Ling et al., 2008; Huttemann et al., 2011). The participation of cytochrome c in apoptosis is dependent on its detachment from the inner mitochondrial membrane followed by its translocation through the outer mitochondrial membrane to attain the cytosol. In the cytosol, cytochrome c engages the Apoptotic protease-activating factor-1 (APAF1), and composes the apoptosome (Liu et al., 1996; Kluck et al., 1996; Kluck et al., 1997, Yang et al., 1997).

2. Generation of reactive species in mitochondria

The use of molecular oxygen as final acceptor of electrons removed from the biological fuels was an evolutionary acquisition that resulted in a significant improvement of the energy yield, a crucial event for the rise of complexes organisms. However, a low percentage of molecular oxygen consumed in mitochondrial respiratory chain is not completely reduced to water generating reactive oxygen species (ROS). Mitochondria clearly represent a primary source of ROS in most aerobic mammalian cells (Turrens et al., 1985). The mitochondrial generation of ROS occurs at electron transport chain as a secondary product of mitochondrial respiratory chain (Murphy, 2009). The primary ROS produced in this process is the superoxide anion ($O_2^{\bullet-}$) resulted from a single electron reduction of molecular oxygen by the electrons leaked from the substrate of respiratory chain. In the mitochondrial matrix a superoxide dismutase (MnSOD) transforms superoxide into a more stable form: hydrogen peroxide (H_2O_2). The rate of H_2O_2 production in isolated mitochondria when in state 4 of respiration is 0.6–1.0 nmol/mg mitochondrial protein/min.(Turrens et al., 1985) but this range was considered over estimated and the superoxide production in normal respiring mitochondria could be around 0.1 nM H_2O_2/mg mitochondrial protein/min. Besides the components of the respiratory chain, other mitochondrial complexes also can generate superoxide, such as: dihydrolipoamide dehydrogenase-containing FAD-linked pyruvate, α-ketoglutarate dehydrogenase complexes (Starkov et al., 2004), as well as the flavoenzymes α-glycerophosphate dehydrogenase. In the respiratory chain, two sites have been found to be responsible for the vectorial generation of $O_2^{\bullet-}$. Mitochondria respiring with complex I/III substrates release superoxide anion into the matrix while complex II/III substrates release superoxide anion into the intermembrane space. In complex I the production of $O_2^{\bullet-}$ probably occurs via autooxidation of the reduced flavin mononucleotide (Turrens & Boveris, 1980) and in complex III the partial reduction of molecular oxygen to $O_2^{\bullet-}$ occurs in the Q-cycle via the semiquinone intermediate (Zhang et al., 1998). The vectorial synthesis of superoxide anion indicates that the resultant H_2O_2 formed can act as a mitochondrial second messenger for both nuclear and mitochondrial genomes. This signaling system could be a requirement for appropriate nuclear and mitochondrial gene expression and metabolome modulation. By this point view, the prooxidant formation of the $O_2^{\bullet-}$ /H_2O_2 second messenger system is essential for the normal physiological function of the metabolome and the random molecular damage promoted by $O_2^{\bullet-}$ /H_2O_2 has been rebutted. However, the physiological function of reactive species is dependent of a fine regulation mechanism warranted by the balance between the generation and decomposition of ROS. Oxidative stress occurs when cells have an imbalance of production and decomposition of ROS that results in damages of biomolecules such as lipids, DNA and proteins. Therefore, H_2O_2 may generate the hydroxyl radical ($HO^{\bullet}$), the most reactive and damaging oxygen species, through the Fenton reaction catalyzed by transition metals (Halliwell & Gutteridge, 1990, Eq. 2).

$$Fe^{2+} + H_2O_2 \rightarrow {}^{\bullet}OH + OH^- + Fe^{3+} \qquad (2)$$

Hydroxyl radical can also be generated in the Haber–Weiss reaction with the superoxide radical as shown in Eq. 3 (Valko et al., 2004)

$$O_2^{\bullet-} + H_2O_2 \rightarrow O_2 + {}^{\bullet}OH + OH^- \qquad (3)$$

Superoxide ion can also contribute for the generation of hydroxyl radical by recycling Fe^{3+} to Fe^{2+} (Eq. 4)

$$Fe^{3+} + O_2^{\bullet-} \rightarrow Fe^{2+} + O_2 \qquad (4)$$

Mitochondria is also a source of reactive nitrogen species (RNS) derived also from a signaling molecule, the nitric oxide ($NO^{\bullet}$) (Moncada & Higgs, 1993; Denninger, 1999, Radi et al., 2002). Nitric oxide is generated by the family of nitric oxide synthases (NOS). The NOS family synthesizes $NO^{\bullet}$ using L-arginine as a substrate and NADPH as reducing agent and the reaction is favored by the presence of Ca^{2+} ions and sulfhydryl groups.

Cells of the immune system produce both the superoxide anion and nitric oxide during the oxidative burst triggered by inflammation processes.

Under these conditions, nitric oxide and the superoxide anion may react to produce significant amounts of a highly oxidative molecule, peroxynitrite anion (ONOO-), able to promote DNA fragmentation, lipid oxidation and protein nitrosylation (Valko et al., 2007; Ghafourifar & Cadenas, 2005). Reaction (5) has one of the highest rate constants known for reactions of $NO^{\bullet}$. Thus, $NO^{\bullet}$ toxicity is linked to its ability to combine with superoxide anion.

$$NO^{\bullet} + O_2^{\bullet-} \rightarrow ONOO^- \qquad (5)$$

To assure the cell redox balance, the evolutionary acquisition of aerobic O_2-dependent metabolism was accompanied by a highly conserved antioxidant enzymatic apparatus that works in a concerted way to promote decomposition of $O_2^{\bullet-}$ (Cu-Zn/Mn superoxide dismutase and cytochrome c) and H_2O_2 (catalase and glutathione peroxidase) as well as to repair oxidative damage of proteins (thioredoxine, glutaredoxine, thioredoxine reductase and others). In addition to the enzymatic apparatus, low molecular molecules contribute also to the redox cell balance by acting as free radical trapping. CoQ is included among these antioxidant molecules and, unlike cytochrome c, its action is not restricted to the mitochondria.

The present chapter is concerned with the state of art of the more recent findings about the antioxidant role played by the two mobile electron carriers present in the mitochondrial respiratory chain of higher organism cells: CoQ and cytochrome c and these findings are described herein.

3. Antioxidant properties of CoQ and cytochrome c

3.1 CoQ

Antioxidants are molecules able to inhibit the oxidation of other molecules by eliminating free radicals or by decreasing their formation. In biological systems the high effectiveness of antioxidant system is fundamental due to the constant generation of free radicals inside the organism at several sites that potentially may cause oxidative damage and consequently loss of function of protein, lipids and nucleic acids (Halliwell & Gutteridge, 2007). The effectiveness of antioxidants against oxidative damage in biological environments is directly related to their chemical structure.

The role of CoQ_{10} in biological energy conversions as a redox component of the mitochondrial electron transport chain is well-described. Despite its ability to generate free radicals acting as pro-oxidant, as discussed before, Mellors & Tappel (1966) proposed an antioxidant role for CoQ showing that the reduced and oxidized forms of CoQ were able to prevent heme-catalyzed lipid peroxidation. Up today an increasing number of works has been conducted to understand the mechanisms of CoQ antioxidant action and the *in vivo* situations in which this property is achieved.

Due to the relatively high hydrophobicity, CoQ is partitioned into the lipid bilayers and can play the antioxidant role toward the impairment of lipid oxidation. Impairment of lipid oxidation is biologically important to the maintenance of the membrane integrity and to prevent the oxidation of lipoproteins (Ingold et al, 1993). Lipid oxidation is an oxidative chain reaction that is triggered by the abstraction of a hydrogen atom ($H^\bullet$) from allylic carbon of an unsaturated chain of a phospholipid (LH) by reactive species such as hydroxyl radical ($^\bullet OH$) resulting in a carbon centered free radical ($L^\bullet$). After intramolecular rearrangement $L^\bullet$ may reacts with another LH or with molecular oxygen generating peroxyl radicals ($LOO^\bullet$) that react with another LH resulting in lipid hydroperoxides (LOOH). Such lipid-derived peroxides suffer homolytic cleavage in the presence of metals in Fenton type reactions. The reactions involved in the initiation and propagation of the lipid peroxidation are summarized below.

$$LH +^\bullet OH \rightarrow L^\bullet + H_2O \tag{6}$$

$$L^\bullet + O_2 \rightarrow LOO^\bullet \tag{7}$$

$$LOO^\bullet + LH \rightarrow LOOH + L^\bullet \tag{8}$$

$$LOOH + Fe^{2+} \rightarrow LO^\bullet + OH^- + Fe^{3+} \tag{9}$$

Molecules that are able to react with these intermediate radical species to form less reactive radicals are considered chain breaking antioxidants, such as vitamin E. One of the mechanisms by which coenzyme Q exerts its antioxidant action inhibiting the lipid oxidation of membranes is by reacting directly with lipid-derived radicals transferring $H^{\bullet}$ and generating the ubisemiquinone radical ($CoQH^{\bullet}$) as shown in Eq. 10.

$$CoQH_2 + L^{\bullet} / LOO^{\bullet} \rightarrow CoQH^{\bullet} + LH / LOOH \qquad (10)$$

Besides the inhibition of the lipid oxidation of membranes, CoQ also protects DNA from oxidation induced by H_2O_2 plus metal ions. This process seems to be important especially for mitochondrial DNA oxidation since such damage is not easily repaired (Bentinger, 2010). It was demonstrated in human lymphocytes that incubation with CoQ results in increase of resistance to H_2O_2 damage and less damage by exposure to oxygen (Litarru & Tiano, 2007). Endogenous CoQ also plays an important role in the protection of protein oxidation. The sensitivity of different proteins to oxidative stress varies to a great extent, depending on their structure, composition and localization (Bentinger et al, 2007) but the protective effect of CoQ is probably mediated by a direct scavenger mechanism. Also, the reactive ferrylmyoglobin formed by the reaction of myoglobin with H_2O_2 can use $CoQH_2$ to be converted to metmyoglobin and oxymyoglobin in a mechanism that allows the hemeprotein to neutralize peroxides that can be harmful to cells (Mordente et al, 1993; Guo et al, 2002; Litarru & Tiano, 2007).

Although several studies have compared the antioxidant efficiency of vitamin E *versus* CoQ, there is no other biological function described for vitamin E differently of CoQ that participates in the energetic metabolism. This raises some evolutionary questions addressed by Beyer (1994) whether the antioxidant role of CoQ is merely a coincidence of its structure or it was selected on the basis of the advantage to organism against oxidative stress. Furthermore, there is more CoQ than vitamin E in tissues and in mitochondria (Joshi et al, 1963; Mellors & Tappel, 1966; Ingold et al, 1993). Due to its hydrophobicity, the antioxidant efficiency of $CoQH_2$ is influenced by the polarity of the environment and it is dependent on the accessibility to the free radicals. In aqueous media, ubiquinol is only about 10% as effective as a chain breaking antioxidant like Vitamin E, possibly because the intramolecular hydrogen bonding between the hydroxyl and methoxy groups in CoQ, and also to the electron withdrawing inductive effects of the methoxy groups that stabilize the phenolate relative to the phenoxyl radical (Ingold et al, 1993). These considerations are important to the interpretation of the *in vitro* analysis of the antioxidant activity of CoQ in aqueous systems and apolar environments, such as liposomes and membranes of organelles and cells. In aqueous media, the reactivity of a free radical generated by oxidant agents is modulated by the proton concentration as a function of the pK_a of the radical group. Thus, the reactivity of $CoQH^{\bullet}$ radicals generated by the reaction with lipid derived free radicals is lower within the phospholipid bilayer.

By using liposomes as membrane model, Frei et al (1990) showed that $CoQH_2$, but not its oxidized form CoQ, scavenged free radicals and inhibited lipid peroxidation with similar efficiency than vitamin E. Also, the simultaneous addition of $CoQH_2$ and vitamin E resulted in oxidation of the quinone sparing vitamin E. In this *in vitro* system, ascorbate or GSH were not able to recycle the oxidized CoQ (Frei et al, 1990). The antioxidant activity of the reduced form $CoQH_2$ is due to its behavior as a phenolic antioxidant inhibiting not only the lipid peroxidation but also regenerating vitamin E, preventing DNA and protein oxidation, and reducing ferrylmyoglobin (James et al, 2004; Littarru & Tiano, 2007; Roginsky et al, 2009). Several groups have been concerned to study the mechanisms of the antioxidant action of CoQ. It was proposed that $CoQH_2$ inhibits lipid peroxidation by decreasing the production of lipid peroxyl radicals ($LOO^\bullet$) and reducing perferryl radicals. $CoQH_2$ could eliminate $LOO^\bullet$ directly acting as a primary scavenger of free radicals (Crane, 2001). Thus, $CoQH_2$ can exert its antioxidant action inhibiting lipid peroxidation directly by acting as a chain breaking antioxidant and indirectly by recycling vitamin E (James et al, 2004; Cuddihy et al, 2008). It was showed that α-tocopherol recycling in mitochondrial membranes is directly dependent on the CoQ/α-tocopherol molar ratio (Lass & Sohal, 2000) and that such recycle process also occurs in vivo (Lass et al, 1999).

$$LOO^\bullet + Vit\ E \rightarrow LOOH + Vit\ E^\bullet \tag{11}$$

$$Vit\ E^\bullet + CoQH_2 \rightarrow Vit\ E + CoQH^\bullet \tag{12}$$

CoQ significantly increases the rate of vitamin E regeneration in membranes a process also observed in low density lipoproteins, presumably by CoQ content present in the blood (Crane, 2001). Alternatively, reduced coenzyme Q could react directly with superoxide and hydroxyl radicals as a free radical scavenger and interfere with the initiation of lipid peroxidation (Beyer, 1990). Differently of others antioxidant coumpounds, CoQ inhibits both the initiation and propagation of lipid and protein oxidation. (Bentinger et al, 2010). In fact, it is probable that this antioxidant is considerably more efficient than that exhibited by vitamin E (Turunen et al, 2003). The reactivity of CoQ and vitamin E with different radicals, including the reaction rate constants, was reviewed by James et al (2004).

In mitochondria it is proposed that the respiratory chain enzymes and other dehydrogenases are able to recycle CoQ to the reduced form able to protect membranes against oxidation. There are at least three enzymes responsible to maintain CoQ_{10} in its reduced form: NADH cytochrome b5 reductase, NADH/NADPH oxidoreductase (also called DT-diaphorase) and NADPH coenzyme Q reductase (Turunen et al, 2004). Mitochondrial DT-diaphorase, a two-electron quinone reductase, seems to have a crucial participation in the antioxidant action of CoQ due to its maintenance in the reduced form $CoQH_2$ (Cadenas, 1995). Differently from NADH and succinate dehydrogenases, which are able to generate the partially reduced coenzyme Q ubisemiquinone, the DT-diaphorase is unique since it can directly reduce CoQ *via* 2 electron transfer without intermediate formation of the semiquinone (Beyer et al, 1996). CoQ can also be reduced by the mitochondrial respiratory chain (Genova et al, 2003; Bentinger et al, 2007). Another mechanism that may contribute to the antioxidant activity of CoQ is the interaction of superoxide dismutase with $CoQH_2$ and DT diaphorase resulting in inhibition of coenzyme autoxidation (Beyer, 1992). Besides the energetic role of coenzyme Q as mobile electron carrier, the antioxidant activity of $CoQH_2$ is important to decrease the oxidative modification of mitochondrial CoQ pool associated to the impairment of the electron transport in the respiratory chain observed during the lipid oxidation of mitochondrial membranes (Forsmark-Andrée et al, 1997). It was showed that the antioxidant effects of CoQ in microsomes and mitochondria are also mediated by vitamin E recycling (Kagan et al, 1990). Recently, it was demonstrated that the enzymes: lipoamide dehydrogenase, thioredoxin reductase and glutathione reductase can also reduce CoQ (Olsson et al, 1999; Xia et al, 2001; Xia et al, 2003).

The high antioxidant efficiency of CoQH$_2$ depends on several factors, including its localization into the membranes, hydrophobicity, the efficiency as scavenger of free radicals and recycling antioxidant cellular systems. Mitochondria are directly implicated with oxidative stress conditions, due to the constant generation of superoxide anions (O$_2\bullet$-) by the respiratory chain, a process which is normally counterbalanced by the antioxidant defense system composed of superoxide dismutase, gluthathione peroxidase and reductase, GSH and NAD(P)H. However, in mitochondrial dysfunctions, the excessive formation of O$_2\bullet$-, and consequently of hydrogen peroxide (H$_2$O$_2$), leads to the generation of the extremely reactive hydroxyl radical ($\bullet$OH) by means of the Fenton-Haber-Weiss reaction (Sies, 1997). Stress oxidative is thought to be involved in the ethyology of many human diseases (Brookes et al, 2004) but also in cell signaling (Linnane et al, 2007) and endogenous and exogenous antioxidants are crucial to modulate theses processes. CoQ and vitamin E addition in cultured cells attenuated ROS production, lipid peroxidation, mitochondrial dysfunction, and cell death induced by amitriptyline (Cordero et al, 2009). In Langendorff preparations of isolated heart, pretreatment with CoQ protected coronary vascular reactivity after isquemia/reperfusion radical scavenger activity (Whitman et al, 1997). CoQ was also able to ameliorate cisplatin-induced acute renal injury in mice (Fouad et al, 2010).

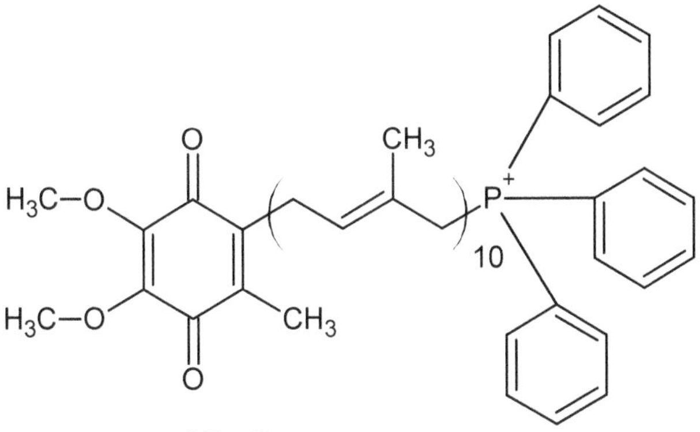

MitoQ

Fig. 5. Chemical structure of triphenylphosphonium-substituted coenzyme Q, MitoQ.

Many *in vitro* and *in vivo* studies showed that CoQ, mainly in its reduced state, may act as an antioxidant protecting membranes from oxidative damage (Beyer, 1990). Besides the antioxidant activity, CoQ participates as cofactor of uncoupling proteins and modulates gene expression associated to cell signaling, metabolism, transport, etc (Linnane et al, 2002). The hydrophobicity of CoQ allows its easy insertion into the mitochondrial inner membrane where it is converted to the reduced form by reductases (Beyer et al, 1996). Although the antioxidant activity of CoQ in biological membranes the relative high hydrophobicity disfavors its use as a therapeutic agent (Kelso et al, 2001). As an alternative, mitochondrial-targeted ubiquinone analogs, including Mito Q, and ubiquinone analogs with a decrease number of carbons in the side chain compared with CoQ10 were developed (Geromel et al, 2002). It was showed that the CoQ analogue decylubiquinone, but not CoQ, decreased ROS production associated to the inhibition of the MPT (mitochondrial permeability transition) and cell death in HL60 cells. Such effect is due to the antioxidant action of decylubiquinone either preventing ROS formation or scavenging ROS generated by cytochrome *bc*1

(Armstrong et al, 2003). MitoQ (Fig. 5) is an orally active antioxidant developed by the pharmaceutical industry to potentially treat several diseases. This compound retains the antioxidant activity of CoQ10 and the triphenylphosphonium cation (TPP$^+$) substituent directs this agent to mitochondria (Tauskela, 2007).

In cultured cells MitoQ was able to accumulate into mitochondria and act against oxidative stress (Murphy & Smith, 2007; James et al, 2005). It was also demonstrated a protective effect of MitoQ in a sepsis model by decreasing the oxidative stress and protecting mitochondria against damage as well as by suppressing proinflammatory cytokine release (Lowes et al, 2008). It was proposed that the antioxidant action of MitoQ may be useful in the treatment of diseases associated to the impairment of mitochondrial Complex I (Plecitá-Hlavatá et al, 2009). On the other hand, it was recently showed that MitoQ may be prooxidant and present proapoptotic action due its quinone group that may participates in redox cycling and superoxide production (Doughan & Dikalov, 2007). Thus, the study of the mechanisms of antioxidant action and other effects of CoQ and derivatives must be considered for the development of quinone-based therapeutic strategies.

3.2 Cytochrome c

Similarly to that was described for CoQ, cytochrome c may also contribute to the generation and trapping of prooxidant species. It has been described that besides the participation in the respiratory chain and apoptosis, cytochrome c exhibits also a prooxidant peroxidase activity and an antioxidant superoxide oxidase activity. However, a whole view of the roles played by cytochrome c in cells leads to the conclusion that the respiratory and pro-apoptotic activities of this protein intrinsically contribute also to the cell redox balance.

It was demonstrate that loss of cytochrome c by mitochondria oxidizing NAD$^+$-linked substrates results in respiratory inhibition associated to a significant increase of ROS production (Davey et al., 1998; Gnaiger et al., 1998; Rossingnol et al., 2000) The depletion of cytochrome c results in respiratory inhibition and maintains reduced the electron carriers upstream the hemeprotein with consequent increasing of the NAD(P)H levels. However, the terminal segment of the respiratory chain is more active than the proximal one in such way that only mild respiratory inhibition has been observed in cells undergoing apoptosis accompanied by cytochrome c release and increased production of ROS. Therefore, only almost total cytochrome c depletion could significantly promote respiratory inhibition and enhance of ROS production at complex I (Davey et al., 1998; Gnaiger et al., 1998; Rossingnol et al., 2000; Kushnareva et al., 2002) Considering the role played by cytochrome c in apoptosis, literature data have correlated this event to an increased peroxidase activity of the hemeprotein. In comparison with pentacoordinated hemeproteins such as myoglobin and horseradish peroxidase, in the native form, cytochrome c reacts very slowly with peroxides (Radi et al., 1991). However, the peroxidase activity of cytochrome c can be favored in conditions leading to loss of the heme iron sixth coordination position with the sulfur atom of Met80 or the replacement of Met80 by other amino acid lateral chains (Nantes et al., 2000; Rodrigues et al., 2007; Nantes et al., 2001; Zucchi et al., 2003). A condition that can strongly favor the peroxidase activity of cytochrome c is the association with negatively charged membranes (Rytömaa, et al, 1992. Ott, et al., 2002. Mugnol, et al, 2008. Rytömaa & Kinnunen, 1994, Rytömaa, M & Kinnunen, 1995, Kawai, et al, 2005; Kagan, et al., 2005).

According to Kagan et al., (2004) the amount of cardiolipin in the outer side of the inner mitochondrial membrane can be increased in a proapoptotic condition and favor the peroxidase activity of cytochrome c. In this scenario, the peroxidase activity of cytochrome c on cardiolipin should be involved in its detachment from the inner mitochondrial membrane to attain the cytosol and trigger apoptosis. In addition, the reaction of cytochrome c with lipid-derived carbonyl compounds results in the production of triplet excited species able to generate $O_2(^1\Delta g)$ by energy transfer to molecular oxygen (Foote, 1968, Nantes et al., 1996; Estevam et al., 2004; Groves, 2006).

Considering the peroxidase activity of cytochrome c culminates with its detachment from the inner mitochondrial membrane leading to the death of cells with unbalanced redox processes, the pro-apoptotic activity of cytochrome c might be included, if not as antioxidant, but as a protective role of this protein for the whole organism.

However, the protective antioxidant activity of cytochrome c can also be exerted in a preventive rather than a destructive way. The elimination of superoxide ion by SOD generates hydrogen peroxide. As discussed before, hydrogen peroxide is a signaling molecule but its accumulation in cells should be prevented to avoid undesirable reaction with transition metal ions and the consequent generation of hydroxyl radical. Hydrogen peroxide can react with Fe^{III} respiratory cytochrome c and convert it to high valence species (oxoferryl forms) that are highly prooxidant species. The high valence species of cytochrome c can attack lipids and trigger a radical propagation leading to oxidative damages of mitochondrial membranes. Therefore, the cellular antioxidant apparatus includes catalase and GPx (glutathione peroxidase) that are responsible for hydrogen peroxide reduction. Fe^{III} cytochrome c competes with SOD for one electron reduction by superoxide ion. The reduction of cytochrome c by superoxide ion is more efficient than SOD to prevent oxidative stress because, by this way, the electron is devolved to the respiratory chain, does not generates hydrogen peroxide and further prevents the generation of high valence species of the hemeprotein. In an apparent paradox but consistent with the competition with cytochrome c, over expression of SOD1 has been related to an increase of the oxidative stress (Goldsteins et al., 2008). Cytochrome c can efficiently acts as a true antioxidant by scavenging $O_2^{\bullet-}$ without producing secondary and potentially harmful ROS (Pereversev et al., 2003).

Also, the reduction of cyt c heme iron by $O_2^{\bullet-}$ impairs peroxidase activity on hydrogen peroxide and the consequent generation of radicalar and excited prooxidant species. As discussed before, even the conditions favoring the peroxidase activity of cytochrome c should not be considered exclusively harmful and damaging events since they culminate with detachment of cytochrome c from the inner mitochondria membrane to participate in the apoptosis in cytosol. It is important to note that the participation of cytochrome c in oxidative and nitrosative stress can also promote damages in the hemeprotein (Estevam et al, 2004; Rodrigues et al., 2007), including impairment of the proapoptotic activity (Suto et al., 2005). However, the association of cytochrome c with unsaturated lipid bilayers is shown to prevent these oxidative damages and preserve the apoptotic activity (Estevam et al, 2004; Rodrigues et al., 2007).

The contribution of cytochrome c for hydrogen peroxide elimination is probably not restricted to the peroxidase mechanism and superoxide ion trapping. It has been proposed the reduction of hydrogen peroxide by Fe^{II} cytochrome c in a mechanism named as electron-leak pathway.

At this point it is important to consider the role played by testicular cytochrome c. Reactive oxygen species generated in the respiratory chain are responsible for damages in biomolecules such as DNA, lipids and proteins of sperm that culminate with loss of cell viability and infertility. Sperm are particularly susceptible to the undesirable effects of ROS because their high content of polyunsaturated fatty acids present in the plasma membrane and a low concentration of ROS scavenging enzymes in the cytoplasm (Jones et al, 1979, Huttemann et al., 2011; Sharma et al, 1999; Liu et al., 2006). Mammalian germ cells express two types of cytochrome c during their development: the somatic cytochrome c and a testis specific cytochrome c that shares 86,5% identity with the somatic counterpart. During meiosis, the expression of somatic cytochrome c declines and testis cytochrome c becomes the predominant form in sperm. Liu et al, reports that testis cytochrome c is three fold more efficient than the somatic one in the catalysis of H_2O_2 reduction and is also more resistant to be degraded by the side products of this reaction. In line with the proposal that apoptosis is also an antioxidant protective mechanism, testis cytochrome c exhibited higher apoptotic activity in the well established apoptosis measurement system using *Xenopus* egg extract. Therefore, testis cytochrome c can protect sperm from the damages caused by H_2O_2 as well as promote the elimination of sperm whose DNA was damaged. Taken together the electron-leak pathway and apoptosis, probably related to a peroxidase activity are the contribution of testis cytochrome c for the biological integrity of sperm produced by mammalian cells.

Therefore, a delicate balance controls both antioxidant and prooxidant activities of cytochrome c with repercussions on both bioenergetics and cell death. In this regard, it is noteworthy that cytochrome c import to mitochondria, synthesis and activities underlying life and death fates for cells are regulated by signaling mechanisms and involves thiol redox balance, allosteric regulation and chemical modifications including nitration and phosphorylation. Cytochrome c is a nuclear-coded protein that is imported by mitochondria as apoprotein and, in the intermembrane space, is converted to the holoprotein by the covalent ligation of the heme group to cystein residues 14 and 17, a process catalyzed by the enzyme heme lyase (Dumont et al, 1991). The addition of the heme group confers redox properties for cytochrome c and enables it to participate, as terminal oxidant agent, in the thiol redox cascade involved in the import and assembly of TIMs (transporters of the inner mitochondrial membrane) (Chacinska et al, 2004, Riemer et al., 2011; Allen et al., 2005). The participation of cytochrome in the respiratory chain as electron carrier is also controlled by allosteric and covalent modification mechanisms. ATP has been characterized as a downregulator of the electron-transfer activity of cytochrome c. The mechanism may involve changes of both charge and structure of cytochrome c and is consistent with the adjustment of respiratory chain activity to the energy demand of cell signaled by the ATP/ADP ratio. Recent findings have strongly shown that the well-known mechanism of protein phosphorylation operates also in the control of proteins responsible for the oxidative phosphorylation. The technique of cytochrome c isolation in the presence of nonspecific phosphatase inhibitors enabled the identification of tissue-specific sites of cytochrome c and evidenced the activities of this protein is under the control of this specific cell signaling mechanism mainly operating in higher organisms (Huttemann et al., 2011). Previously, it was demonstrated that redox reaction of cytochrome c with a model aldehyde, diphenylacetaldehyde, is under the control of the protonation of two tyrosine residues (Rinaldi et al, 2004). More recent findings established the phosphorylation of cytochrome c tyrosine residues is involved in the control of the transmembrane potential that in health conditions should not attain the maximal to avoid increase of ROS generation (Yu et al., 2008; Zhao et al., 2010). The consequences of cytochrome c nitration in biological systems have been investigated and demonstrated that a small structural change promoted by nitration of tyrosine 74 does not preclude ligation of cytochrome c to Apaf-1 but this cytochrome c form became unable to activate caspases (Garcia-Heredia et al., 2010).

Similarly to CoQ, the more recent findings about cytochrome c biological functions operating under the control of cell signaling mechanisms led to investigations concerning the therapeutical use of cytochrome c properties by administration of exogenous proteins as well as by controlling its phosphorylation in pathological conditions (Huttemann et al, 2011; Piel et al., 2007, 2008).

Fig. 6 summarizes the antioxidant activity of cytochrome c.

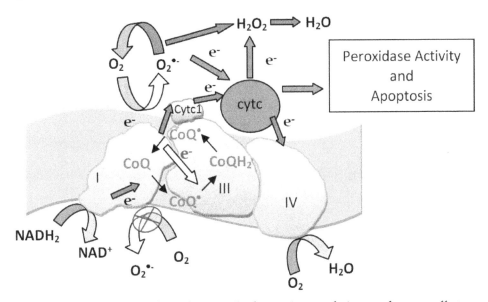

Fig. 6. Antioxidant activity of cytochrome c. In the respiratory chain, cytochrome c affects the generation of $O_2^{\bullet-}$ and H_2O_2 by making the electron transfer of the respiratory chain more fluent (green arrows). The hemeprotein also eliminates the generated $O_2^{\bullet-}$ and H_2O_2 through a cytochrome c mediated electron-leak pathway (red arrows). Further, the peroxidase activity of cytochrome c associated to apoptosis is also a protective mechanism for the whole organism as in the case of testis cytochrome c.

4. Conclusion

The evolutionary acquisition of the O_2-dependent aerobic metabolism resulted in a highly more efficient use of the energetic fuels and a cell signaling mechanism based on reactive species. Concerning the ROS, the primary species produced in mitochondria is $O_2^{\bullet-}$ from which both the signaling molecule, H_2O_2, and the highly deleterious derivative, hydroxyl radical are generated. Therefore, a very efficient antioxidant apparatus was also evolved to assure cell redox balance and repair of oxidative damages. The antioxidant apparatus encompasses enzymes able to decompose reactive species (SOD, catalase) and repair oxidative damages (thioredoxin, glutaredoxine) and free radical trapping (ascorbic acid, tocopherol, lipoic acid). More recently, cytochrome c was included in the category of antioxidant enzymatic apparatus due to its capacity to oxidize superoxide ion and devolve the electron to the respiratory chain as well as by the capacity to reduce hydrogen peroxide. The electron transport in the respiratory chain can also be considered an antioxidant activity of cytochrome c because it contributes for the fluency of electron transport. The antioxidant activity of CoQ is based on the direct and indirect trapping of free radicals and it is not restricted to mitochondria but exerted in the whole cellular and extra-cellular media. The beneficial antioxidant activity of CoQ has been studied with the aim to develop an antioxidant therapy by the use of CoQ analogous and derivatives. Figure 7 summarizes the antioxidant activity of mobile electron carriers of the respiratory chain.

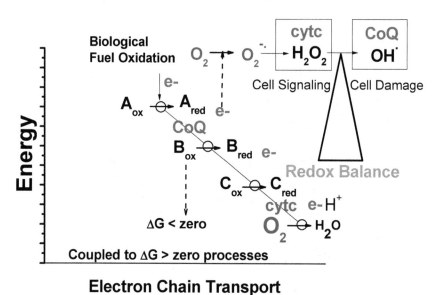

Electron Chain Transport

Fig. 7. Antioxidant activity of mobile electron carriers of the respiratory chain. The aerobic oxidation of biological fuels by using molecular oxygen as final acceptor of electrons in an electron transport chain allowed an efficient mechanism of withdrawing energy from biological fuels concomitant with the generation of reactive species for cell signaling but also able to promote cell damage. The redox cell balance is achieved by prevented the accumulation of reactive species without prejudicing the signaling function. Cytochrome c contributes to the maintenance of the adequate levels of hydrogen peroxide in cells by means of fluency of electron transport in the respiratory chain, oxidation of superoxide ion and reduction of hydrogen peroxide and CoQ by means of direct and indirect trapping of free radicals.

5. Acknowledgments

Authors are grateful to Fundação de Amparo à Pesquisa do Estado de São Paulo – FAPESP – for the financial support.

6. References

Abriata, L.A., Cassina, A., Tórtora, V., Marín, M., Souza, J.M., Castro, L., Vila, A.J., Radi, R. (2009). Nitration of solvent-exposed tyrosine-74 on cytochrome c triggers heme iron-methionine-80 bond disruption: nuclear magnetic resonance and optical spectroscopy studies. *J Biol Chem*, Vol. 284, No. 1, (January 2009), pp.(17–26), ISSN 1083-351X

Allen, S., Balabanidou, V., Sideris, D.P., Lisowsky, T., Tokatlidis, K. (2005). Erv1 mediates the Mia40-dependent protein import pathway and provides a functional link to the respiratory chain by shuttling electrons to cytochrome c. *J Mol Biol*, Vol. 353, No. 5, (November 2005), pp.(937–944), ISSN 0022-2836

Armstrong, J.S., Whiteman, M., Rose, P., & Jones, D.P. (2003). The Coenzyme Q10 analog decylubiquinone inhibits the redox-activated mitochondrial permeability transition: role of mitochondrial complex III. *J Biol Chem*, Vol. 278, No. 49, (December 2003), pp. (49079-49084), ISSN 1083-351X

Bennoun, P. (1982). Evidence for a respiratory chain in the chloroplast. *Proc Natl Acad Sci USA*, Vol. 79, No. 14, (July 1982), pp.(4352-4356), ISSN 1091-6490

Bentinger, M; Brismar, K & Dallner, G. (2007). The Antioxidant Role of Coenzyme Q. *Mitochondrion*, Vol. 7, (March, 2007), pp. (41-50), ISSN 1567-7249

Bentinger, M; Tekle, M; Dallner, G. (2010). Coenzyme Q – Biosynthesis and Functions. *Biochemical and Biophysical Research Communications*, Vol. 396, No. 1, (May, 2010), pp. 74-79, ISSN 0006-291X

Beyer, R. E. (1994). The relative essentiality of the antioxidative function of coenzyme Q--the interactive role of DT-diaphorase. *Mol Aspects Med*, Vol. 15, No.1 , (1994), pp. (117-129), ISSN 0098-2997

Beyer, R. E., Segura-Aguilar, J., Di Bernardo, S., Cavazzoni, M., Fato, R., Fiorentini, D., Galli, M. C., Setti, M., Landi, L., & Lenaz, G. (1996). The role of DT-diaphorase in the maintenance of the reduced antioxidant form of coenzyme Q in membrane systems. *Proc. Natl. Acad. Sci. U. S. A.*, Vol. 93, No. 6, (March 1996), pp. (2528-2532), ISSN 1091-6490

Beyer, R.E. (1992). An analysis of the role of coenzyme Q in free radical generation and as an antioxidant. *Biochem Cell Biol*, Vol. 70, No. 6, (June 1992), pp. (390-403), ISSN 1208-6002

Beyer, RE. (1990). The Participation of Coenzyme Q in Free Radical Production and Antioxidation. *Free Radical Biology and Medicine,* Vol. 8, No. 6, (February, 1990), pp. (545-565), ISSN 08915849

Boyer, P.D. (1993). The binding change mechanism for ATP synthase-some probabilities and possibilities. *Biochim Biophys Acta,* Vol. 1140, No. 3, (January 1993), pp.(215-250), ISSN 0005-2728

Brookes, PS; Yoon, Y; Robotham, JL; Anders, MW & Sheu, SS. (2004). Calcium, ATP, and ROS: A MitochondrialLove-hate Triangle. *American Journal Physiology-Cell*, Vol. 287, No. 4, (October, 2004), pp. (817-833), ISSN 0363-6143

Cadenas E. (1995). Antioxidant and Prooxidant Functions of DT-diaphorase in Quinone Metabolism. *Biochemical Pharmacology*, Vol. 49, No. 2, (January, 1995), pp. (127-140), ISSN 0006-2952

Castresanal J., M. Lubben, (1994). Evolution of cytochrome oxidase, an enzyme older than atmospheric oxygen. *The EMBO J.*, Vol. 13, No. 11, (June 1994), pp.(2516-2525), ISSN 0261-4189

Chacinska, A., Pfannschmidt, S., Wiedemann, N., Kozjak, V., Sanjuán Szklarz, L.K., Schulze-Specking, A., Truscott, K.N., Guiard, B., Meisinger, C., Pfanner, N. (2004). Essential role of Mia40 in import and assembly of mitochondrial intermembrane space proteins. *The EMBO J.*, Vol. 23, No. 19, (September 2004), pp.(3735-3746), ISSN 0261-4189

Chew, G.T., Hamilton, S.J., Watts, G.F. (2007). Therapeutic regulation of endothelial dysfunction in type 2 diabetes mellitus. *Diab Vasc Dis Res*, Vol. 4, No. 2, (June 2007), pp.(89-102), ISSN 1752-8984

Cordero, MD; Moreno-Fernández, AM; Gomez-Skarmeta, JL; de Miguel, M; Garrido-Maraver, J; Oropesa-Avila, M; Rodríguez-Hernández, A; Navas, P & Sánchez-Alcázar, JA. (2009). Coenzyme Q10 and alpha-Tocopherol Protect Against Amitriptyline Toxicity. *Toxicology and Applied Pharmacology*, Vol. 235, No. 3, (March, 2009), pp. (329-337), ISSN 0041-008X

Crane, F.L., Hatefi, Y., Lester, R.L., & Widmer, C. (1957). Isolation of a quinone from beef heart mitochondria. *Biochim Biophys Acta*, Vol.25 , No.1 , (July 1957), pp. (220-221), ISSN 0006-3002

Crane, FL. (2001). Biochemical Functions of Coenzyme Q_{10}. *The Journal of the American College of Nutrition*, Vol. 20, No. 6, (December, 2001), pp. (591-598), ISSN 1541-1087

Cross, A.R., Jones, O.T. (1991). Enzymic mechanisms of superoxide production. *Biochim Biophys Acta*, Vol. 1057, No. 3, (May 1991), pp.(281-298), ISSN 0005-2728

Cuddihy, SL; Ali, SS; Musiek, ES; Lucero, J; Kopp, SJ; Morrow, JD & Dugan, LL. (2008). Prolonged α-Tocopherol Deficiency Decreases Oxidative Stress and Unmasks α-Tocopherol-dependent Regulation of Mitochondrial Function in the Brain. *The Journal of Biological Chemistry*, Vol. 283, No. 11, (March, 2008), pp. (6915-6924), ISSN 1083-351X

Davey, G.P., Peuchen, S., Clark, J.B. (1998). Energy thresholds in brain mitochondria. Potential involvement in neurodegeneration. *J Biol Chem*, Vol. 273, No.21, (May 1998), pp.(12753-12757), ISSN 1083-351X

Denninger, J.W., Marletta, M.A. (1999). Guanylate cyclase and the (NO)-N-/cGMP signaling pathway. *Biochim Biophys Acta*, Vol. 1411, No. 2-3, (May 1999), pp.(334-350), ISSN 0005-2728.

Doughan, A.K., & Dikalov, S.I. (2007). Mitochondrial redox cycling of mitoquinone leads to superoxide production and cellular apoptosis. *Antioxid Redox Signal*, Vol. 9, No. 11, (November 2007), pp. (1825-1836), ISSN 1557-7716

Duchen, MR. (1999). Contributions of mitochondria to animal physiology: From homeostatic sensor to calcium signalling and cell death. *J Physiol*, Vol. 516, No. 1, (April 1999), pp.(1-17), ISSN 1469- 7793

Dudkina, N.V., Sunderhaus, S., Boekema, E.J., Braun, H.P. (2008). The higher level of organization of the oxidative phosphorylation system: mitochondrial supercomplexes. *J Bioenerg Biomembr*, Vol. 40, No. 5, (October 2008), pp.(419-424), ISSN 1573-6881

Dumont, M.E., Cardillo, T.S., Hayes, M.K., Sherman, F. (1991). Role of cytochrome c heme lyase in mitochondrial import and accumulation of cytochrome c in Saccharomyces cerevisiae. *Mol Cell Biol*, Vol. 11, No. 11, (November 1991), pp.(5487–5496), ISSN 1098-5549

Echtay, K.S., Winkler, E., Frischmuth, K., Klingenberg, M. (2001). Uncoupling proteins 2 and 3 are highly active H(+) transporters and highly nucleotide sensitive when activated by coenzyme Q (ubiquinone). *Proc Natl Acad Sci USA*, Vol. 98, No. 4, (February 2001), pp.(1416–21), ISSN 1091-6490.

Erecińska, M., Wilson, D.F. (1982). Regulation of cellular energy metabolism. *J Memb Biol*, Vol. 70, No. 1, pp.(1-14), ISSN 1432-1424

Estevam, M.L., Nascimento, O.R., Baptista, M.S., Di Mascio, P., Prado, F.M., Faljoni-Alario, A., Zucchi, Mdo.R., Nantes, I.L. (2004). Changes in the spin state and reactivity of cytochrome c induced by photochemically generated singlet oxygen and free radicals. *J Biol Chem*, Vol. 279, No. 38, (September 2004), pp.(39214–39222), ISSN 1083-351X.

Feher, G. (1970). *Electron Paramagnetic Resonance with Applications to Selected Problems in Biology* (1 ed.) Gordon & Breach Science Publishers , ISBN 0677026757, New York.

Fernández-Vizarra, E., Tiranti, V., Zeviani, M. (2009). Assembly of the oxidative phosphorylation system in humans: What we have learned by studying its defects. *Biochim Biophys Acta*, Vol. 1793, No. 1, (January 2009), pp.(200–211), ISSN 0005-2728.

Fitton, V., Rigoulet, M., Ouhabi, R., Guérin, B. (1991). Mechanistic stoichiometry of mitochondrial oxidative phosphorylation. *Biochemistry*, Vol. 30, No. 32, (August 1991), pp.(3576–3582), ISSN 1178-6264

Foote, C.S. (1968). Mechanisms of photosensitized oxidation. There are several different types of photosensitized oxidation which may be important in biological systems. *Science*, Vol. 162, No. 857, (November 1968), pp.(963–970), ISSN 1095-9203

Forsmark-Andrée, P; Lee, CP; Dallner, G & Ernster, L. (1997). Lipid Peroxidation and Changes in the Ubiquinone Content and the Respiratory Chain Enzymes of Submitochondrial Particles. *Free Radical Biology and Medicine*, Vol. 22, No. 3, (August, 1997), pp. (391-400), ISSN 08915849

Fouad, AA; Al-Sultan, AI; Refaie, SM & Yacoubi MT. (2010). Coenzyme Q10 Treatment Ameliorates Acute Cisplatin Nephrotoxicity in Mice. *Toxicology*, Vol. 274, No. 1-3, (July- August, 2010), pp. (49-56), ISSN 0300-483X

Frei, B., Kim, M.C., & Ames, B.N. (1990). Ubiquinol-10 is an effective lipid-soluble antioxidant at physiological concentrations. *Proc Natl Acad Sci U S A*, Vol. 87, No. 12, (June 1990), pp. (4879–4883), ISSN 1091-6490

Gale, P.H., Arison, B.H., Trenner, N.R., Page, A.C. Jr., & Folkers, K. (1963). Coenzyme Q. 36. Isolation and characterization of coenzyme Q10 (H-10). *Biochemistry*, Vol.2, (Jan-Feb 1963), pp. (196-200), ISSN 0006-2960

García-Heredia, J.M., Díaz-Moreno, I., Nieto, P.M., Orzáez, M., Kocanis, S., Teixeira, M., Pérez-Payá, E., Díaz-Quintana, A., De la Rosa, M.A. (2010). Nitration of tyrosine 74 prevents human cytochrome c to play a key role in apoptosis signaling by blocking caspase-9 activation. *Biochim Biophys Acta*, Vol. 1797, No.6-7, (June-July 2010), pp.(981–993), ISSN 0005-2728

Genova, ML; Pich, MM; Biondi, A; Bernacchia, A; Falasca, A; Bovina, C; Formiggini, G; Parenti, CG & Lenaz, G. (2003). Mitochondrial Production of Oxygen Radical Species and the Role of Coenzyme Q as an antioxidant. *Experimental Biology and Medicine*, Vol. 228, No. 5, (May, 2003), pp. (506-513), ISSN 1535-3702

Geromel, V., Darin, N., Chretien, D., Benit, P., DeLonlay, P., Rotig, A., Munnich, A., & Rustin, P. (2002). Coenzyme Q(10) and idebenone in the therapy of respiratory chain diseases: rationale and comparative benefits. *Mol. Genet. Metab*, Vol. 77, No. 1-2, (September-October 2002), pp. (21–30), ISSN 1096-7192

Ghafourifar P., Cadenas, E. (2005). Mitochondrial nitric oxide synthase. *Trends Pharmacol Sci*, Vol. 26, No. 4 pp.(190–195), ISSN 0165-6147

Giorgio, M., Trinei, M., Migliaccio, E., Pelicci, P.G. (2007). Hydrogen peroxide: a metabolic by-product or a common mediator of ageing signals? *Nat Rev Mol Cell Biol*, Vol. 8, No. 9, (September 2007), pp.(722–728), ISSN 1471-0072

Gnaiger, E., Lassnig, B., Kuznetsov, A., Rieger, G., Margreiter, R. (1998). Mitochondrial oxygen affinity, respiratory flux control and excess capacity of cytochrome c oxidase. *J Exp Biol*, Vol. 201, No. 8, (April 1998), pp.(1129-1139), ISSN 1477-9145.

Goldsteins, G., Keksa-Goldsteine, V., Ahtoniemi, T., Jaronen, M., Arens, E., Akerman, K., Chan, P.H., Koistinaho, J. (2008). Deleterious role of superoxide dismutase in the mitochondrial intermembrane space. *J Biol Chem*, Vol. 283, No. 13, (March 2008), pp.(8446–8452), ISSN 1083-351X.

Gómez-Díaz, C., Rodríguez-Aguilera, J.C., Barroso, M.P., Villalba, J.M., Navarro, F., Crane, F.L., Navas, P. (1997). Antioxidant ascorbate is stabilized by NADH-coenzyme Q10 reductase in the plasmamembrane. *J Bioenerg Biomembr*, Vol. 29, No. 3, (June 1997), pp.(251–257), ISSN 1573-6881

Groves, J.T. (2006) High-valent iron in chemical and biological oxidations. *J Inorg Biochem*, Vol. 100, No. 4, (April 2006), pp.(434–447), ISSN 0162-0134

Guo, Q; Corbett, JT; Yue, G; Fann, YC; Qian, SY; Tomer, KB & Mason, RP. (2002). Electron Spin Resonance Investigation of Semiquinone Radicals Formed from the Reaction of Ubiquinone 0 with Human Oxyhemoglobin. *The Journal of Biological Chemistry*, Vol. 277, No. 8, (February, 2002), pp. (6104-6110), ISSN 1083-351X

Halliwell, B., & Gutteridge, J.M.C. (2007) *Free radicals in biology and medicine* (4 ed.), Oxford University Press, ISBN 019856869X, New York.

Halliwell, B., Gutteridge, J.M.C. (1990). Role of free radicals and catalytic metal ions in human disease: an overview. *Methods Enzymol*, Vol. 186, pp.(1–85), ISSN 0076-6879

Hatefi, Y. (1985). The mitochondrial electron transport and oxidative phosphorylation system. *Ann Rev Biochem*, Vol. 54, pp.(1015–1069), ISSN 0066-4154

Hüttemann, M., Pecina, P., Rainbolt, M., Sanderson, T.H., Kagan, V.E., Samavati, L., Doan, J.W., Lee, I. (2011). The multiple functions of cytochrome c and their regulation in life and death decisions of the mammalian cell: From respiration to apoptosis. *Mitochondrion*, Vol. 11, No. 3, (May 2011), pp.(369–381), ISSN 1567-7249

Ingold, KU; Bowry, VW; Stocker R & Walling, C. (1993). Autoxidation of Lipids and Antioxidation by alpha-Tocopherol and Ubiquinol in Homogeneous Solution and in Aqueous Dispersions of Lipids: Unrecognized Consequences of Lipid Particle Size as Exemplified by Oxidation of Human Low Density Lipoprotein. *Proceedings of the National Academy of Sciences*, Vol. 90, No. 1, (January, 1993), pp. (45-49), ISSN 0027-8424

James, A.M., Smith, R.A., & Murphy, M.P. (2004). Antioxidant and prooxidant properties of mitochondrial Coenzyme Q. *Arch Biochem Biophys*, Vol. 423, No. 1, (March 2004), pp. (47-56), ISSN 0003-9861

James, AM; Cocheme, HM; Smith RA &, Murphy, MP. (2005). Interactions of Mitochondria-targeted and Untargeted Ubiquinones with the Mitochondrial Respiratory Chain and Reactive Oxygen Species. Implications for the Use of Exogenous Ubiquinones as Therapies and Experimental Tools. *The Journal of Biological Chemistry*, Vol. 280, No. 22, (June, 2005), pp. (21295–21312), ISSN 1083-351X

Jones, R., Mann, T., Sherins, R. (1979). Peroxidative breakdown of phospholipids in human spermatozoa, spermicidal properties of fatty acid peroxides, and protective action of seminal plasma. *Fertil Steril*, Vol. 31, No. 5, (May 1979), pp.(531–537), ISSN 1556-5653

Joshi, V.C., Jayaraman, J., & Ramasarma, T. (1963). Tissue concentrations of conenzyme Q, ubichromenol and tocopherol in relation to protein status in the rat. *Biochem J*, Vol. 88, No. 25, (July 1963), pp. (25-31), ISSN 0264-6021

Kagan, V.E., Borisenko, G.G., Tyurina, Y.Y., Tyurin, V.A., Jiang, J., Potapovich, A.I., Kini, V., Amoscato, A.A., Fujii, Y. (2004). Oxidative lipidomics of apoptosis: redox catalytic interactions of cytochrome c with cardiolipin and phosphatidylserine. *Free Rad Biol Med*, Vol. 37, No. 12, (December 2004), pp.(1963–1985), ISSN 0891-5849

Kagan, V.E., Tyurin, V.A., Jiang, J., Tyurina, Y.Y., Ritov, V.B., Amoscato, A.A., Osipov, A.N., Belikova, N.A., Kapralov, A.A., Kini, V., Vlasova, I.I., Zhao, Q., Zou, M., Di, P., Svistunenko, D.A., Kurnikov, I.V., Borisenko, G.G. (2005). Cytochrome c acts as a cardiolipin oxygenase required for release of proapoptotic factors. *Nat Chem Biol*, Vol. 1, No. 4, (August 2005), pp.(223–232), ISSN 1552-4469

Kagan, V; Serbinova, E & Packer, L. (1990). Antioxidant Effects of Ubiquinones in Microsomes and Mitochondria are Mediated by Tocopherol Recycling. *Biochemical and Biophysical Research Communications*, Vol. 169, No. 3, (June, 1990), pp. (851-857), ISSN 0006-291X

Kawai, C., Pessoto, F.S., Rodrigues, T., Mugnol, K.C., Tórtora, V., Castro, L., Milícchio, V.A., Tersariol, I.L., Di Mascio, P., Radi, R., Carmona-Ribeiro, A.M., Nantes, I.L. (2009). pH-sensitive binding of cytochrome c to the inner mitochondrial membrane. Implications for the participation of the protein in cell respiration and apoptosis. *Biochemistry*, Vol. 48, No. 35, (September 2009), pp.(8335-8342), ISSN 1178-6264

Kawai, C., Prado, F.M., Nunes, G.L., Di Mascio, P., Carmona-Ribeiro, A.M., Nantes, I.L. (2005). pH-dependent interaction of cytochrome c with mitochondrial mimetic membranes. *J Biol Chem*, Vol. 280, No. 41 , (October 2005), pp.(34709–34717), ISSN 1083-351X

Kelso, G. F., Porteous, C. M., Coulter, C. V., Hughes, G., Porteous, W. K., Ledgerwood, E. C., Smith, R. A., & Murphy, M. P. (2001). Selective targeting of a redox-active ubiquinone to mitochondria within cells: antioxidant and antiapoptotic properties. *J. Biol. Chem* , Vol. 276, No. 7, (February 2001), pp. (4588–4596), ISSN 1083-351X

Kluck, R.M., Bossy-Wetzel, E., Green, D.R., Newmeyer, D.D. (1997). The release of cytochrome c from mitochondria: a primary site for Bcl-2 regulation of apoptosis. *Science*, Vol. 275, No. 5303, (February 1997), pp.(1132–1136), ISSN 1095-9203

Kluck, R.M., Martin, S.J., Hoffman, B.M., Zhou, J.S., Green, D.R., Newmeyer, D.D. (1997). Cytochrome c activation of CPP32-like proteolysis plays a critical role in a Xenopus cell free apoptosis system. *EMBO J*, Vol. 16, No. 15, (August 1997), pp.(4639–4649), ISSN 0261-4189

Kushnareva, Y., Murphy, A.N., Andreyev, A. (2002). Complex I-mediated reactive oxygen species generation : modulation by cytochrome c and NAD(P)u oxidation-reduction state. *Biochem J,* Vol. 368, No. 2, (December 2002), pp.(545-553), ISSN 1470-8728

Lass, A & Sohal, RS. (2000). Effect of coenzyme Q_{10} and α-Tocopherol Content of Mitochondria on the Production of Superoxide Anion Radicals. *The FASEB Journal,* Vol. 14, No. 1, (January, 2000), pp. (87-94), ISSN 1530-6860.

Lass, A; Forster, MJ & Sohal RS. (1999). Effects of Coenzyme Q10 and alpha-Tocopherol Administration on Their Tissue Levels in the Mouse: Elevation of Mitochondrial alpha-Tocopherol by Coenzyme Q10. *Free Radical Biology and Medicine,* Vol. 26, No. 11-12, (June, 1999), pp. (1375-1382), ISSN 0891-5849

Linnane, AW; Kopsidas, G; Zhang, C; Yarovaya, N; Kovalenko, S; Papakostopoulos, P; Eastwood, H; Graves, S & Richardson, M. (2002). Cellular Redox Activity of Coenzyme Q10: Effect of CoQ10 Supplementation on Human Skeletal Muscle. *Free Radical Research,* Vol. 36, No. 4, (April, 2002), pp. (445-453), ISSN 1071- 5762

Linnane, AW; Kios, M & Vitetta, L. (2007). Coenzyme Q(10)--Its Role as a Prooxidant in the Formation of Superoxide Anion/Hydrogen Peroxide and the Regulation of the Metabolome. *Mitochondrion,* Vol. 7, (June, 2007), pp. (51-61), ISSN 1567-7249

Litarru, GP & Tiano, L. (2007). Bioenergetic and Antioxidant Properties of Coenzyme $Q_{10.}$ *Molecular Biotechnology,* Vol. 37, No. 1, (August, 2007), pp. (31-37), ISSN 1073-6085

Liu, X., Kim, C.N., Yang, J., Jemmerson, R., Wang, X. (1996). Induction of apoptotic program in cell-free extracts: requirement for dATP and cytochrome c. *Cell,* Vol. 86, No. 1, pp.(147-57), ISSN 0092-8674

Liu, Z., Lin, H., Ye, S., Liu, Q., Meng, Z., Zhang, C., Xia, Y., Margoliash, E., Rao, Z., & Liu, X. (2006). Remarkably high activities of testicular cytochrome c in destroying reactive oxygen species and in triggering apoptosis. *Proc Natl Acad Sci USA,* Vol. 103, No. 24, (June, 2006), pp.(8965-8970), ISSN 1091-6490

Lowes, D.A., Thottakam, B.M., Webster, N.R., Murphy, M.P., & Galley, H.F. (2008). The mitochondria-targeted antioxidant MitoQ protects against organ damage in a lipopolysaccharide-peptidoglycan model of sepsis. *Free Radic Biol Med,* Vol.45, No. 11, (December 2008), pp. (1559-1565), 0891-5849

Mellors, A., & Tappel, A.L. (1966). Quinones and quinols as inhibitors of lipid peroxidation. *Lipids,* Vol.1, No. 4, (July 1966), pp. (282-284), ISSN 1558-9307

Mellors, A., & Tappel, A.L. (1966). The inhibition of mitochondrial peroxidation by ubiquinone and ubiquinol. *J Biol Chem.,*Vol. 241, No. 19, (October 1966), pp. (4353-4356), ISSN 1083-351X

Mitchell, P. Chemiosmotic coupling in oxidative and photosynthetic phosphorylation. (1966) *Biol Rev Camb Philos Soc,* Vol. 41, No. 3, (August 1966), pp.(445-502), ISSN 1469-185X

Moncada, S., Higgs, A. (1993). The L-arginine–nitric oxide pathway. *N Engl J Med,* Vol. 329, No. 27, (December 1993), pp.(2002-2012), ISSN 1533-4406

Moore, T. (1959). Ubiquinone and vitamin E. *Nature,* Vol. 184, (August 1959), pp.(607-608), ISSN 0028-0836

Mordente, A; Matorana GE; Santini SA; Miggiano, GA; Petitti, T; Giardina, B; Battino, M & Littarru, GP. (1993) Antioxidant Effect of Coenzyme Q on Hydrogen Peroxide-Activated Myoglobin. *Clin Investig,* Vol. 71, No. 8, (1993), pp. (92-96), ISSN 0941-0198

Mugnol, K.C., Ando, R.A., Nagayasu, R.Y., Faljoni-Alario, A., Brochsztain, S., Santos, P.S., Nascimento, O.R., Nantes, I.L. (2008). Spectroscopic, structural, and functional characterization of the alternative low-spin state of horse heart cytochrome c. *Biophys J*, Vol.94, No. 10, (May 2008), pp.(4066–4077), ISSN 1470-8728

Murphy, M.P. (2009). How mitochondria produce reactive oxygen species. *Biochem J* Vol. 417, No. 1, pp.(1–13), ISSN 1470-8728

Murphy, MP &.Smith, RA. (2007). Targeting Antioxidants to Mitochondria by Conjugation to Lipophilic Cations. *Annual. Review of Pharmacology and Toxicology*, Vol. 47, (February, 2007), pp. (629–656), ISSN 0362-1642

Nantes, I.L., Bechara, E.J.H., Cilento, G. (1996). Horseradish peroxidase-catalyzed generation of acetophenone and benzophenone in the triplet state. *Photochem Photobiol*, Vol. 63, No. 6, (June 1996), pp.(702–708), ISSN 1751-1097

Nantes, I.L., Faljoni-Alário, A., Nascimento, O.R., Bandy, B., Gatti, R., Bechara, E.J. (2000) Modifications in heme iron of free and vesicle bound cytochrome c by tert-butyl hydroperoxide: a magnetic circular dichroism and electron paramag-netic resonance investigation. *Free Radic Biol Med*, Vol. 28, No. 5, (March 2000), pp.(786–796), ISSN 0891-5849

Nantes, I.L., Mugnol, K.C.U. (2008) Incorporation of Respiratory Cytochromes in Liposomes: An Efficient Strategy to Study the Respiratory Chain. *J Liposs Res,*Vol.18, No. 3, pp.(175-194), ISSN 1532-2394

Nantes, I.L., Zucchi, M.R., Nascimento, O.R., Faljoni-Alario, A. (2001). Effect of heme iron valence state on the conformation of cytochrome c and its association with membrane interfaces. A CD and EPR investigation. *J Biol Chem*, Vol. 276, No. 1, (January 2001), pp.(153–158), ISSN 1083-351X

Olsson, J.M., Xia, L., Eriksson, L.C., & Björnstedt, M. (1999). Ubiquinone is reduced by lipoamide dehydrogenase and this reaction is potently stimulated by zinc. FEBS Lett, Vol. 448, No. 1, (April 1999), pp. (190-192), ISSN 0014-5793

O'Reilly, C.M., Fogarty, K.E., Drummond, R.M., Tuft, R.A., Walsh, J.V.Jr. (2003). Quantitative analysis of spontaneous mitochondrial depolarizations. *Biophys J*, Vol. 85, No. 5, (November 2003), pp.(3350–3357), ISSN 1470-8728

Ott, M., Robertson, J.D., Gogvadze, V., Zhivotovsky, B., Orrenius, S. (2002). Cytochrome c release from mitochondria proceeds by a two-step process. *Proc Natl Acad Sci*, Vol. 99, No. 3, (February 2002), pp.(1259–1263), ISSN 1091-6490

Ow, Y.-L.P., Green, D.R., Hao, Z., & Mak, T.W. (2008). Cytochrome c: functions beyond respiration. *Nature Rev Mol Cell Biol*, Vol. 9, (July, 2008), pp.(532–542), ISSN 1471-0072

Papucci, L., Schiavone, N., Witort, E., Donnini, M., Lapucci, A., Tempestini, A., Formigli, L., Zecchi-Orlandini, S., Orlandini, G., Carella, G., Brancato, R., Capaccioli, S. (2003). Coenzyme q10 prevents apoptosis by inhibiting mitochondrial depolarization independently of its free radical scavenging property. *J Biol Chem* Vol. 278, No. 30, (July 2003), pp.(28220–28228), ISSN 1083-351X

Pereverzev, M.O., Vygodina, T.V., Konstantinov, A.A., Skulachev, V.P. (2003). Cytochrome c, an ideal antioxidant. *Biochem Soc Trans*, Vol. 31, No. 6, (December 2003), pp.(1312–1315), ISSN 0300-5127

Piel, D.A., Deutschman, C.S., Levy, R.J. (2008). Exogenous cytochrome C restores myocardial cytochrome oxidase activity into the late phase of sepsis. *Shock*, Vol. 29, No. 5, (May 2008), pp.(612–616), ISSN 1073-2322

Piel, D.A., Gruber, P.J., Weinheimer, C.J., Courtois, M.R., Robertson, C.M., Coopersmith, C.M., Deutschman, C.S., Levy, R.J. (2007). Mitochondrial resuscitation with exogenous cytochrome c in the septic heart. *Crit Care Med*, Vol. 35, No. 9, (September 2007), pp.(2120–2127), ISSN 1530-0293

Plecitá-Hlavatá, L; Jezek, J & Jezek, P. (2009). Pro-oxidant Mitochondrial Matrix-Targeted Ubiquinone MitoQ10 Acts as Anti-oxidant at Retarded Electron Transport or Proton Pumping Within Complex I. *The International Journal of Biochemistry and Cell Biology*, Vol. 41, No. 8-9, (August-September, 2009), pp. (1697-1707), ISSN 1357-2725

Radi, R., Cassina, A., Hodara, R. (2002). Nitric oxide and peroxynitrite interactions with mitochondria. *Biol Chem*, Vol. 383, No. 3-4, (March-April 2002), pp.(401-409), ISSN 1437-4315

Radi, R., Turrens, J.F., Freeman, B.A. (1991). Cytochrome c-catalyzed membrane lipid peroxidation by hydrogen peroxide. *Arch Biochem Biophys*, Vol. 288, No. 1, (July 1991), pp.(118–125), ISSN 0003-9861

Riemer, J., Fischer, M., Herrmann, J.M. (2011). Oxidation-driven protein import into mitochondria: Insights and blind spots. *Biochim Biophys Acta*, Vol. 1808, No. 3, (March 2011), pp.(981–989), ISSN 0005-2728

Rinaldi, T.A., Tersariol, I.L., Dyszy, F.H., Prado, F.M., Nascimento, O.R., Di Mascio, P., Nantes, I.L. (2004). Protonation of two adjacent tyrosine residues influences the reduction of cytochrome c by diphenylacetaldehyde: a possible mechanism to select the reducer agent of heme iron. *Free Rad Biol Med*, Vol. 36, No. 6, (March 2004), pp.(802-810), ISSN 0891-5849

Rodrigues, T., de França, L.P., Kawai, C., de Faria, P.A., Mugnol, K.C., Braga, F.M., Tersariol, I.L., Smaili, S.S., Nantes, I.L. (2007). Protective role of mitochondrial unsaturated lipids on the preservation of the apoptotic ability of cytochrome C exposed to singlet oxygen. *J Biol Chem*, Vol. 282, No. 35, (August 2007), pp.(25577-25587), ISSN 1083-351X

Roginsky, V.A., Tashlitsky, V.N., & Skulachev, V.P. (2009). Chain-breaking antioxidant activity of reduced forms of mitochondria-targeted quinones, a novel type of geroprotectors. *Aging*, Vol.1, No. 5, (May 2009), pp. (481-489), ISSN 1945-4589

Rossingnol, R., Malgat, M., Mazat, J.P., & Letellier, T. (2000). Tissue variation in the control of oxidative phosphorylation : implication for mitochondrial diseases. *Biochem J*, Vol. 347, No. 1, (November 1999), pp.(41-43), ISSN 0264-6021

Rytömaa M., & Kinnunen, P.K.J. (1995) Reversibility of the binding of cytochrome c to liposomes. Implications for lipid-protein interactions. *J Biol Chem*, Vol. 270, No. 7, (February 1995), pp.(3197-3202), ISSN 0021-9258

Rytömaa, M., & Kinnunen, P.K.J. (1994). Evidence for two distinct acidic phospholipid-binding sites in cytochrome c. *J Biol Chem*, Vol. 269, (January 1994), pp.(1770–1774), ISSN 0021-9258

Rytömaa, M., Mustonen, P., & Kinnunen, P.K. (1992). Reversible, nonionic, and pH-dependent association of cytochrome c with cardiolipin-phosphatidylcholine liposomes. *J Biol Chem*, Vol. 267, (November 1992), pp.(22243–22248), ISSN 0021-9258

Schmelzer, C., Lindner, I., Vock, C., Fujii, K., Döring, F. (2007). Functional connections and pathways of coenzyme Q10-inducible genes: an in-silico study. *IUBMB Life,* Vol. 59, No. 10, (October 2007), pp.(628–633), ISSN 1521-6551

Sharma, R.K., Pasqualotto, F.F., Nelson, D.R., Thomas, A.J.Jr., & Agarwal, A. (1999). The reactive oxygen species−total antioxidant capacity score is a new measure of oxidative stress to predict male infertility. *Hum Reprod,* Vol. 14, No. 11, (November 1999), pp.(2801–2807), ISSN 1460-2350

Sies, H. (1997). Oxidative stress: Oxidants and Antioxidants. *Experimental Physiology,* Vol. 82, No. 2, (March, 1997), pp. (291-295), ISSN 1469-445X

Solaini, G., Sgarbi, G., Lenaz, G., & Baracca, A. (2007). Evaluating mitochondrial membrane potential in cells. *Biosci Rep,* Vol. 27, (June 2007), pp.(11–21), ISSN 1573-4935

Starkov, A.A., Fiskum, G., Chinopoulos, C., Lorenzo, B.J., Browne, S.E., Patel, M.S., & Beal, M.F. (2004) Mitochondrial alpha-ketoglutarate dehydrogenase complex generates reactive oxygen species. *J Neurosc,* Vol. 24, (September 2004), pp.(7779–7788), ISSN 0270-6474

Suto, D., Sato, K., Ohba, Y., Yoshimura, T., & Fujii, J. (2005). Suppression of the pro-apoptotic function of cytochrome c by singlet oxygen via a haem redox state-independent mechanism. *Biochem J,* Vol. 392, (December 2005), pp.(399–406), ISSN 0264-6021

Swartz, H.M., Bolton, J.R., & Borg, D.C. (1972). *Biological Applications of Electron Spin Resonance.* Wiley, ISBN 9780471838708, New York.

Tauskela, J.S. (2007). MitoQ - a mitochondria-targeted antioxidant. *IDrugs,* Vol.10, No. 6, (June 2007), pp. (399-412), ISSN 2040-3410

Tuominen, E.K., Wallace, C.J., & Kinnunen, P.K. (2002). Phospholipid-cytochrome *c* interaction: evidence for the extended lipid anchorage. *J Biol Chem,* Vol. 277, No. 11, (March 2002), pp.(8822-8826), ISSN 0021-9258

Turrens, J.F., & Boveris, A. (1980). Generation of superoxide anion by the NADH dehydrogenase of bovine heart mitochondria. *Biochem J,* Vol. 191, no. 2, (November 1980), pp.(421–127), ISSN 0264-6021

Turrens, J.F., Alexandre, A., & Lehninger, A.L. (1985). Ubisemiquinone is the electron donor for superoxide formation by complex III of heart mitochondria. *Arch Biochem Biophys,* Vol. 237, No. 2, (March 1985), pp.(408–414), ISSN 0003-9861

Turunen, M; Olsson, J & Dallner, G. (2004). Metabolism and Function of Coenzyme Q. *Biochimica et Biophisica Acta,* Vol. 1660, No.1-2, (January, 2004), pp. (171-199), ISSN 0005-2736

Valko, M., Izakovic, M., Mazur, M., Rhodes, C.J., & Telser, J. (2004). Role of oxygen radicals in DNA damage and cancer incidence. *Mol Cell Biochem,* Vol. 266, No. 1-2, (November 2004), pp.(37–56), ISSN 1098-5549

Valko, M., Leibfritz, D., Moncol, J., Cronin, M.T., Mazur, M., & Telser, J. (2007). Free radicals and antioxidants in normal physiological functions and human disease. *Int J Biochem Cell Biol,* Vol. 39, No. 1, (August 2007), pp.(44–84), ISSN 1357-2725

Whitman, GJ; Niibori, K; Yokoyama, H; Crestanello, JA; Lingle, DM & Momeni R. (1997). The Mechanisms of Coenzyme Q10 as Therapy for Myocardial Ischemia Reperfusion Injury. *Molecular Aspects of Medicine,* Vol. 18, No. (December, 1997), pp. (195-203), ISSN 0098-2997

Wittig, I., Braun, H.P., & Schägger, H. (2006). Blue native PAGE. *Nat Protoc*, Vol. 1, (June 2006), pp.(418–428), ISSN 1754-2189

Wolf, D.E., Hoffman, C.H., Trenner, N.R., Arison, B.H., Shunk, C.H., Linn, B.D., McPherson, J.F., & Folkers, K. (1958). Coenzyme Q. I. Structure studies on the coenzyme Q group. *J Am Chem Soc*, Vol. 80, No. 17, pp.(4752), ISSN 0002-7863

Xia, L., Björnstedt, M., Nordman, T., Eriksson, L.C., & Olsson, J.M. (2001). Reduction of ubiquinone by lipoamide dehydrogenase. An antioxidant regenerating pathway. Eur J Biochem., Vol. 268, No. 5, (March 2001), pp. (1486-1490), ISSN 1432-1033

Xia, L., Nordman, T., Olsson, J.M., Damdimopoulos, A., Björkhem-Bergman, L., Nalvarte, I., Eriksson, L.C., Arnér, E.S., Spyrou, G., & Björnstedt, M.(2003). The mammalian cytosolic selenoenzyme thioredoxin reductase reduces ubiquinone. A novel mechanism for defense against oxidative stress. J Biol Chem., Vol. 278, No. 4, (January 2003), pp. (2141-2146), ISSN 1083-351X

Xiong, J., & Bauer, C.E. (2002). Complex evolution of photosynthesis. *Annu Rev Plant Biol*, Vol. 53, (June 2002), pp.(503–521), ISSN 1543-5008

Yang, J., Liu, X., Bhalla, K., Kim, C.N., Ibrado, A.M., Cai, J., Peng, T.I., Jones, D.P., & Wang, X. (1997). Prevention of apoptosis by Bcl-2: release of cytochrome *c* from mitochondria blocked. *Science*, Vol. 275, No. 5303, (February 1997), pp.(1129–1132), ISSN 1095-9203

Yu, H., Lee, I., Salomon, A.R., Yu, K., & Hüttemann, M. (2008). Mammalian liver cytochrome c is tyrosine-48 phosphorylated in vivo, inhibiting mitochondrial respiration. Biochim. Biophys. Acta. Vol. 1777, No. 7-8, (July-August 2008), pp. (1066–71), ISSN 0006-3002

Zamzami, N., Maisse, C., Metivier, D., & Kroemer, G. (2007). Measurement of membrane permeability and the permeability transition of mitochondria. Methods. Cell. Biol., Vol. 80, (February 2007), pp. (327–340), ISSN 1941-7322

Zhang, L., Yu, L., & Yu, C.-A. (1998). Generation of superoxide anion by succinate-cytochrome c reductase from bovine heart mitochondria. J. Biol. Chem., Vol. 273, No. 51, (December 1998), pp. (33972-33976), ISSN 0021-9258

Zhao, X., Leon, I.R., Bak, S., Mogensen, M., Wrzesinski, K., Hojlund, K., & Jensen, O. N. (2011). Phosphoproteome analysis of functional mitochondria isolated from resting human muscle reveals extensive phosphorylation of inner membrane protein complexes and enzymes. Mol. Cell. Proteomics., Vol. 10, No. 1, (September 2010), pp. (1-14), ISSN 1535-9476

Zucchi, M.R., Nascimento, O.R., Faljoni-Alario, A., Prieto, T., & Nantes, I.L. (2003). Modulation of cytochrome c spin states by lipid acyl chains: a continuous-wave electron paramagnetic resonance (CW-EPR) study of haem iron. Biochem. J., Vol. 370, No. 2, (November 2002), pp. (671-678), ISSN 0264-6021

3

Bioenergetics Theory of Aging

Alexander G. Trubitsyn
Institute of Biology and Soil Sciences,
Far East Division Russian Academy of Sciences,
Vladivostok,
Russia

1. Introduction

The average lifespan of people in developed countries has tripled since ancient times while its maximum longevity (about 120 years) has remained invariable. The strategic goal of gerontology is to exceed this limit, i.e. to develop remedies which would allow the living of an indefinitely long life. However there have not been any significant advances in solving this problem so far. There is still no answer to even the fundamental question: what is the primary cause of degradation for all of an organism's functions (otherwise known as aging)? Actually, there are too many answers to this question: over 300 aging theories have been developed, and each of them provides a different response (Medvedev, 1990), although the majority of these theories now have only historical importance. Theories of aging are traditionally divided into two alternative groups. First, stochastic theories claim that there are no specific aging genes and that an organism's deterioration is the result of damaging factors. Second, and by way of contrast, programmed-aging theories assert that longevity is predetermined by a genetic program. Stochastic theories have dominated the discussion since gerontology became a branch of science, and the idea that aging is programmed has not yet received wide recognition, even though there is a lot of empirical evidence supporting it. There are several factors which impede the wide recognition of this idea. First, there is no evidence that longevity is under the control of natural selection; and second, there is no convincing mechanism the programmed of aging. Adherents of this view currently search for longevity genes in a practically blind or ad hoc fashion (Holzenberger et al., 2003; Kenyon, 2010). Many such genes have been found for various organisms, ranging from unicellular creatures to mammals, but it is still unclear what processes they control (Anisimov, 2003).

There are several different theories which are currently under consideration and which are based on reliable, proven evidence: i) the free radical theory which claims that aging is caused by an increased damage rate in cell structures due to an increased generation-rate of reactive oxygen species (ROS) by their own mitochondria; ii) the protein error theory which states that the primary cause is the age-dependent retardation of the protein synthesis rate; iii) the replicative senescence theory which argues that an age-dependent organism's senility is caused by the limitation of cell proliferation. There is also reliable evidence in support of other theories which are not as popular, for instance the immunological theory and several versions of neuroendocrinal theories.

The goal of this report is: (1) to show that despite the beliefs of the supporters of the stochastic theories, longevity is controlled by natural selection, i.e. specific aging genes exist; (2) such genes program a lowering of the bioenergetics level (degradation of Gibbs energy, ΔG). In turn, such degradation results in an age-dependent increase in the ROS generation rate, a decrease in the protein synthesis rate, and a limitation of cell division. These three phenomena form the basis for a large number of secondary destructive processes which result in the degradation of all physiological organisms' functions, i.e. the causes of aging. The very idea that bioenergetics exerts an impact upon aging is not a novelty. Hasty and Vijg (2002) have recently stated in theory that proper energy-saving could support a living system indefinitely. B.N. Ames (2004) has remarked that mitochondrial bioenergetics supports the metabolism's cell processes and that its attenuation can result in the age-dependent degradation of all of an organism's physiological functions. And indeed, life as a phenomenon is characterised by a number of physical and chemical processes driven by the power of the bioenergetics machine. A gradual decrease in bioenergetics level can cause the degradation of all vital processes. However they also believe that the cause of age-dependent bioenergetics attenuation is to be identified with the mechanism postulated by the free-radical theory. The following fact seems to reject the assumption of the direct programming of bioenergetics attenuation: one of main bioenergetics parameters is the mitochondrial membrane potential $\Delta\psi$. In vitro tests have shown that the superoxide ($O_2^{\cdot -}$) generation rate in the electron transport chain decreases as $\Delta\psi$ decreases. Consequently, in the process of bioenergetics attenuation the ROS level should decrease, but the tests show its increase in all tissues. And only that version of the vicious cycle brought forward by the free-radical theory can explain this paradox. Another mechanism which we have already suggested explains the increase in the number of reactive oxygen species during programmed bioenergetics attenuation (Trubitsyn, 2006). The bioenergetics mechanism of aging under consideration represents the integration of several of the author's articles published earlier (Trubitsyn, 2006, 2006a, 2009, 2010, 2011).

2. The increase in the level of reactive oxygen species is predetermined by programmed bioenergetics decay

2.1 Introduction
The free-radical theory of aging (the theory of oxidative stress, the oxidative damage theory and the mitochondrial theory of aging) was proposed by D. Harman (1956) in the middle of the 20th century and its improved version continues to dominate discussion. Its supporters claim that there are no specific aging genes because longevity cannot be controlled by natural selection (Kirkwood, 2002, 2008; Medawar, 1952). According to their view, age-dependent organism degradation results from the damage to cell structures by the ROS that are generated by mitochondria (Trifunovic & Larsson, 2008). This theory fascinates researchers by virtue of its simplicity and clarity. Indeed, it has been established that as an organism gets older, the ROS generation by the mitochondrial respiratory chain (electron transport chain) increases and the amount of damage to cell structures increases as well. The conclusion is obvious and the method for preventing aging is equally so: the neutralisation ROS by antioxidants. The age-dependent increase in the ROS generation rate is assumed to be just that: the ROS generated by mitochondria produce injury to its own mitochondrial DNA (mtDNA), which results in a defect in the respiratory chain. This, in turn, increases the rate of ROS production and as a result a vicious cycle arises.

When the free radical theory of aging appeared, it stirred up a brisk discussion which continues to this day. Empirical data has shown that there is no appreciable loss in the respiratory chain's functions during aging (Barrientos et al., 1996; Rasmussen et al., 2003). This is also supported by experiments indirectly related to the electron transport chain. For example, research into intra-cellular organelle transfers has shown that the mitochondria of old donors recover their functional activity completely when transferred to ρ0 HeLa cells (HeLa cells free of mtDNA) (Hayashi et al., 1994; Isobe et al., 1998). Mitochondrial dysfunctions are also eliminated when HeLa cell nuclei are transferred into the cells of old donors (Isobe et al., 1997). The authors concluded that nuclear factors are responsible for age-related mitochondrial deficiency. In addition, the conclusion that the age-dependent accumulation of mtDNA mutations is modulated by the nuclear genome was also made by Yao et al. (2007). The discussion has become especially vigorous over the last decade. On the one hand, based on this theory, it was claimed that "aging is no longer an unsolved problem in biology" (Hayflick, 2007; Holliday, 2006). On the other hand, R.M. Howes (2006) has declared that the "overly exuberant and exaggerated past expectations and claims of the free-radical theory have been quieted by extensive randomised, double-blind, controlled human studies. A half century of data demonstrates its lack of predictability and it has not been validated by the scientific method. Widespread use of antioxidants has failed to quell the current pandemic of cancer, diabetes, and cardiovascular disease or [even] to stop or reverse the aging process." His position is also supported by G. Bjelakovic et al. (2008) who have collected a great deal of data to show that antioxidants neither result in the beneficial effects expected nor do they increase life expectancy (in the best-case scenario). Gems & Doonan (2009) asked a question in a recent review: "Is the theory really dead, or does it just need to be modified?" Actually, there is more than ample evidence against the aging mechanism postulated by this theory than there is evidence in support of it.

2.2 Schema of the mitochondrial bioenergetics machine

Mitochondria generate about 90 percent of the energy in any eukaryotic cell. Therefore, only the mitochondrial bioenergetics machine will be considered here. Any energy system can be quantitatively described by its propellant power (F) and by its effect (A): F = kA: this is the force and the work in mechanical engineering (k is the friction) and the electromotive force and current in electrical engineering (k is the resistance). It is the free-energy change (Gibbs energy, ΔG) and current in chemical thermodynamics (in bioenergetics in particular). Such terms as the bioenergetics level and the level of energy production are used in bioenergetics to express the propellant power. To make it clear, let us recall that the ΔG of macroergic (high-energy) coenzymes that function in the bioenergetics machine (ATP, NAD, NADP, GSH, etc.) is determined by the value of their concentrations ratio of the reduced form to the oxidised one and by the temperature. For ATP, for instance, $\Delta G = \Delta G^0 - RT \ln[ATP]/[ADP][P_i]$, where ΔG^0 is the standard Gibbs energy that is measured with everything at 1 molar concentration: $[ATP] = [ADP] = [P_i] = 1M$; R is the gas constant; T is the absolute temperature. The more negative Gibbs energy there is, the higher the energy potential the bioenergetics machine generates. As follows from the above expression for ΔG, the concentrations ratio of the reduced to oxidised forms of macroergic coenzymes ($[ATP]/[ADP]$, $[NADH]/[NAD^+]$, etc.) is the only variable which determines the energy potential for warm-blooded animals.

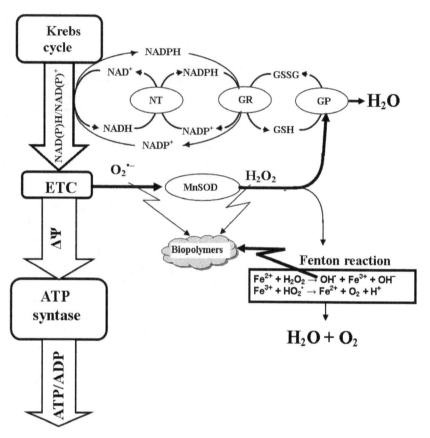

Bioenergetics machine. The primary motive power, NADH/NAD+, is created in the Krebs cycle. The mitochondrial membrane potential, Δψ, is created by the electron flow from NADH to oxygen through the electron transport chain (ETC). ATP-synthase phosphorylates ADP into ATP at the expense of Δψ. *Scavenging mechanism.* The superoxide radical ($O_2^{\cdot-}$) produced by ETC is transformed into hydrogen peroxide, H_2O_2, by manganese superoxide dismutase (MnSOD). H_2O_2 is then decomposed into H_2O and O_2 mainly through the reaction that is catalyzed by glutathione peroxidase (GP) and partially through the Fenton reaction; the last produces an extremely aggressive hydroxyl radical. The glutathione peroxidase activity mainly predetermines the rate of the scavenging process. This activity is sustained by the energy provided by glutathione (GSH) oxidation. The thus GSSG formed is reduced again into GSH at the expense of the oxidation of NADPH in a reaction that is catalysed by glutathione reductase (GR). The NADP+ formed is reduced in turn at the expense of the oxidation of NADH in the reaction catalysed by nicotinamide nucleotide transhydrogenase (NT). The NAD+ formed is reduced by the reactions of the Krebs cycle. The NADP+ can also be directly reduced in the isocitrate dehydrogenase reaction of the Krebs cycle. The chain of these redox reactions is the electrons' pipeline from the Krebs cycle to glutathione peroxidase. *The mechanism of ROS increase.* The programmed bioenergetics decline leads to a proportional decrease in GP activity, which increases the H_2O_2 level. As hydrogen peroxide is a substrate for the Fenton reaction, this augments the H_2O_2 flow through the Fenton reaction, which elevates the content of free radicals. Thus, a decline in the bioenergetics level is followed by an increase in the total amount of reactive oxygen species and its aggressiveness.

Fig. 1. Scheme explaining the mechanism of the ROS increase under the bioenergetics decline.

Researchers divide energy-metabolism reactions into a different number of functional blocks depending upon their purpose. For example, Ainscow and Brand (1998) have divided it into nine blocks connected to each other by five intermediates. To solve the problem under consideration, the bioenergetics machine may be divided into three blocks (the Krebs cycle, the electron transport chain and ATP-synthase) connected by two intermediates ([NADH]/[NAD$^+$] and $\Delta\psi$ (Fig.1)). According to this scheme, the output potential ([ATP]/[ADP]) is generated in three stages. At the first stage, the primary electromotive force, [NADH]/[NAD$^+$], is created by reducing NAD$^+$ to NADH. This serves as the propellant power for stage two where electrons are transferred from NADH to oxygen via the electron transport chain, generating the mitochondrial membrane potential $\Delta\psi$. At the third stage, $\Delta\psi$ is the electromotive force for ATP-synthase which generates the output potential. If there are no excessive loads (in stage four or close to it) then the [NADH]/[NAD$^+$] change results in a proportional change in $\Delta\psi$ and in [ATP]/[ADP].

2.3 ROS-scavenging mechanisms

During the aerobic metabolism, a small number of the electrons that flow from NADH via the respiratory chain react with oxygen directly reducing oxygen to superoxide anion ($O_2^{\bullet-}$ or $HO_2^{\bullet}$) (Demin et al., 1998; Scandalios, 2002a) which can damage cell biopolymers. Cells have a protective system that can be conditionally divided into three functional lines of defence: preventative mechanisms, ROS-scavenging mechanisms, and emergency-response mechanisms. The preventative mechanisms either prevent $O_2^{\bullet-}$ generation or oxidise superoxide back into O_2 at its location of generation (Brand, 2000; Skulachev, 2001). The emergency-response mechanisms are actuated when the ROS amount exceeds a critical level and when the cumulative effect of other mechanisms cannot improve the situation. However ROS not only damage biopolymers but it also plays an important role in the regulation of transcription factors, growth factors and other intracellular signal systems (Brigelius-Flohe et al., 2003; Cerimele et al, 2005; Rhee, 1999; Scandalios, 2002). The cell needs ROS, but their concentration should be maintained at a safe level. Therefore, there is a dedicated ROS-scavenging mechanism to maintain the ROS homeostasis. This mechanism performs the $O_2^{\bullet-}$ detoxification through a two-stage process (Fig. 1). At first, the manganese-containing mitochondrial superoxide dismutase (MnSOD) transforms superoxide into hydrogen peroxide (H_2O_2) (Jonas et al., 1989; Scandalios, 2002a) which is then decomposed by catalase and peroxidases. Most H_2O_2 is decomposed in cytosol by catalase and in the mitochondrial matrix by the glutathione and thioredoxin systems (catalase is absent in the mitochondrial matrix) (Wei et al., 2001). The glutathione system consists of glutathione peroxidase (GP) and glutathione reductase (GR). The GP potency is maintained due to the oxidation of glutathione (GSH) which is converted into its disulphide form (GSSG). Next, the GR catalyses the reduction of the oxidised glutathione at the expense of NADPH oxidation (Arai et al., 1999; Jo et al., 2001; Iantomasi et al., 1993). The NADP$^+$ thus formed is reduced again to NADPH in the isocitrate dehydrogenase reaction of the Krebs cycle (Jo et al., 2001). There is an analogous system – the thioredoxin system – which functions in parallel with the glutathione system and which also consists of thioredoxin peroxidase (TP) and thioredoxin reductase (TR) (Jo et al, 2001; Nordberg & Arner, 2001). Similarly, the TP potency is maintained by the oxidation of thioredoxin which is then reduced by TR, also at the expense of NADPH oxidation (Lewin et l., 2001). For the sake of simplicity, this parallel system is not shown in Fig. 1. The reaction that is catalysed by these peroxidases is simple: H_2O_2 takes two electrons from the glutathione (thioredoxin) and two protons from the environment and then decays into two water molecules: $H_2O_2 + 2e^- + 2H^+ = 2H_2O$. Only GP and TP catalyse this reaction directly; the other reactions are a pipeline by which energy is transferred from the Krebs cycle to glutathione peroxidase with thioredoxin peroxidase providing their activity (Iantomasi et al., 1993). The activity of any energy-dependent chemical reaction depends upon the energy supply (Westerhoff & van Dam, 1987). Therefore, the more the $NADPH/NADP^+$ ratio is generated in the Krebs cycle, the higher the GP and TP activity, and vice versa. It was shown experimentally that bioenergetics attenuation results in decrease of the scavenging mechanism's activity (Jo et al., 2001). It should be also noted that the ROS-scavenging mechanism can to some extent adapt to changes in the ROS level: the cell responds to a higher ROS concentration by a higher synthesis rate for MnSOD and glutathione-system enzymes (Meewes et al., 2001). An increase in the gene expression of those enzymes is mediated by the transcription nuclear factor-κB that is activated under excessive amounts of ROS (Scandalios, 2002a; Schreck et al., 1991).

2.4 Fenton reaction

There is additional the ferrous-ion catalysed means of hydrogen peroxide decomposition, which is called the Fenton reaction. In its simplest form, the Fenton chemistry is a chain mechanism of certain reactions in which H_2O_2 breaks up into water and oxygen and where Fe^{2+} is regenerated (Dunford, 2002):

$$Fe^{2+} + H_2O_2 \rightarrow Fe^{3+} + {}^{\bullet}OH + HO^- \tag{1}$$

$$Fe^{2+} + {}^{\bullet}OH \rightarrow Fe^{3+} + HO^- \tag{2}$$

$${}^{\bullet}OH + H_2O_2 \rightarrow HO_2{}^{\bullet -} + H_2O \tag{3}$$

$$HO_2{}^{\bullet} + Fe^{2+} \rightarrow Fe^{3+} + HO_2{}^- \tag{4}$$

$$HO_2{}^{\bullet} + Fe^{3+} \rightarrow Fe^{2+} + O_2 + H^+ \tag{5}$$

As distinct from the glutathione system, the iron decomposes H_2O_2 due to its ability to undergo cyclic oxidation and reduction. However, such redox activity of iron can generate free radicals capable of causing a wide range of biological injuries (Liu et al. 2003). The hydroxyl radicals (${}^{\bullet}OH$) formed during the Fenton reaction are true chemical predators: indeed, the reactivity of ${}^{\bullet}OH$ is so great that, if they are formed in living systems, they will react immediately with whatever biological molecule is in their vicinity, producing secondary radicals of variable reactivity (Halliwell & Gutteridge 1984; Yu & Anderson 1997). Among $O_2{}^{\bullet-}$, H_2O_2 and ${}^{\bullet}OH$, only the hydroxyl radical can directly cause double-stranded DNA breaks (Aruoma 1994).

2.5 The mechanism of age-dependent increase in ROS level

The Fenton reaction actually shunts the ROS-scavenging mechanism. As a result, H_2O_2 molecules are decomposed both by the glutathione system and the Fenton reaction. As the two ways of hydrogen peroxide decomposition compete for the substratum, the fraction of

H_2O_2 which can produce $^\bullet OH$ (Q_r) is predetermined by both the activity of the Fenton reaction (A_f) and that of glutathione peroxidase (A_g): $Q_r = A_f / (A_f + A_g)$. Thus, the lower the level of the activity of glutathione peroxidase and thioredoxin peroxidase, the higher the level of ROS production.

As has been mentioned, a decrease in the energy metabolism rate should, in theory, result in a lowering of the $O_2{}^{\bullet-}$ generation rate. Indeed, this is just what happens. However the concentration of the other ROS does not only depend upon the $O_2{}^{\bullet-}$ generation rate: the programmed age-dependent delay in the bioenergetics level results in a decrease in GP and TP activity. This raises the concentration of their substrate, H_2O_2. Since hydrogen peroxide is a substrate for the Fenton reaction as well, it augments the current through this reaction. As a result, the total amount of ROS and their aggressiveness increases despite a decrease in the $O_2{}^{\bullet-}$ generation rate.

2.6 Conclusion

Accordingly, the leading cause of the age-dependent increase in the amount of ROS and its aggressiveness is a programmed attenuation of cellular bioenergetics rather than a progressive accumulation of mutations in mtDNA due to the creation of a vicious cycle.

3. The age-dependent attenuation of bioenergetics underlies a decrease in the general level of protein synthesis

3.1 Introduction

A different popular aging theory, the protein-error theory, is based on the indisputable fact that the bulk protein synthesis slows down during aging (Rattan, 1996, 2009; Ryazanov & Nefsky, 2002). According to the theory, such retardation results in a decreasing protein turnover rate which causes the accumulation of defective macromolecules. S.I.S. Ratton (1996), who has investigated this process in detail, reports that "the implications and consequences of slower rates of protein synthesis are manifold, including a decrease in the availability of enzymes for the maintenance, repair and normal metabolic functioning of the cell, an inefficient removal of inactive, abnormal and damaged macromolecules in the cell, the inefficiency of the intracellular and intercellular signalling pathways, and a decrease in the production and secretion of hormones, antibodies, neurotransmitters and the components of the extra cellular matrix." The reason behind a slower protein synthesis rate is seen in the stochastic accumulation of molecular damage and the progressive failure of maintenance and repair (Rattan, 2009). It entails damage to fragments of the translation mechanism: "a decline in the efficiency and accuracy of ribosomes, an increase in the levels of rRNA and tRNA, and a decrease in the amounts and activities of elongation factors" (Rattan, 2006). At the same time, there is empirical evidence which allows for the explanation of the slowing down of overall protein synthesis by the attenuation of cellular bioenergetics.

3.2 The mechanism for the decrease in the level of cell protein synthesis

It was D.A. Young (1969) who discovered a relationship between the protein synthesis rate and the bioenergetics level for the first time. When conducting experiments on glucocorticoid hormones, he noticed that the rate of amino acids inclusion into a growing polypeptide chain depends upon the entry of carbohydrates (glucose, pyruvate, and lactate) into cells. An assumption was made that this effect is connected with the ATP generation rate. It was shown thereafter that the protein synthesis rate depends upon the ADP/ATP and GDP/GTP ratios rather than on the absolute ATP value (Hucull et al., 1985; Mendelsohn et al., 1977; Young, 1970). In these tests, minor changes in the nucleotide diphosphate / nucleotide triphosphate ratio resulted in a significant effect on the range corresponding to a physiological energy level. The authors came to the conclusion that the ADP/ATP and/or GDP/GTP ratios are a physiological regulator of the protein synthesis rate.

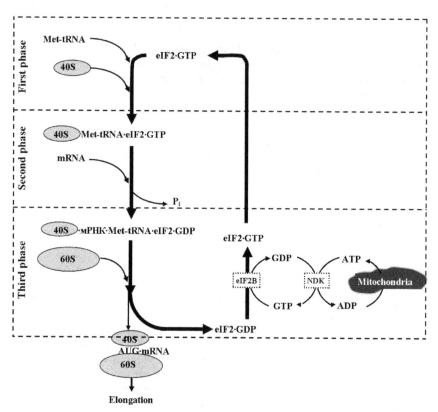

The initiation of translation can be divided into three phases. Phase one: the initiator methionine transport RNA (Met-tRNA) binds with the pre-existing binary complex eIF2·GTP and the 40S ribosomal subunit to provide the pre-initiation complex 40S·Met-tRNA·eIF2·GTP. Phase two: the pre-initiation complex binds to messenger RNA (mRNA). When the pre-initiation complex stops at the initiation codon of the mRNA, the GTP molecule is hydrolysed to GDP, inorganic phosphorus (Pi) is liberated and the energy of oxidation is spent on bond formation. This powers the ejection of the factors bound to the 40S ribosomal subunit in the third phase. The continuity of the initiation of these events requires the recycling of initiation factor molecules. eIF2 is released as an inactive binary complex with GDP and requires a guanine nucleotide exchange factor, eIF2·B, to catalyse regeneration of the eIF2 GTP. Energy support of regeneration is carried out at the expense of GTP oxidation. The GDP formed is then reduced at the expense of ATP oxidation in a reaction catalysed by nucleoside diphosphate kinase (NDK). The ADP formed is in turn reduced to ATP in the mitochondrial bioenergetics machine. The programmed bioenergetics decline decreases the eIF2 recirculation rate and thus reduces the general level of protein synthesis.

Fig. 2. The simplified scheme for the initiation of translation and its connection with bioenergetics.

The molecular mechanism of protein synthesis is currently well-understood and has been detailed in a number of reviews (Pain, 1996; Rattan, 2009). It was shown that the protein synthesis rate for eukaryotes is controlled at the translation level (Hucul, et al., 1985; Kimball et al., 1998). Among three translation stages (initiation, elongation and termination), the regulatory stage is the initiation (Hucul, et al., 1985; Kimball et al., 1998). The goal of this stage is the sequential binding of first the 40s and then the 60s ribosomal subunit to a messenger RNA molecule. At least 12 recirculation eukaryotic initiation factors (eIF) are involved in this stage. The initiation process can be divided into three phases (Fig. 2): (1) the association of the Met-tRNA initiator and several initiation factors with the 40s ribosomal subunit so as to form the pre-initiation complex; (2) the binding of this complex to a messenger RNA (mRNA) molecule, and (3) the addition of the 60s ribosomal subunit to assemble an 80s ribosome at the initiation codon.

The first initiation phase starts with the binding of the Met-tRNA initiator to a pre-existing double complex eIF2·GTP. When this preinitiation complex binds to mRNA at the second phase, GTP is oxidised to form GDP, and the oxidation energy is used to create bonds, with inorganic phosphorus being released. At the third stage, when the goal has been reached, the preinitiation complex disintegrates into separate initiation factors; these factors are then recycled to catalyse further initiation events. eIF2 is released as a binary complex with GDP, which is stable but not functionally active, i.e. it is unable to bind to a new Met-tRNA. A guanine nucleotide exchange factor, eIF2B, is required to catalyse the regeneration of the eIF2·GTP. Energy for such regeneration is provided by ATP oxidation to form ADP and the ADP is then reduced in the bioenergetics machine. Thus, the total protein synthesis level is originally regulated by the eIF2 recirculation rate which, in turn, depends upon the cellular bioenergetics value.

If the GDP-to-eIF2·GDP reduction is interrupted, the protein synthesis in the cell is blocked (Clemens, 1994). The natural mechanism protecting an organism in various stressful situations is based on this phenomenon: the phosphorylating of α-subunit eIF2 by different specific protein kinases blocks the reaction of the GDP-to-GTP exchange, which results in a complete protein synthesis termination in the cell followed by apoptosis (Clemens, 1994; Clemens et al., 2000). Such specific protein kinases are expressed in the cell when emergencies occur, such as an occurrence of the double-stranded replicative form of viral RNA (Jeffrey et al. 2002; Pain, 1996; Robert et al., 2006), irreparable damage of the genetic apparatus (Zykova et al., 2007; Jeffrey et al, 2002), acute shortage in amino acids (Clemens et al., 2001; Harmon et al., 1984), and malignant cell transformation (Clemens, 1994, Mendelsohn et al., 1977). Under normal physiological conditions when there are no specific protein kinases, the GDP-to-GTP exchange rate in the eIF2·GDP complex (and, consequently, the total protein synthesis rate) is regulated by the cellular bioenergetics (Hucull et al., 1985).

3.3 Conclusion
Programmed bioenergetics decline is the original cause of overall protein synthesis decrease rather than the stochastic accumulation of molecular damage.

4. The Hayflick limit is caused by the age-related decrease in the bioenergetics level

4.1 Introduction
Tissue senility is the most visible phenomenon and one of the most harmful phenomena of organism aging. Its cause was determined half a century ago (Hayflick & Moorhead, 1961):

higher eukaryotic cells do not divide infinitely, and, after a certain number of doublings, they enter a nondividing but viable state. Human fibroblasts, for example, are able to divide 53 ± 6 times over 302 ± 27 days and be in a stationary state for 305 ± 41 more days (Bayreuther et al., 1988). This limitation of division, the Hayflick limit, underlies the replicative aging theory, which is recognized to be one of the most striking modern aging theories (Anisimov, 2003). The main postulate of this theory is that, due to accumulation of old nondividing cells, tissue renewing homeostasis is violated, which causes their degradation (Hornsby, 2002; Itahana et al., 2004; Yegorov & Zelenin, 2003).

4.2 Modern views on the cause of cell proliferation limitation

A convincing mechanism of termination of old cells division was predicted theoretically by A.M. Olovnikov in 1971 and then confirmed experimentally (Greider & Blackburn, 1985). Vertebrates' chromosome ends from the DNA 3'-end have repeating nucleotide sequences — telomeres. They prevent fusion of chromosome ends, protect DNA from nuclease digestion, and participate in doubled chromosome disjunction in mitosis. In embryonic cells telomeres are synthesized by a special enzyme telomerase, which most somatic cells do not have. Because of the necessity of RNA-primer during DNA reduplication initiation, the telomere ends of somatic cells chromosomes are shortened with every cycle. As a result, after a certain number of doublings, the telomere end is depleted and divisions are terminated due to chromosome erosion (Itahana et al., 2004). This mechanism was confirmed by numerous empirical facts: 90–95% of potentially immortal cancer cells possess telomerase activity and the telomere end of their chromosomes is not shortened; suppression of telomerase activity in these cells causes shortening of the telomere end and division termination, i.e., aging; and restoration of telomerase activity makes them potentially immortal again. Therewith, facts contradictory to this conception were accumulated. The most convincing of them were obtained by a research group led by Blasco (Blasco et al., 1997). They obtained mice zygotes lacking a telomerase gene but with full-sized initial chromosome telomere ends. Mice developed from these zygotes were not only viable, but also fertile. This initial telomere length was sufficient to maintain normal viability of six mouse generations. In the first generation, for example, mice passed through youth and maturity successfully and died in old age having 80% of telomeres in reserve. Only in the fifth and sixth generations did anomalies caused by chromosome telomere end depletion appear. These data were confirmed by another group of authors led by Herrera (Herrera et al., 1999). They obtained an analogous mouse line, but with a shortened initial telomere end, and repeated the experiments of Blasco et al. These mice were viable for only four generations, and anomalies in late generations were related with depletion of telomeres in cells of tissues with the most intensive proliferation (Lee et al., 1998). By the present time, researchers of the telomere mechanism incline to the conclusion that loss of the telomere end indeed leads to chromosome erosion and cell death, but cell proliferation termination during normal physiological cell aging happens earlier than this critical moment and a cell that has expended all its proliferative potential still contains a significant telomere reserve. The telomere mechanism serves as an additional barrier on the road to reproduction of malignant cells (Itahana et al., 2004). The conclusion that there is nonparticipation of the telomere apparatus in the mechanism of termination of old cells' division could have been drawn from the very beginning. It followed from the results of the initial Hayflick experiments that, after a certain number of doublings, a cell enters a nondividing, but viable, state, and there is no sense in discussing viability if division termination due to chromosome erosion is accepted. Therefore, the question of the Hayflick limit's nature is without answer. Apparently, an alternative reason for this phenomenon should be looked for in the mechanism of cell division.

4.3 The reason for termination of proliferation of old cells

The cell division cycle (proliferative cycle) is divided into four phases (Sherr, 1994): G1, S, G2, and M (Fig.1).

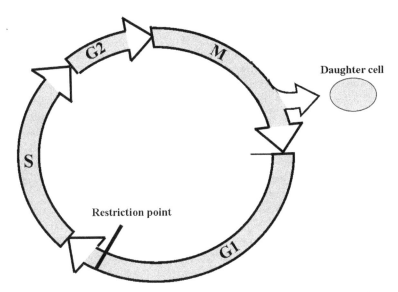

Cycle of cell reduplication is divided into 4 phases: G1 (gap 1), S (synthesis), G2 (gap 2), and M (mitosis).

Fig. 3. Phases of proliferative cell cycle.

In the G1-phase, precursor molecules necessary for DNA reduplication and doubling of all cell structures in the following division are synthesized. In the following S-phase, DNA is reduplicated, and after a short G2-phase, a cell enters M. Numerous studies have showed that all non dividing cells stay in the G1-phase. If a cell has passed through this phase, then it will pass through other phases automatically with almost equal speed. As far as entry of old cells into irreversible proliferative quiescence is concerned, we will be interested only in events occurring in the G1-phase. Control of the cell division rate is performed by endogenous and exogenous (for a cell) regulatory factors that are stimulators and inhibitors of proliferation. As an example of such regulation, the data of one of the first works in this field (Leof, et al., 1982) accurately reflecting the essence of the phenomenon are shown. The effect of different growth factors on mouse fibroblasts was studied. It was shown that, right after mitosis completion, a cell enters a proliferative quiescence state between the M and G1-phases (the G0-phase). To be removed from this state, a cell needed an external proliferative signal from platelet derived growth factor (PDGF). No structural or biochemical changes in a quiescent cell happened without the signal, and it remained insensitive to other proliferative stimuli. This primary stimulus is a competence factor. After a cell has received this signal, the biochemical reactions for new division cycle preparation begin, stopping a period of time. For further development of biochemical events, epidermal growth factor (EGF), not PDGF, was necessary. The addition of EGF caused continuation of biochemical and structural changes, but after some time a new halt at the G1/S-transition occurred, which was called the restriction point R. The passage of the last several hours of the G1-phase happened only under somatomedin C stimulation (Sm-C). The last two factors were called the first and the second progression factors. All tissue cells are stimulated by its growth factors. In addition to growth factors, passing through a cell cycle is regulated by a large group of inhibitors (Sherr & Roberts, 1999; Sherr, 2000). Cells can leave a cycle and move to a quiescent state. There are three types of quiescence. (1) Irreversible quiescence, or the terminal differentiation state, in which cells lose growth factors' receptors and become incapable of returning to a proliferative cycle (for example, neural, secretory, and muscular). (2) Temporal quiescence necessary for a cell to function within one or another tissue. This occurs if a cell does not receive the necessary proliferative stimulus from growth factors or there are exogenous inhibitors in the environment that void their proliferative signal. Such cells retain the integrity of their receptor apparatus, and, in appropriate conditions, they are able to come back to a cycle (for example, hepatocytes, fibroblasts, and others). (3) Proliferative quiescence of old cells that spend all their proliferative potential is similar to temporal quiescence. Cells retain their receptor apparatus and the integrity of all structures necessary for proliferation, although division does not occur.

The first experiments to determine the reasons for termination of proliferation of old cells were performed by Rittling et al. (1986). They studied 11 biochemical reactions happening sequentially in the G1-phase in young and old cells. It was shown that, in old cells, all reactions occur in the same way as in young cells, but old cells stop at the restriction point and deepen in quiescence, not reflecting the proliferative stimulus by the second progression factor. If after some period of time these cells are stimulated by proliferative factors again, they will pass through all the stages of preparation to transfer to the S-phase and will come back to a proliferative quiescence. The authors concluded that, in old cells that have expended all their proliferative potential, the restriction point becomes impassable.

Events happening in the restriction point are studied intensely, mainly by researchers of carcinogenesis. Their interest is due to the fact that malignant cells pass this point without stopping, while a delay of the cycle of dividing normal postembryonic cells here is obligatory and, for old cells, as has already been mentioned, this point becomes an insuperable barrier. To date significant success in studying biochemical events in this point has been achieved.

The main regulators of reactions occurring in the division cycle are cyclin-dependent kinases (Cdks). They are the controllers of all events: the determine the order of reactions, their duration, and their intensity (Sherr, 1996). The function of cyclin-dependent kinases is simple: de novo synthesized gene-regulating proteins of a division cycle E2F leaves a translational conveyor, figuratively speaking, in a package. This package is retinoblastoma protein (Rb). Until these proteins are bound with Rb, they are inactive. Cyclin-dependent kinases phosphorylate Rb protein, and, after that, regulatory proteins are released and activate genes necessary for the division cycle (Sherr, 2000; Frolov & Dyson, 2004). Cdks themselves can be in an active or inactive state. Regulation of cyclin-dependent kinases' activity is quite complex (Morgan,1995) but it is enough to know two principal moments to uncover the discussed topic: (1) Cdk is activated when it is conjugated with a specific cyclin (which is evident from its name) and (2) an active Cdk-cyclin complex can be deactivated again if it is conjugated with a specific inhibitor of cyclin-dependent kinases. To date eight types of cyclin-dependent kinases marked with the numbers Cdk1, Cdk2, etc.; ten types of cyclins marked with the Latin letters cyclin A, cyclin B, etc.; and a large group of Cdk inhibitors that have individual number labels and represent several families have been found in mammals. Three Cdks (2, 4, and 6); the cyclins D, E, and A; and the inhibitors of INK4 (p15[ink4b], p16[ink4a], p18[ink4c], and p19[ink4d]) and CIP/KIP families (p21[cip1], p27[kip1], and p57[kip2]) regulate passage through G1 (Sherr & Roberts 1999; Sherr, 2000). INK4 inhibitors specifically interact with Cdk4 and 6 and function in the G1-phase until the restriction point, and CIP/KIP interact with all Cdks. Research on G1-phase events has increased greatly in the past decade. New biochemical participants and ways in which they interact have been found. Information about them can be found in several reviews and original papers (Bockstaele et al., 2009; Larrea et al., 2008; Rahimi & Leoff, 2007; Sherr, 2000). Here only the main events minimally sufficient for understanding of termination of old cells division will be discussed. Leaving out the details, the G1-phase passing scheme discussed in (Sherr, 1996; Sherr & Roberts, 1999) can be shown in the following way (Fig. 4). The level of inhibitor p27 in quiescent cells is high, which prevents the reaction for division preparation. In response to mitogen stimulation, cyclin D is expressed and the active complex cyclin D-Cdk4 is formed, as are phosphorylates Rb. As a result gene-regulating E2F proteins are released and phosphorylated Rb is degraded. Then E2F proteins activate enzyme genes necessary for DNA reduplication in the S-phase and cyclin E, Cdk2, and E2F genes. Released cyclin E and Cdk2 form an active E-Cdk2 complex, which began interacting with p27, phosphorylating Rb, and activating regulatory protein genes. It is important that the cyclin E-Cdk2 complex activates the genes of their components, i.e., it reproduces itself. As a result a positive feedback loop is formed and promotes rapid p27 removal and E2F proteins and S-phase proteins' avalanche-like increase, which allows a cell to pass through a restriction point. With this E2F increased expression induces synthesis of inhibitor p53, which terminates the E2F expression unnecessary in the S-phase. However, this and the following cycle reactions are outside the discussed topic. Two research groups simultaneously and independently drew a considerable specification of the character of interaction of p27 with active cyclin E-Cdk complex (Vlach et al, 1997; Sheaf et al., 1997). Until their works it was considered that p27 and active cyclin E-Cdk2 complex interaction had a single consequence — complex inactivation. They performed a study of the kinetics of the molecular interactions of these compounds and showed that not only does the inhibitor inactivate the complex, but the complex can also attack an inhibitor phosphorylating it on threonine 187.

Figuratively speaking, there is a competition for survival between inhibitor p27 and the cyclin E-Cdk2 complex. Its outcome is determined by the reaction energy supply: with a high ATP level, the cyclin E-Cdk2 complex has an advantage. It phosphorylates p27; after that, this inhibitor becomes a target for ubiquitin-dependent proteolytic machinery and is destroyed. If the bioenergetics level becomes lower than a certain value, then even p27 inactivates cyclin E-Cdk2. As a result a positive feedback loop of E2F synthesis and S-phase transition are blocked. An ability to inactivate its inhibitor belongs only to cyclin E-Cdk2 complex and was not found in other analogous complexes.

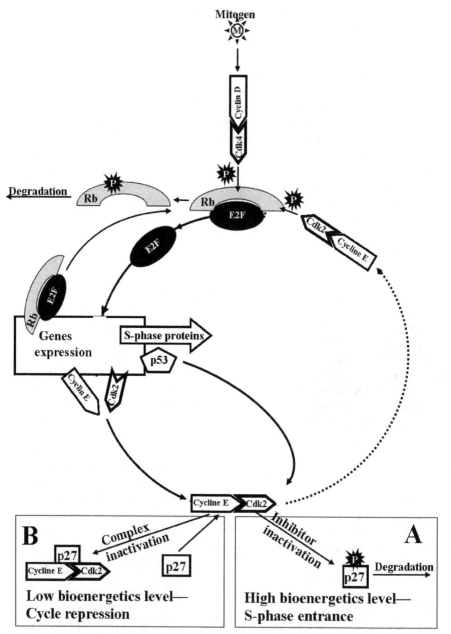

Fig. 4. Simplified scheme of control of passing through a restriction point.

In response to mitogen stimulation, an active cyclin D-Cdk4 complex is synthesized, which phosphorylates Rb protein. As a result gene-regulating E2F proteins are released and phosphorylated Rb degrades. E2F proteins activate genes of proteins essential for DNA reduplication in the S-phase, cyclin E and cyclin-dependent kinase 2 (Cdk2) genes, as well as E2F itself. Combined cyclin E and Cdk2 form an active complex which interacts with inhibitor of cyclin-dependent kinases p27. Two consequences are possible. A. If the bioenergetics level is within the physiological norm, then Cdk2 activated by cyclin E phosphorylates p27. Then the inhibitor becomes a target for degradation. After that cyclin E-Cdk2 phosphorylates Rb and additional gene-regulating proteins are released. As E2F activates cyclin E, Cdk2, and E2F genes, then there is a positive feedback loop promoting rapid p27 removal and an avalanche-like increase of S-phase proteins, which allows a cell to pass through a restriction point. At the same time, increased expression of E2F induces inhibitor p53 synthesis, which inactivates cyclin the E-Cdk2 complex and terminates unnecessary S-phase E2F expression. B. If the bioenergetics level is below a certain critical level, then p27 forms a tight bond with the cyclin E-Cdk2 complex and inactivates it. As a result an increased expression of S-phase proteins does not occur, the p27 level remains high, and entry into the S-phase becomes impossible.

These data can explain the results of the abovementioned research by Rittling et al. The central event of the G1-phase restriction point of the cell cycle is triggering of a self-accelerating cascade of reactions controlled by the cyclin E-Cdk2 complex. This is an essential condition of inhibitor p27 removal and accumulation of all precursors for DNA reduplication and cell division. It is satisfied only with a normal physiological level of bioenergetics. When bioenergetics in old cells decreases until some threshold level, cyclin E-Cdk2 loses its ability to inactivate p27 and itself becomes a target. As a result inhibitor removal stops and S-phase transition becomes impossible. All this information can be summarized in the following way: cyclin-dependent kinase inhibitor p27 prevents passage through the restriction point. There is a special pump for its removal in a cell. Its work efficiency depends on the energy supply. During the programmed decrease of cell bioenergetics, below a certain threshold level, it stops inhibitor removal and cell division becomes impossible.

It should be mentioned that a critical level is achieved after a certain number of divisions. Thus, bioenergetics decrease and the lifespan depend not on the calendar time of an organism's existence, but from the number of past divisions in its critical tissues, i.e., the amount of the past divisions is a biological clock. An organism counts time on proliferative clock.

5. Longevity is under control of natural selection

5.1 Introduction
Several lines of evidence show that genes exert strong controls on longevity and patterns of aging (Carey, 2003; Holzenberger et al., 2003; Kenyon, 2010; Vaupel, 2003). Therefore, the specific genes that program longevity and the selective pressure that would lead the genes to the development during evolution are to exist (Bredesen, 2004; Mitteldorf, 2004; Skulachev, 2001). The most of evolutionists, nevertheless, deny the possibility that longevity is under the control of natural selection (Medawar, 1952; Kirkwood, 2002). In 1952 P.B.Medawar has shown that life expectancy is not under control of individual (Darwinian) natural selection. He has noticed that animals in habitat never live till an old age and perish from the various external reasons at youngish age; therefore the natural selection cannot differentiate them by the longevity sign. Hence the specific genes programming aging cannot exist. This conception dominates till now.

The aim of this section was to show the mechanism by which natural selection controls species-specific longevity. The ecological approach was used to solve the problem. As known from ecological laws, the intrinsic population growth rate (r_{in}), the length of the generation (T), and the net reproductive rate (R_0) are interconnected by dependence, according to the following formula: $r_{in} = lnR_0/T$ (MacArthur & Connell, 1966). It is shown here that during evolution the r_{in} value is stabilized by interpopulation (group) natural selection (not individual selection) at the level which corresponds to environmental pressure in the ecological niche of the species. This leads to the conclusion that species-specific longevity and fertility are under the control of natural selection and depend inversely on each other.

5.2 Population size oscillations and extinction risk

The state of a population's size over the long-term is a measure of population welfare. Stability or an increase in size testifies to the well-being of the community, but a decrease indicates that the population is under risk of extinction. A practical determination of this criterion represents a difficult problem because biological systems are dynamic. Successive changes in biological systems are termed "disturbance" (White & Pickett, 1985). Disturbances are inherent in all biological communities and occur on a wide range of quantitative, spatial, and temporal scales (Pickett & White, 1985). The size of any population determined by observation is in fact its value at an instantaneous time cut-off (Southwood, 1981). Population number can change in time by hundreds, thousands, and in some species, even by millions (Nicholson, 1954). Population size oscillations are forced by varying environmental factors, such as the infections, the availability of food, the number of predators and parasites, etc. The mean population size, population number averaged over some period of observation, is a much more informative characteristic. Based on theoretical averaging over a prolonged time interval, this parameter is considered to be the dynamically equilibrium size (N_{eq}). However, the fate of a population depends on its minimal size (N_{min}), i.e., the lowest value which a population reaches in the process of oscillations. N_{min} depends on both N_{eq} and swings in the population size. The minimal population size is a genetic bottleneck that is an evolutionary event in which the population is often reduced by several orders of magnitude (Leberg, 1992; Richards & Leberg, 1996). Populations are potentially immortal, but each of them is always subject to the risk of extinction due to minimum viable population size (Green, 2003; Tracy & George, 1992). The last is the smallest population size that will persist over some specified length of time with a specified probability (Hedrick & Gilpin, 1996). If a population size is reduced below this value, even if for a moment in time, then the population becomes doomed to extinction during future generations due to genetic drift (Cherry & Wakeley, 2003; Gilpin & Soule, 1986). Therefore, the extinction risk is maximal in the N_{min} state because a significant part of a population is prevented from reproducing. This increases genetic drift, as the rate of the drift is inversely proportional to the population size (Frankham, 1996; Lande, 1993; Shaffer, 1981).

5.3 The interpopulation natural selection

The mechanism of interpopulation natural selection is simple: "Small populations can fluctuate out of existence quite rapidly" (Leigh, 1975). In other words, preferred extinction of populations having less N_{min} is the essence of interpopulation natural selection. Natural selection, as a whole, consists of two stages. During the first stage, the classical Darwin-Wallace individual selection rejects organisms which are less adapted to the given environment. As the members of the population serve as an environmental factor for each individual, attributes can arise that are useful only to their carriers but neutral or harmful for the other individuals. Such attributes become harmful for the community, but they are supported by individual natural selection. Longevity and a number of psychological attributes, for example, are a concern to them (Gadgil, 1975). The interpopulation selection takes such attributes under control as the populations are units of natural selection in the second stage (Levins, 1962; Wilson, 1973). If any attribute decreases N_{eq}, or increases the amplitude of oscillations and spreads in the population, then the population perishes as a whole. In contrast, the attributes that increase N_{min} promote population survival, which is an evolutionary mechanism for developing characteristics that are useless or even harmful for individuals, but beneficial for the community (e.g., altruism, care of posterity, and bravery). To determine the direction of evolution for a specific attribute of a species' populations, it is necessary to assess the dependence of N_{min} from a quantitative expression of this attribute. The pressure of group selection is always directed to an increase in N_{min}.

5.4 Evolution mechanism of longevity and fertility

To solve the problem under consideration, we need to consider the dynamics of populations of an abstract species of vertebrates with overlapping generations. In so doing, we shall determine the dependence on N_{min} from the intrinsic population growth rate (r_{in}) at a various value of environment pressure in the species' ecological niche, remembering thus that

$$r_{in} = lnR_0 / T \qquad (1)$$

A change in size of population, dN/dt, depends on the difference between birth and death rates:

$$dN / dt = dN_b / dt - dN_m / dt. \qquad (2)$$

Accordingly, population size influences birth and death rates:

$$dN_b / dt = bN \text{ and } dN_m / dt = mN \qquad (3)$$

where b and m are density-dependent are birth and death rates, respectively.

Population size does not influence the birth and death rates directly, but through changes in environmental parameters. When the population size increases, food resources are exhausted, the number of predators and parasites grow, infections are increased, and living space per capita declines. All this raises the level of environmental pressure upon a population. As a result, the birth rate decreases but the death rate increases (Fig. 5):

$$b = b_{in} - aN; \ m = m_{in} + jN \qquad (4)$$

where b_{in} and m_{in} are intrinsic are birth and death rates that are realized, provided that N is negligible; a and j are environmental pressures on the birth and death rates respectively. Substituting b and m from equation (4) in equation (2), and taking into account equation (3), it follows that:

$$dN / dt = (b - m)N = (b_{in} - m_{in})N - (a + j)N^2 \qquad (5)$$

Having designated $b_{in} - m_{in} = r_{in}$ and $a + j = p$, equation (5) can be rewritten as

$$dN / dt = N(r_{in} - pN) \qquad (6)$$

where r_{in} is the intrinsic population growth rate (*time* $^{-1}$), and p is the environmental pressure (*time* $^{-1}N^{-1}$).

The dynamics of any population is complicated by feedback among population size and the environmental pressure. The environmental conditions vary after changes in population size with some time delay. Let, for example, population size grow from N_{min} at time t_0 to N_{eq} at time t. At once, as the population size reaches N_{eq}, the environmental pressure remains at the level that existed at the moment, $t\text{-}\tau$. The time delay, τ, is the time necessary for the breeding of parasites and predators and a reduction of food resources and vital space per capita. As a result, the population size proceeds to increase to the equilibrium point N_{eq}, and reaches the point, $N_{max} > N_{eq}$. As this state is unstable, the population size is reduced and, for the same reason, passes the N_{eq} point and falls to $N_{min} < N_{eq}$; this is the nature of auto-oscillations about N_{eq} (Macfadien, 1963; May, 1973). Being forced out of the equilibrium condition, a population enters an auto-oscillation regimen and the amplitude of the oscillations can serve as a criterion of population responsiveness to environmental variability.

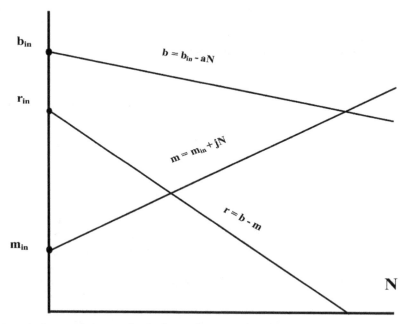

The increase in the population number leads to a decrease in b and an increase in both m and r relative to their intrinsic values, b_{in}, m_{in}, and r_{in}. The slopes of the line depend on the environmental pressure on the birthrate (a) and mortality (j).

Fig. 5. Dependence of birth (b) and death (m) rates and the population growth rate (r) from the population number (N).

With the delay effect, equation (6) becomes:

$$dN_t / dt = N_t(r_{in} - pN_{(t-\tau)}) \qquad (7)$$

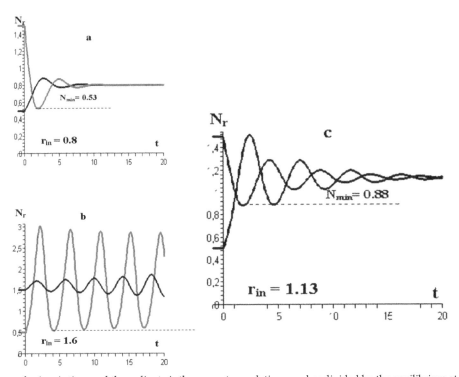

Exes: abscissa is time and the ordinate is the currant population number divided by the equilibrium at time t_0 ($N_r = N_t/N_{eq}$). The consequences of two situations in the environment are modulated (two curves): (1) a favorable coincidence that raises the outbreak of the population number ($N_0 > N_{eq} = 1.5$) and (2) coincidence of severe conditions that causes population depression ($N_0 < N_{eq} = 0.5$). (a) Populations having small values of r_{in} are reduced to the minimal value after outbreaks of numbers (red curve): coincidences of favorable circumstances in the environment threaten the existence of such populations to a greater extent. (b) Populations having large values of r_{in} reach the lowest number after a state of decay (red curve): coincidences of unfavorable circumstances threaten the existence of such populations to a greater extent. (c) There is an optimal r_{in} value under the present environmental pressure when favorable and unfavorable environment cataclysms are followed by an equal aftereffect. The value of N_{min} is maximal under this r_{in}; this r_{in} value is maintained by interpopulation selection.

Fig. 6. The influence of the intrinsic population growth rate on parameters of population size oscillations.

It can be seen that the dynamics of the population are determined by three parameters: τ and p, are factors of the habitat, but r_{in} is an intrinsic characteristic of the population. Each of the factors influences oscillation characteristics. Parameter τ is the regeneration time of density-dependent environmental factors. As the environmental pressure is a complex value, then τ is also a multifactorial distributed characteristic of the environment (Schley &

Gourley, 2000). However, it can be accepted as a discrete characteristic at solving many tasks analogous to our problem (May, 1981; Schley & Gourley, 2000). The numerical solution of equation (7) shows that τ influences the amplitude of the population size oscillations: the greater the τ, the greater the amplitude of oscillations. Species that are under $\tau < 0.3$ have the least variability; perturbed size of its populations monotonously return to the equilibrium state. In the range $0.3 < \tau < 1.6$, an oscillatory return to an equilibrium number occurs. The further τ increases cause continuous oscillations. If $\tau > 2.2$, then populations become non-viable; the smallest external disturbance provokes increasing oscillations that decrease N_{min} to nil. It is apparent that within an ecological niche, in the overwhelming majority of vertebrate species the τ value is limited by 0.4-1.5. Therefore, we shall accept in further calculations that this parameter of the ecological niche of the abstract species under consideration is equal to 1.

The numerical solution of equation (7) shows that the variation of r_{in} influences both the N_{eq} value and the amplitude of oscillations that predetermines changes in N_{min} (Fig. 6). The dependence of N_{min} from r_{in}, calculated with other parameters unchanged ($p = 1$; $\tau = 1$), is shown in Fig. 7. The curve of dependence $N_{min}(r_{in})$ has a maximum under certain $\acute{r}_{in}$. As mentioned above, the selective pressure is always directed to an increase in N_{min}. In the case in point, the directions of selective pressure are opposite from larger and smaller r_{in} values. Hence, it appears that the intrinsic rate of population growth is stabilized by group selection on the level which corresponds to the maximal N_{min} value:

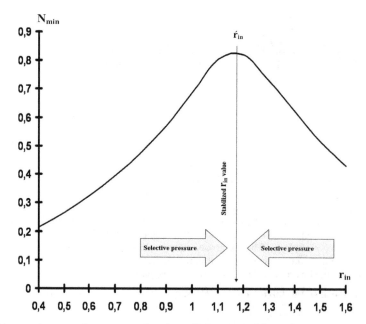

Under constant environmental parameters ($p = 1$; $\tau = 1$) the curve of dependence $N_{min}(r_{in})$ has a maximum under a certain $\acute{r}_{in}$. As the extinction risk is inversely proportional to N_{min}, then selective pressure pushes r_{in} of populations of species to this value.

Fig. 7. Scheme of stabilization of the intrinsic population growth rate by interpopulation selection.

if the environmental pressure is constant, then any population of a species deviating from this value will have a greater extinction risk. As $r_{in} = lnR_0/T$, then the length of the generation and the net reproduction rate are stabilized by interpopulation natural selection. On a long-term temporal scale, the environmental pressure becomes constant.

However, in the course of evolution, it can gradually vary during a change of parameters of an ecological niche under influence, for example, changes of climate. The calculated dependence of the stabilized $\acute{r}_{in}$ value from p shows that variation in environmental pressure causes a change of the stabilized intrinsic population growth rate: the greater the p value, the greater the $\acute{r}_{in}$.

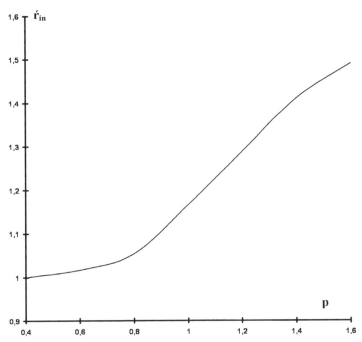

An increase in environmental press in the ecological niche of a species in the course of evolution causes growth of the intrinsic population growth rate and vice versa.

Fig. 8. The dependence of a stabilized intrinsic population growth rate ($\acute{r}_{in}$) from environmental pressure (p).

Thus, the value of the intrinsic rate of population growth is under natural selection control and it is predetermined by environmental pressure within the ecological niche of the species.

The conclusion that the intrinsic rate of population growth is stabilized by interpopulation natural selection can be made logically without resorting to mathematical calculations. Population size oscillations are inevitable because of stochastic variations in the environment and are harmful as they increase the extinction risk. The intrinsic rate of the population growth influences population responsiveness to environmental fluctuations. When a population is in a state of size reduction, the maximal r_{in} is preferably for oscillation damping. Under these conditions, a decrease in population size in any given half cycle of the oscillation will be minimal as the high rate of breeding serves as a brake for the decrease.

But, such r_{in} values will become threatening when the opposite phase begins as it intensifies the increase in population size. According to the theory of risk spread, the greater the extension of a population on top, the deeper it falls in foot. The same intensification of amplitudes is provoked by an inverse extreme value of the intrinsic population growth rate. A natural population cannot have such an extreme or any arbitrary of r_{in} value. There is an optimal value of the intrinsic rate of population growth which ensures minimal possible population oscillation (Fig. 6). That value is sustained by interpopulation natural selection because deflection of the r_{in} to any side from the value increases the population extinction risk; the above-stated mathematical calculations alone have demonstrated this.

Let's look now what in fact is hidden behind the intrinsic population growth rate. According to equation (1), these are two population characteristics: 1) the net reproduction rate and 2) the length of the generation, neither of which can be programmed by the genome directly. In a general sense, the length of the generation is the time from which the individuals are born to the time most offspring, on average, are produced for a population. The concept of "post-reproductive age" is applicable to the full only to post-industrial man and his/her pets. Animals of post-reproductive age are rare in natural habitats (Medawar, 1952). Analyses of cohort life tables of natural populations show that the length of the generation is actually equal to the mean survival of the population age-groups. Thus the average longevity in the habitat is under natural selection control.

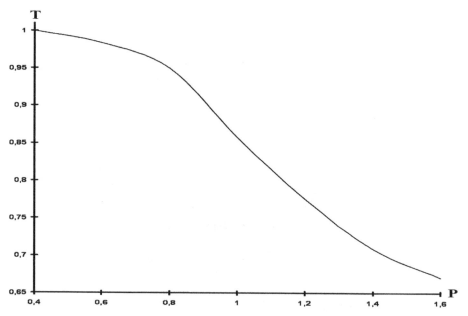

The increase in environmental pressure in a species ecological niche results in a longevity decrease.

Fig. 9. Dependence of longevity (T) on environmental pressure in the ecological niche of a species.

The variation in the net reproduction rate corresponds to the variation in fertility of the population members. To understand it, we should recollect that $R_0 = N_T/N_0$, where N_0 is the initial population size and N_T is the population size one generation later. It is apparent that if R_0 is increasing in evolution, the fertility is growing, and vice versa.

Thus, longevity and fertility are actually under the control of interpopulation natural selection. The selective pressure acts on both components of r_{in} simultaneously, but the distribution of forces are unequal for different species because of environmental factor specificity. As a result, only a correlation between longevity and fertility exist in nature. This evolution mechanism of longevity is apparently applicable only for vertebrata. Invertebrates, by virtue of their huge variety, can have others, and various, evolution mechanisms determine species specific longevity.

6. The mechanism of aging

According to the above considerations, the mechanism of programmed aging is represented as follows (Fig. 10): the genetic program controls the only function – bioenergetics decline. The latter causes the increase in the ROS level, a lowering of the protein synthesis rate, the cessation of cells dividing and some other processes; every one of them in turn spawns a number of secondary harmful processes. As the number of cells dividing (proliferative time) increases, these destructive phenomena in an organism's tissues augment progressively, which gradually leads to the organism's destruction.

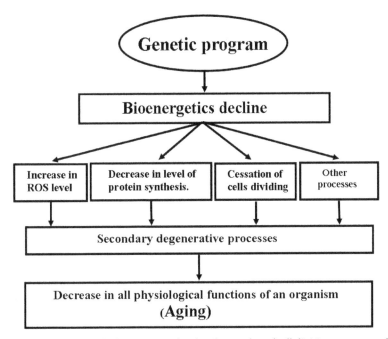

The genetic program decreases the bioenergetics level as the number of cell divisions augments. This results in the increase in the ROS level, the lowering of the protein synthesis rate, the cessation of cells dividing and some other injurious processes. In turn, each one of them spawns a number of secondary harmful processes which leads to a decrease in all of the physiological functions of an organism, i.e. aging.

Fig. 10. Scheme of the bioenergetics mechanism of aging.

7. Conclussion

Gerontology has entered the 21 Century with significant empirical baggage but without a theory capable of generalising the data and discovering the general regularities of the aging process. Instead, as mentioned, more than 300 different theories have been developed. There is still no consistent opinion as to what the primary driving force of aging actually is. The majority of researchers are convinced that there are no genes for aging and that stochastic factors underlie the aging. Those who trust in the programmed theory assume that almost every process influencing aging is governed by its own genes, i.e. aging is multifactorial. According to this, genes of aging exist and they control the sole driving force of aging – proliferative-dependent bioenergetics decline. It can be shown that this programmed process underlies, whether expressly or by implication, any theory of aging based on real phenomenon. This can provide the basis for the creation of a united theory of aging.

The present situation in modern gerontology does not suggest any hope of the achievement of the abovementioned strategic aim: numerous efforts to elaborate a remedy for senescence based on stochastic theories have yielded no result. The strict restriction of food (calorie restriction) is the only trick that has been developed which authentically increases species' maximal lifespan. The mechanism for this phenomenon is not yet understood, but it is easily explained by the bioenergetics theory: the lack of food detains cell division which in turn leads to a lag of the proliferative clock relative to calendar time. The programmed theories do not much promise success because of the large number of genes that operate in the ageing process. A decrease in the level of bioenergetics is apparently programmed by only a few genes. The analysis of the evolutionary plasticity of fruit fly populations has shown that longevity is programmed by no more than by three genes (Mylnikov, 1997).

One relevant inference to be made of the theory stated above is that the manipulation of any secondary phenomenon generated by the decline in bioenergetics cannot give effect to an increase in the maximal lifespan. A means to operate bioenergetics has to be found - it is the only way towards healthy and unlimited longevity. This is complicated problem but it can be solved in the near future: the bioenergetics machine is already studied well enough, the regulator of energetical homeostasis is visible, and the potent arsenal of experimental techniques is created. The period depends mainly on facilities for research.

8. References

Ainscow, E.K. & Brand, M.D. (1998). Control analysis of systems with reaction blocks that 'cross-talk'. *Biochemica et Biophysica Acta*, Vol. 1366, pp. 284-290, ISSN 0006-3002.

Anisimov, V.N. (2001). Life span extension and cancer risk: myths and reality *Experimental Gerontology*, Vol. 36, Issue 7, pp. 1101-1136, ISSN 0531-5565

Anisimov, V.N. (2003). Priorities in Fundamental Research in Gerontology: Russian Contribution. *Advances in Gerontology*, Vol. 12, pp. 9-27, ISSN 1561-9125.

Arai, M., Imai, H., Koumura, T., Yoshida, M., Emoto, K., Umeda, M., Chiba, N. & Nakagawa, Y. (1999). Mitochondrial phospholipid hydroperoxide glutathione peroxidase plays a major role in preventing oxidative injury to cell. *Journal of Biological chemistry*, Vol. 274, Issue 8, pp. 4924-4933, ISSN 0021-9258.

Barrientos, A., Casademont, J., Rotig, A., Miro, O., Urbano-Marquez, A., Rustin, P. & Gardellach, F. (1996). Absence of relationship between the level of electron transport chain activities and aging in human skeletal muscle. *Biochemical and Biophysical Research Communication*, Vol. 229, Issue 2, pp. 536-539, ISSN 0006-291X.

Bayreuther, K., Rodemann, P., Hommel, R., Dittmann, K., Albiez, M. & Francz, P.I. (1988). Human Skin Fibroblasts in Vitro Differentiate Along a Terminal Cell Lineage. *Proceedings National Academy of Sciences USA*, Vol.85, No.14, pp. 5112-5116. ISSN 0027-8424.

Bjelakovic, G., Nikolova, D., Gluud, L.L., Simonetti, R.G. & Gluud, C. (2008). Antioxidant supplements for prevention of mortality in healthy participants and patients with various diseases. *Cochrane Database of Systematic Reviews*, Issue 2, Art. No. CD 007176, ISSN 1469-493X .

Blasco, M.A., Lee, H.W., Hande, M.P., Samper, E., Lansdorp, P.M., DePinho, R.A. & Greider, C.W. (1997). Telomere Shortening and Tumor Formation by Mouse Cells Lacking Telomerase RNA. *Cell*, Vol. 91, No. 1, pp. 25–34, ISSN 0092-8674.

Bockstaele, L., Bisteau, X., Paternot, S., Roger, P.P. (2009). Different Regulation of Cycline-dependent Kinase 4 (CDK4) and CDK6, Evidence That CDK4 Might Not Be Activated by CDK7, and Design of CDK6 Activating Mutation. *Molecular and Cell Biology*, Vol. 29, No. 15, pp. 4188–4200, ISSN 0270-7306.

Brand, M.D. (2000). Uncoupling to survive? The role of mitochondrial inefficiency in aging. *Experimental Gerontology*, Vol. 35, pp. 811-820, ISSN 0531-5565.

Bratic, I. & Trifunovic, A. Mitochondrial energy metabolism and ageing. (2010). *Biochimica et Biophysica Acta*, Vol. 1797, pp. 961–967, ISSN 0006-3002.

Bredesen, D.E. (2004). Non-existent aging program: haw dos it work? *Aging Cell*, Vol.3, No.5, pp. 255-259, ISSN: 1474-9718

Brigelius-Flohe, R., Banning, A. & Schnurr, K. (2003). Selenium-dependent enzymes in endothelial cell function. *Antioxidants and Redox Signaling*, Vol. 5, No. 2, pp. 205-215, ISSN 1523-0864.

Carey, J.R. (2003). Life span: a conceptual overview. In *Life span: evolutionary, ecological, and demographic perspectives*. Carey, J.R. & Tuljapurkar, S., eds., pp. 1-18, Population council, ISBN: 0878341110, New York.

Cerimele, F., Battle, T., Lynch, R., Frank, D.A., Murad, E., Cohen, C., Macaron, N., Sixbey, J., Smith, K., Watnick, R.S., Eliopoulos A., Shehata, B. & Arbiser, J.L. (2005). Reactive oxygen species signaling and MAPK activation distinguish Epstein-Barr Virus (EBV)-positive versus EBV-negative Burkitt's Lymphoma. *Proceedings National Academy of Sciences USA*, Vol. 102, No.1, pp. 175-179, ISSN 0027-8424.

Cherry, J.L. & Wakeley, J. (2003). A diffusion approximation for selection and drift in a subdivided populations. *Genetics*, Vol. 163, No. 1, pp. 421-428, ISSN 0016-6731.

Clemens, M.J. (1994). Regulation of eukaryotic protein synthesis by protein kinases that phosphorylate initiation factor eIF-2. *Molecular Biology Reports*, Vol. 19, pp. 201-10, ISSN 0301-4851

Clemens, M.J., Bushell, M., Jeffrey, I.W., Pain, V.M. & Morley, S.J. (2000). Translation initiation factor modifications and the regulation of protein synthesis in apoptotic cells. *Cell Death & Differentiation*. Vol. 7, No. 7, pp. 603-615, ISSN 1350-9047.

Clemens, M.J., Pain, V.M., Wong, S-T. & Henshaw, E.C. (1982). Phosphorylation inhibits guanine nucleotide exchange on eukaryotic initiation factor 2. *Nature*, Vol. 296. pp. 93-95, ISSN 0028-0836.

Demin, O.V., Kholodenko, B.N. & Skulachev, V.P. (1998). A model of O_2^- generation in the complex III of the electron transport chain. *Molecular and cell biology*, Vol. 184, No 1-2, pp. 21-33, ISSN 0270-7306.

Dunford, H.B. (2002). Oxidations of iron(II)/(III) by hydrogen peroxide: from aquo to Enzyme. *Coordination Chemistry Reviews*, Vol. 233-234, pp. 311-318, ISSN 0010-8545

Frankham, R. (1996). Relationship of genetic variation to population size in wildlife. *Conservation Biology*, Vol. 10, No. 6, pp. 1500-1508, ISSN 0888-8892.

Frolov, M.V. & Dyson, N.J. (2004). Molecular Mechanism of E2F-dependent Activation and pRB-mediated Repression. *Journal of Cell Sciences*, Vol. 117, No. 1, pp. 2173-2181, ISSN 0021-9533.

Gadgil, M. (1975). Evolution of social behavior through interpopulation selection. *Proceedings of National Academy of Science USA*, Vol. 72, No. 3, pp. 1199–1201, ISSN 0027-8424.

Gems, D. & Doonan, R. (2009). Antioxidant defense and aging in C. elegans. Is the oxidative damage theory of aging wrong? *Cell Cycle*, Vol. 8, No. 11, pp. 1681-1687, ISSN 1538-4101.

Gilpin, M.E. & Soulé, M. E. (1986). Minimum viable populations: processes of species extinction. In *The science of scarcity and diversity*. Soule, M.E., ed., pp.19-34. Conservation biology: Sinauer Associates, ISBN: 0878937951, Sunderland.

Green, D.M. (2003). The ecology of extinction: population fluctuation and decline in amphibians. *Biological conservation*, Vol. 111, Issue 3, pp. 331-343, ISSN 0006-3207.

Greider, C. W. & Blackburn, E. H. (1985). Identification of a Specific Telomere Terminal Transferase Activity in Tetrahymena Extracts. *Cell*, Vol. 43, No. 1, pp. 405–413, ISSN 0092-8674.

Halliwell, B. & Gutteridge, J.M.C. (1984). Oxygen toxicity, oxygen radicals, transition metals and disease. Biochemical Journal, Vol. 219, No.1, pp. 1-14, ISSN 0006-2936.

Harman, D. (1956) Aging: a theory based on free radical and radiation therapy. *Journal of Gerontology*, Vol. 11, pp. 298-300.

Harmon, C.S., Proud, C.G. & Pain, V.M. (1984). Effect of starvation, diabetes and acute insulin treatment on a regulation of polypeptide-chain initiation in rat skeletal muscle. *Biochemical Journal*, Vol. 223, pp. 687-696, ISSN 0006-2936.

Hasty, P. & Vijg J. (2002). Genomic priorities in aging. *Science*, Vol. 296, No. 5571, pp.1250-1251, ISSN 0036-8075.

Hayashi, J-I., Ohta, S., Kagawa, Y., Kondo, H., Kaneda, H., Yonekawa, H., Takai, D. & Miyabayashi , S. (1994). Nuclear but not mitochondrial genome involvement in human age-related mitochondrial dysfunction. Functional integrity of mitochondrial DNA from aged subjects. *Journal of Biological Chemistry*, Vol. 269, Issue 9, pp. 6878-6883, ISSN 0021-9258.

Hayflick, L. & Moorhead, P. S. (1961). The Serial Cultivation of Human Diploid Cell Strains. *Experimental Cell Research*, Vol.25, No.15, pp. 585–621, ISSN 0014-48-24.

Hayflick, L. (2007). Biological aging is no longer an unsolved problem. *Annals of the New York Academy of Sciences*, Vol. 1100, pp. 1-13, ISSN 1749-6632.

Hedrick, P.W. & Gilpin, M.E. (1996). Genetic effective size of a metapopulation. In *Metapopulation biology: ecology, genetics and evolution*, Hanski, I. & Gilpin, M.E., eds, pp. 165-191, Academic Press, ISBN 0123234468, New York.

Herrera, E., Samper, E., Martin-Caballero, J., Flores, J.M., Lee, H-W. & Blasco, M.A. (1999). Disease States Associated With Telomerase Deficiency Appear Earlier in Mice With Short Telomeres. *EMBO Journal*, Vol.18, pp. 1950-2960, ISSN 0261-4189.

Holliday, R. (2006). Aging is no longer an unsolved problem in biology. *Annals of the New York Academy of Sciences*, Vol. 1067, pp. 1-9, ISSN 1749-6632

Halliwell, B. & Gutteridge, J.M.C. (1984). Oxygen toxicity, oxygen radicals, transition metals and disease. *Biochemical Journal*, Vol. 219, No.1, pp. 1-14, ISSN 0006-2936.

Holzenberger, M., Dupont, J., Ducos, B., Leneuve P., Geloen, A., Even, P.C., Cervera, P. & Le Bouc, Y. (2003). IGF-1 receptor regulates lifespan and resistance to oxidative stress in mice. *Nature* 421, 182-187, ISSN 0047-6374.

Hornsby, P.J. (2002). Cellular Senescence and Tissue Aging in Vivo. *Journal of Gerontology series A: Biological and Medical Sciences*, Vol. 57, No. 7, pp. B251-B256, ISSN 1079-5006.

Howes, R.M. (2006). The free radical fantasy. A panoply of paradoxes. *Annals of the New York Academy of Sciences*, Vol. 1067, pp. 22-26, ISSN 1749-6632

Hucull, J.H., Henshaw, E.S. & Young, D.A. (1985). Nucleotide diphosphate regulation of overall rates of protein biosynthesis acting at the level of initiation. *The Journal of Biological chemistry*, Vol. 260, No. 29, pp. 15585-15591, ISSN 0021-9258.

Jeffrey, I.W., Bushell, M., Tilleray, V.J., Morley, S. & Clemens, M.J. (2002). Inhibition of protein synthesis in apoptosis: Differential requirements by the tumor necrosis factor α family and a DNA-damaging agent for caspases and the Double-stranded RNA-dependent Protein Kinase. *Cancer Research*, Vol. 62, No. 8, pp. 2272 – 2280, ISSN 0008-5472.

Jo, S.H., Son, M.K., Koh, H.J., Lee, S.M., Song, I.H., Kim, Y.O., Lee, Y.S., Jeong, K.S., Kim, W.B., Park, J.W., Song, B.J. & Huh, T.L. (2001). Control of mitochondrial redox balance and cellular defense against oxidative damage by mitochondrial NADP$^+$-dependent isocitrate dehydrogenase. *Journal of Biological chemistry*, Vol. 276, No. 19, pp. 16168-16176, ISSN 0021-9258.

Jonas, S.K., Riley, P.A. & Willson, R.L. (1989). Hydrogen peroxide cytotoxicity. Low-temperature enhancement by ascorbate or reduced lipoate. *Biochemical Journal*, Vol. 264, No.3, pp. 651-655, ISSN 0006-2936.

Iantomasi, T., Favilli, F., Marraccini, P., Stio, M., Treves, C., Quattrone, A., Capaccioli, S., Vincenzini, M.T. & Quatrone, A. (1993). Age and GSH metabolism in rat cerebral-cortex, as related to oxidative and energy parameters. *Mechanism of Ageing and Development*, Vol. 70, No. 1-2, pp. 65-82, ISSN 0047-6374.

Isobe, K., Kishino, S., Inoue, K., Takai, D., Hirawake, H., Kita, K., Miyabayashi, S. & Hayashi, J-I. (1997). Identification of inheritance modes of mitochondrial diseases by introduction of pure nuclei from mtDNA-less HeLa cells to patient-derived fibroblasts. *The Journal of Biological chemistry*, Vol. 272, No. 19, pp. 12606-12610. ISSN 0021-9258.

Isobe, K., Ito, S., Hosaka, H., Iwamura, Y., Kondo, H., Kagawa, Y. & Hayashi, J-I. (1998). Nuclear-recessive mutations of factors involved in mitochondrial translation are responsible for age-related respiration deficiency in human skin fibroblasts. *The Journal of Biological chemistry*, Vol. 273, No. 8, pp. 4601-4606. ISSN 0021-9258.

Itahana, K., Campisi, J., Goberdhan, P. & Dimri, G.P. (2004). Mechanisms of Cellular Senescence in Human and Mouse Cells. *Biogerontology*, Vol.5, No.1, pp. 1–10. ISSN 1389-5729

Kenyon, C.J. (2010). The genetics of ageing. *Nature*, Vol.464, pp.504-512, ISSN 0028-0836.

Kimball, S.R., Fabian, J.R., Pavitt, G.D., Hinnebusch, A.G. & Jefferson, L.S. (1998). Regulation of guanine nucleotide exchange through phosphorylation of eukaryotic initiation factor eIF2. Role of the a- and δ-subunits of eIF2B. *The Journal of Biological chemistry*, Vol. 273, No. 21, pp. 12841–12845, ISSN 0021-9258.

Kirkwood, T. B. L. (2002). Evolution of ageing. *Mechanism of Ageing and Development*, Vol. 123, Issue 7, pp. 737-745, ISSN 0047-6374.

Kirkwood, T. B. L. (2008). Understanding ageing from an evolutionary perspective. *Journal of Internal Medicine*, Vol. 263, pp. 117–127, ISSN 0954-6820.

Korshunov, S.S., Skulachov, V.P. & Starkov, A.A. (1997). High protonic potential actuates mechanism of production of reactive oxygen species in mitochondria. *FEBS Letters*, Vol. 416, Issue 1, pp. 15-18, ISSN 0014-5793.

Lande, R. (1993). Risk of population extinction from demographic and environmental stochasticity and random catastrophes. *American Naturalist*, Vol. 142, No. 6, pp. 911-927, ISSN 0003-0147.

Larrea, M. D., Liang, J., Da Silva, T., Hong, F., Shao, S.H., Han, K., Dumond, D. & Slingerland, J.M. (2008). Phosphorilation of p27kip1 Regulates Assembly and Activation of Cycline D1-Cdk4. *Molecular and Cellular Biology*. Vol.28, No.20, pp. 4662–4672. ISSN 0270-7306.

Leberg, P.L. (1992). Effects of population bottlenecks on genetic diversity as measured by allozyme electrophoresis. *Evolution*, Vol. 46, No.2, pp. 477-494, ISSN 0014-3820.

Lee, H.-W., Blasco, M. A., Gotlieb, G. J., Horner, I.I., Grieder, C.W. & DePinho, R.A. (1998). Essential Role of Mouse Telomerase in Highly Proliferative Organs. *Nature*, Vol.392, pp. 569–574. ISSN 0028-0836.

Leigh, E.G. (1975). Population fluctuations, community stability, and environmental variability. In *Ecology and Evolution of Communities*, Cody, M.L. & Diamond, J.M., eds, pp. 51-73, Harvard University Press, ISBN 0674224442, Harvard.

Leof, E. B., Wharton, W., Van Wyk, J. J., & Pledger,W. (1982). Epidermal Growth Factor (EGF) and Somatomedin C Regulate G1 Progression in Competent BALB/c 3T3 Cells. *Experimental Cell Research*, Vol. 141, Issue 1, pp. 107–115. ISSN 0014-4824.

Levins, R. (1962). Theory of fitness in a heterogeneous environment. 1. The fitness set and adaptive function. *American Naturalist*, Vol. 96, No. 891, pp. 361-373, ISSN 0003-0147.

Lewin, M.H., Hume, R., Howie, A.F., Richard, K., Arthur, J. R., Nicol, F., Walker, S. W. & Beckett, G. J. (2001). Thioredoxin reductase and cytoplasmic glutathione peroxidase activity in human foetal and neonatal liver. *Biochemica et Biophysica Acta*, Vol.1526. Issue 3, pp. 237-241, ISSN 0006-3002.

Lihtgow, G.J. & Andersen, J.K. (2000). The real Dorian Gray mouse. *BioEssays*, Vol. 22, No. 5, pp. 410-413, ISSN 0265-9247.

Liu, R., Liu, W., Doctrow, S.R. & Baudry, M. (2003). Iron toxicity in organotypic cultures of hippocampal slices: role of reactive oxygen species. *Journal of Neurochemistry*, Vol. 85, No. 2, pp. 492-502, ISSN 0169-5088.

Lloyd, R.V., Hanna P.M. & Mason, R.P. (1996). The origin of the hydroxyl radical oxygen in the Fenton reaction. *Free Radical Biology and Medicine*, Vol.22, No. 5, pp. 885-888, ISSN 0891-5849.

MacArthur, R.H. & Connell, J.H. (1966). *The theory of populations*. New York: Willey.

Macfadyen, A. (1966). *Animal ecology, aims and methods*. London: Isaac Pitman & Sons.

May, R.M. (1973). Time-delay versus in population model with two and three trophic levels. *Ecology*, Vol. 54, No. 2, pp. 315-325. ISSN 0012-9658.

May, R.M. (1981). Models for single population. In *Theoretical ecology. Principles and applications. Second edition*, May, R.M., ed., pp. 5-29, Blackwell, ISBN 0632007680, London.

Medawar, P.B. (1952). *An unsolved problem of biology*. Levis, London.

Medvedev, Z.A. (1990). An Attempt at a Rational Classification of Theories of Aging, *Biological Reviews*, Vol. 65, Issue 3, pp. 375–398, ISSN 1464-7931.

Meewes, C., Brenneisen, P., Wenk, J., Kuhr, L, Ma, W., Alikoski, J., Poswig, A., Krieg, T. & Scharffetter, K. (2001). Adaptive antioxidant response protects dermal fibroblast from UVA-induced phototoxicity. *Free Radical Biology and Medicine*, Vol. 30, Issue 3, pp. 686-698, ISSN 0891-5849.

Mendelsohn, S.L., Nordeen, S.K. & Young, D.A. (1977). Rapid changes in initiation-limited rates of protein synthesis in rat thymic lymphocytes correlate with energy charge. *Biochememical and Biophysical Research Communication*, Vol. 79, No. 1, pp. 53-60, ISSN 0006-291X.

Mitteldorf, J. (2004). Ageing selected for its own sake. *Evolutionary Ecology Research*, Vol. 6, pp.1–17, ISSN 1522-0613.

Montero, H., Rojas, M., Arias, C.F. & López, S. (2008). Rotavirus Infection Induces the Phosphorylation of eIF2α but Prevents the Formation of Stress Granules. *Journal of Viroljgy*, Vol. 82, No. 3, pp. 1496-1504, ISSN 0022-538X

Morgan, D. O. (1995). Principles of Cdk Regulation, *Nature*, Vol. 374, pp. 131–134, ISSN 0028- 0836.

Mylnikov, S.V. (1997). Genetic determination of rate of aging in some lines Drosopltila melanogaster. *Advanses in Gerontology*, Vol. 1, pp. 50-56, ISSN 1561-9125 (in Russian).

Nicholson, A.J. (1954). An outline of the dynamics of animal populations. *Australian Journal of Zoology*, Vol. 2, No. 1, pp. 9-65, ISSN 0004-959X.

Nordberg, J. & Arner E.S.J. (2001). Reactive oxygen species, antioxidants and the mammalian thioredoxin system. *Free Radical Biology and Medicine*, Vol. 31, No. 11, pp. 1287-1312, ISSN 0891-5849.

Pain, V.M. (1996). Initiation of protein synthesis in eukaryotic cells. *European Journal of Biochemistry* Vol. 236, No. 3, pp. 747-771, ISSN 0014-2956

Pickett, S.T.A. & White P.S. (1985). Patch dynamics: A synthesis. In *The ecology of natural disturbance and patch dynamics*, Pickett, S.T.A. & White, P.S., eds, pp. 371-384, Academic Press, ISBN 0125545215, London.

Rahimi, R. A. & Leoff, E. B. (2007). TGF-beta Signaling: a Tail of Two Responses. *Journal of Cellular Biochemistry*, Vol. 102, No.3, pp. 593–608, ISSN 0730-2312.

Rasmussen, U.F., Krustrup, P., Kjaer, M. & Rasmussen, H.N. (2003). Experimental evidence against the mitochondrial theory of aging. A study of isolated human skeletal muscle mitochondria. *Experimental Gerontology*, Vol.38, No. 8, pp. 877-886, ISSN 0531-5565.

Rattan, S. I. S. (1996). Synthesis, Modifications, and Turnover of Proteins During Aging, *Experimental Gerontjlogy*, Vol.31, No.1-2, pp. 33–47. ISSN 0531-5565

Rattan, S.I.S. (2006). Theories of biological aging: genes, proteins, and free radicals. *Free Radical Research*, Vol. 40, No. 12, pp. 1230-1238, ISSN 1071-5762.

Rattan, S. I. S. (2009). Synthesis, Modifications, and Turnover of Proteins During Aging. In *Protein Metabolism and Homeostasis in Aging*, Tavernarakis N. ed., Landes Bioscience and Springer Science+Business Media. ISBN 1441970010

Rittling, S.R., Brooks, K.M., Cristofalo, V.J. & Baserga, R. (1986). Expression of Cell Cycle Dependent Genes in Young and Senescent W1-38 Fibroblasts. *Procidings of National Academy of Sciences USA*, Vol. 83, pp. 3316–3320, ISSN 0027-8424.

Rhee, S.G. (1999). Redox signaling: hydrogen peroxide as intracellular messenger. *Experimental and Molecular Medicine*, Vol. 31, No. 2, pp. 53-59, ISSN 1226-3613.

Richards, C. & Leberg, P.L. (1996). Temporal changes in allele frequencies and a population's history of severe bottlenecks. *Conservation Biology*, Vol. 10, No. 3, pp. 832-839, ISSN 0888-8892.

Robert, F., Kapp, L.D., Khan, S.N., Acker, M.G., Kolitz, S., Kazemi, S., Kaufman, R.J., Merrick, W.C., Koromilas, A.E., Lorsch, J.R. & Pelletier, J. (2006). Initiation of protein synthesis by hepatitis C virus is refractory to reduced eIF2·GTP ·met-RNA$_i$met ternary complex availability. *Molecular Biology of the Cell*, Vol. 17, pp. 4632-4644, ISSN 1059-1524.

Ryazanov, A.G. & Nefsky, B.S. (2002). Protein turnover plays a key role in aging. *Mechanism of Ageing and Development*, Vol. 123, No. 2-3, pp. 207-213, ISSN 0047-6374.

Scandalios, J.G. (2002). Oxidative stress responses–what have genome-scale studies taught us? *Genome Biology*, Vol. 3. No. 7, pp. 1019-1021, ISSN 1465-6906

Scandalios, J.G. (2002). The rise of ROS. *Trends in Biochemical Sciences*, Vol. 27. P. 483-486. ISSN. 0968-0004.

Schreck, R., Rieber, P. & Baeuerle. P.A. (1991). Reactive oxygen intermediates as apparently widely used messengers in the activation of the NF-kappa B transcription factor and HIV-1. *EMBO Journal*, Vol. 10, No. 8, pp. 2247-2258, USSN 0261-4189.

Shaffer, M.L. (1981). Minimum population sizes for species conservation. *BioScience*, Vol. 31, No. 2, pp. 131-134, ISSN: 0265-9247.

Sheaf, R.J., Groudine, M., Gordon, M., Roberts, J.M., & Clurman, B.E. (1997). Cycline E-CDK2 Is Regulator of p27kip1. *Genes and Development*, Vol.11, No.11, pp. 1464–1478. ISSN 0890-9369.

Sherr, C. J. (1994). G1 Phase Progression: Cycling on Cue. *Cell*, Vol. 79, No.4, pp. 551–555. ISSN 0092-8674.

Sherr, C. J. (1996). Cancer Cell Cycles. *Science*, Vol. 274, pp.1672–1677, ISSN 0036-8075.

Sherr, C.J. & Roberts, J.M. (1999). Cdk Inhibitors: Positive and Negative Regulators of G1 Phase Progression, *Genes and Development*, Vol. 13, No. 12, pp. 1501–1512, ISSN 0890- 9369.

Sherr, C.J. (2000). The Pezcoller Lecture: Cancer Cell Cycles Revisited. *Cancer Research*, Vol. 60, pp. 3689–3695, ISSN 0008-5472.

Schley, D. & Gourley, S.A. (2000). Linear stability criteria for population models with periodically perturbed delays. *Journal of Mathematical Biology*, Vol. 40, No. 6, pp. 500- 524, ISSN 0303-6812.

Skulachev, V.P. (2001). NAD(P)+ decomposition and antioxidant defense of the cell. *FEBS Letters*, Vol. 492. Issue 1, pp. 1-3 ISSN 0014-5793.

Skulachev, V.P. (2001). The programmed death phenomena, aging, and the samurai low of biology. *Experimental Gerontoljgy*, Vol. 36, No. 7, pp. 995-1024, ISSN 0531-5565.

Southwood, T.R.E. (1981). Bionomic strategies and population parameters. In *Theoretical Ecology. Principles and applications*. Second Edition, May, R.M., ed., pp. 30-52, Blackwell, ISBN 0632007680, London.

Tracy, C.R. & George, T.L. (1992). On the determinants of extinction. *American Naturalist*, Vol. 139, No. 1, pp. 102-122, ISSN 0003-0147.

Trifunovic, A. & Larsson, N-G. (2008). Mitochondrial dysfunction as a cause of ageing. *Journal of Internal Medicine*, Vol. 263, No. 2, pp. 167–178, ISSN 0954-6820.

Trubitsyn, A.G., (2006). The Modified Variant of Mitochondrial Theory of Aging, *Advanses in Gerontology*, Vol. 18, pp. 21-28, ISSN 1561-9125, (In Russion).

Trubitsyn, A.G., (2006a). Evolution Mechanism of Species-Specific Lifespan, *Advanses in Gerontology*,Vol. 19, pp. 13-24, ISSN 1561-9125, (In Russion).

Trubitsyn, A.G., (2009). The Mechanism of Phenoptosis: 1. Age-Dependent Decrease of the Overall Rate of Protein Synthesis is Caused by the Programmed Attenuation of Bioenergetics, *Advanses in Gerontology*, VoL. 22, No. 2, pp. 223-227, ISSN 1561-9125, (In Russion).

Trubitsyn, A.G., (2010). Aging as a result of the implementation of the phenoptosis program. *Russian Journal of general chemistry*, Vol. 80, No. 7, 1490-1500, ISSN 1070-3632.

Trubitsyn, A.G. (2011). The Mechanism of Phenoptosis: 2. The Hayflick Limit Is Caused by Programmed Decrease of the Bioenergetics Level. *Advances in Gerontology*, Vol. 1, No. 2, pp. 147–152, ISSN 2079-0570.

Vaupel, J.W. (2003). Post-Darwinian longevity. In *Life span: evolutionary, ecological, and demographic perspectives*, Carey & Tuljapurkar, eds, pp. 258-269, Population Council,. ISBN, New York.

Vlach, J., Hennecke, S., & Amati, B. (1997) Phosphorylation-Depenedent Degradation of the Cyclin- Dependent Kinase Inhibitor p27[Kip1]. *The EMBO Journal*, Vol.16, pp. 5334–5344. ISSN 0261-4189

Wei, Y-H., Lu, C-Y., Wei C-Y., Ma, Y-S., & Lee, H-C. (2001). Oxidative stress in human aging and mitochondrial disease–consequences of defective mitochondrial respiration and impair antioxidant enzymes. *Chinese Journal of Physiology*, Vol.44. No. 1, pp. 1-11, ISSN: 0304- 4920

White, P.S., Pickett, S.T.A. (1985). Natural disturbance and patch dynamics: an introduction. In *The ecology of natural disturbance and patch dynamics*, Pickett, S.T.A. & White, P.S., eds., pp. 3-13, Academic Press, ISBN 0125545215, London.

Westerhoff, H.V., Van Dam, K. (1987). Thermodynamics and Control of Biological Free-Energy Transduction. Elsevier Science Publisher B.V. (Biomedical division), ISBN 0444807837 Amsterdam.

Wilson, E.O. (1973). Group selection and its significance for ecology. *BioScience*, Vol. 23, No. 11, pp. 631-638, ISSN 0265-9247.

Yegorov, E. E. & Zelenin, A. V. (2003). Duration of Senescent Cell Survival in Vitro As a Characteristic of Organism Longevity, an Additional to Proliferative Potential of Fibroblast, *FEBS Letters*. Vol. 541, Issue 1, pp. 6–10. ISSN 0014-5793.

Young, D.A. (1969). Glucocorticoid Action on Rat Thymus Cells. Interrelationships between carbohydrate, protein, and adenine nucleotide metabolism and cortisol effects on these functions *in vitro*. *Journal of Biological chemistry*, Vol. 244, No. 8, pp. 2210-2217, ISSN 0021-9258.

Young, D.A. (1970). Glucocorticoid Action on Rat Thymus Cells. II. Interrelationships between ribonucleic acid and protein metabolism and between cortisol and substrate effects on these metabolic parameters in vitro. *Journal of Biological chemistry*, Vol. 245, pp. 2747-2752. ISSN 0021-9258.

Yao, Y-G., Ellison, F. M., McCoy, J. P., Chen, J. and Young, N. S. (2007). Age-dependent accumulation of mtDNA mutations in murine hematopoietic stem cells is modulated by the nuclear genetic background. *Human Molecular Genetics*, Vol. 16, No. 3, pp. 286–294, ISSN 0964-6906.

Yu, T.W. & Anderson, D. (1997). Reactive oxygen species-induced DNA damage and its modification: A chemical investigation. *Mutation Research*, Vol. 379, Issue 2, pp. 201-210. ISSN 0027-5107

Zykova, T.A., Zhu, F, Zhang, M. Bode, Y, A. & Dong, Z. (2007). Involvement of ERKs, RSK2 and PKR in UVA-induced signal transduction toward phosphorylation of eIF2 (Ser[51]). *Carcinogenesis*, Vol. 28, No. 7, pp.1543-1551, ISSN 0143-3334

4

Sirtuin-Dependent Metabolic Control and Its Role in the Aging Process

Sara Santa-Cruz Calvo,
Plácido Navas and Guillermo López-Lluch*
*Centro Andaluz de Biología del Desarrollo,
Universidad Pablo de Olavide-CSI,
CIBERER- Instituto Carlos III. Carretera de Utrera Km. 1,Sevilla
Spain*

1. Introduction

During last years, the protein family of sirtuins, composed by NAD^+-dependent deacetylases, has emerged as a key factor in aging. From yeast to humans, sirtuins are involved in metabolic changes that induce a higher respiratory capacity accompanied by lower oxidative damage. They are involved in the control of glucose catabolism, fatty acid metabolism, respiratory chain activity in mitochondria and several other metabolic processes including control of antioxidant capacity in cells and tissues (Dali-Youcef et al., 2007; Elliott & Jirousek, 2008; Lomb et al., 2010; Pallas et al., 2008).

As these deacetylases are dependent on the $NAD^+/NADH$ ratio, they can be considered as important sensors of the metabolic status of the cells and probably because this they are one of the main family of proteins involved in the regulation of metabolism in the cell (Li & Kazgan, 2011). Further, their relationship with the AMPK-dependent pathway, that controls respiratory metabolism by inhibiting insulin-dependent signaling, highlights the importance of these proteins in metabolic regulation and especially in insulin-resistance, diabetes and obesity (Canto et al., 2009; Ruderman et al., 2010).

Sirtuins have been involved in aging process and considered important factors in delaying aging process and increase longevity (Guarente, 2000; Tissenbaum & Guarente, 2001). However, very recent studies have questioned the role of these deacetylases in longevity (Burnett et al., 2011; Viswanathan & Guarente, 2011). But their activity in yeast, worms and flies still permits to correlate its function in metabolism and dietary-dependent modulations with aging process (Guarente, 2008). However, to date, in mammals and, especially in humans, their role in longevity is not clear. Whereas in lower organisms only one member has been found, SIR2, in mammals, seven members have been described to date. This fact indicates a higher complexity in interactions, targets and functions in higher animals than in lowers. Further, in mammals, the specific distribution of these deacetylases among the different cell compartments also indicates several local-dependent influences of sirtuins.

*Corresponding Author

Aging can be considered a severe deleterious process that affects all the compartments in cells and also all the tissues and organs in the organism. Apart of the different theories of aging (Jin, 2010), the main common factor is the accumulation of non-metabolizable or degradable molecules into cells and tissues that impair their correct function. In few words, we age because our organism accumulates rubbish and we are unable to eliminate or recycle it. Most of the damaged molecules are the result of a unbalanced metabolism that produces high levels of reactive molecules accompanied by a low capacity of the endogenous antioxidant mechanisms of cells and the recycling mechanisms such as proteasome and autophagy or DNA-damage reparation (Asha Devi, 2009; Fleming & Bensch, 1991; Maynard et al., 2009; Perez et al., 2009; Sohal et al., 1994). As results, oxidized molecules accumulate into cells impairing their physiology at all levels. Then, a balanced and controlled metabolism will improve oxidant/antioxidant relationship and delay the accumulation of oxidized molecules in aged cells and tissues.

The present chapter is focused on the role of the metabolism in aging process and the importance of sirtuins in its control. We will describe the different pathways regulated by sirtuins and how modifications in NAD^+/NADH ratio can affect the activity of these deacetylases. Moreover, we will discuss the possible role of NADH-dependent oxidoreductases in the control of metabolism through these proteins. Furthermore, the role of a known polyphenol, resveratrol, as agonist of sirtuins and caloric restriction in aging and metabolic control will be also revised.

2. Sirtuins, a heterogeneous family of protein deacetylases

Sirtuins are a family of proteins that share a conserved NAD^+-dependent acetyl-lysine deacetylase and ADP-ribosyltransferase activity. They have been related to the regulation of the metabolism and also lifespan being involved in cell survival and apoptosis, cell proliferation and senescence. They are widely located in all the organs and near all the subcellular locations. The seven isoforms found to date in humans localize either in the nucleus, cytoplasm or mitochondria. The use of modified organisms showing increasing gene dosage of sirtuin orthologs in eukaryotes such as yeast, worms or flies have demonstrated that these enzymes are directly involved in lengthening of longevity (Guarente, 2007). Further, the relationship between calorie restriction and longevity indicate that metabolism is directly involved in aging, and then, as sirtuins are involved in the control of metabolism, a direct link between the activity and modulation of these proteins and a longer lifespan seems to be convincing (Balcerczyk & Pirola, 2010).

In contrast with class I, II and IV deacetylases, mainly involved in the control of epigenetic processes (Kuzmichev & Reinberg, 2001), sirtuins are members of the class III characterized to be dependent on NAD^+. These enzymes catalyze the reaction shown in figure 1. They bind to a Nε-acetyl-lysines of the target protein and deacetylate them by using NAD^+ as substrate and producing nicotinamide (NAM) and 2'-O-acetyl-ribose (2'-O-AADPR) as products (Hirsch & Zheng, 2011). In this process, increasing levels of NAD^+ increase the activity of sirtuins whereas higher NAM or NADH levels exert an inhibitory effect (Wolberger, 2007). Further, the expression of sirtuins are also regulated by the ratio NAD^+/NADH since higher mRNA levels have been found when NADH/NAD^+ levels rise (Gambini et al., 2011). Then, they can be considered as metabolic sensors since they can modulate their activity and levels depending on the ratio NAD^+/NADH.

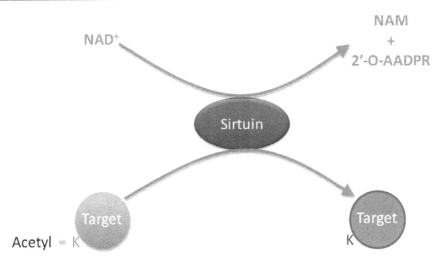

Fig. 1. Deacetylation of K-acetyl residues in targets of sirtuins.

The first member of this family studied in deep was the yeast Sir2. This deacetylase is responsible of silencing chromatin by deacetylation of histones (Blander & Guarente, 2004) and has been related to the increase in longevity in yeast, worms and flies. Apart of its activity reducing the accumulation of chromatin of ribosomal RNA (rRNA) genes in yeast, the prolongevity of Sir2 has been also related to the modulation of mitochondrial function providing benefit to slow aging and associated diseases (Guarente, 2008).

To date, seven sirtuins have been described in mammals. They are designed as SIRT1 through SIRT7. Based on the homology of the 250 aminoacids core domain, the mitochondrial SIRT3, the nuclear-cytosolic SIRT1 and the mainly nuclear SIRT2 show the closest homology to yeast SIR2 (Frye, 2000). However, if we attend to the alignment of the aminoacid sequence of the human members we can see that the identity at the aminoacid sequence is very low among the members of sirtuins family (Table 1) and only deacetylase sirtuin-type domain shows some homology being highly in the NAD+-binding and in the catalytic domains (Figure 2). A possible explanation for these high differences in sequence between the members of sirtuin family in mammals can be found in the plethora of targets that can be recognized by the different members of the family and in their different and selective locations into the cell.

	hSIRT1	hSIRT2	hSIRT3	hSIRT4	hSIRT5	hSIRT6	hSIRT7
hSIRT1	100						
hSIRT2	34.34	100					
hSIRT3	30.94	44.85	100				
hSIRT4	24.55	21.88	26.02	100			
hSIRT5	25.63	22.71	27.66	27.13	100		
hSIRT6	21.56	24.21	28.84	28.43	20.59	100	
hSIRT7	20.97	22.28	23.70	28.14	22.01	36.77	100

Table 1. Pairwise comparison of aminoacid sequences from the human sirtuin members. From BLAST (basic local alignment search tool) analysis of the indicated proteins in figure 2.

```
ySIR2/P06700   NPSNGIFYGPSFTKRESLNARMFLKYYG------------AHKFLDTYLPEDLNSLYIYLIKLLGFEVKDQALIGTINSIVHINSQERVQDLGSAISVTNVEDPLAKKQTVRLIKDLQRAINKVL   232
hSIRT1/Q96EB6  NLYDEDDDDEGEEEEEAAAAAIGYRDNLLFGDEIITNGFHSCESDEEDRASHASSSDWTPRPRIGPYTFVQQHLMIGTDPRTILKDLLPETIPPPELDDMTLWQIVINILSEPFKR-----------   234
hSIRT2/Q8IXJ6  NLFSQTLSLGSQKER------------------------------------------------------------------------------------------LIDELTLEGVARYMQSE-------------   74
hSIRT3/Q9NTG7  GASSVVGSGGSSDK--------------------------------------------------------------------------------------------GKLSLQDVAELIRAR-------------   135
hSIRT4/Q9Y6E7  PASPP---------------------------------------------------------------------------------------------------------LDPEKVKELQRFITLS-----------   54
hSIRT5/Q9NXA8  ARPSS-----------------------------------------------------------------------------------------------------------------SMADFRKFFAKA-----------   50
hSIRT6/Q8N6T7  DPPEE-----------------------------------------------------------------------------------------------------------------LERRVWELARLVWQS--------   44
hSIRT7/Q9NRC8  DDPEE-----------------------------------------------------------------------------------------------------------------LRGKVRELASAVRNA--------   99


ySIR2/P06700   CTRLRLSNFFTIDHFIQKLHTARKIIVLVLTGAGVSTSLGIPDFRS--SEGFYS---KIKHLGLDDPQDVFNYNIFMHDPSVFYNIANMVLPPEKIYSPLHSFIKMLQMKGKLLRNYTQNIDNL   349
hSIRT1/Q96EB6  ---KKRKDINTIEDAVLLQECKKIIVLTGAGVSVSCGIPDFRS--RDGIYARLAVDFPDLPDPQAMFDIEYFRKDPRPFFKFAKEIYPGQFQPSLCHKFIALSDKEGKLLRNYTQNIDTL   350
hSIRT2/Q8IXJ6  ----------------RCRRVICLVGAGISTSAGIPDFRSPSTGLYD--NLEKYHLPYPEAIFEISYFKKHPEPFFALAKELYPGQFKPTICHYFMRLLKDKGLLLRCYTQNIDTL   172
hSIRT3/Q9NTG7  ----------------ACQRVVMVGAGISTPSGIPDFRSPGSGLYS--NLQQYDLPYPEAIFELPFFHNPKPFFTLAKELYPGNYKPNVTHYFLRLLHDKGLLLRLYTQNIDGL   233
hSIRT4/Q9Y6E7  ----------------KRLLVMTGAGISTESGIPDYRSEKVGLYA---RTDRRPIQHGDFVRSAPIRQRYWARNFVGWFQFSSHQENPAHWALSTWEKLGKLYWLVTQNVDAL   148
hSIRT5/Q9NXA8  ----------------KHIVIISGAGVSAESGVPTFRG--AGGYWR--KWQAQDLATPLAFAHNPSRVWEFYHYRREVMGSKEPNAGHRAIAECETRLGKQGRRVVITQNIDEL   145
hSIRT6/Q8N6T7  ----------------SSVVFHTGAGISTASGIPDFRG----------------PHGVWTMEERGLAPKFDTTFES-----ARPTQTHMALVQLERVGLLRFLVSQNVDGL   118
hSIRT7/Q9NRC8  ----------------KYLVVYTGAGISTAASIPDYRG----------------PNGVWTLLQKGRS---------VSAADLSEAEPTLTHMSITRLHEQKLVQHVVSQNCDGL   721


ySIR2/P06700   ESYAGISTDKLVQCHGSFATATCV--TCHWNLPGERIFNKIRNLELPLCPYCYKKRREYFPEGYNNKVGVAASQGSMSERPPYILNSYGVLKPDITFFGEA---LPNKFHKSIREDILEC   464
hSIRT1/Q96EB6  EQVAGIQR---IIQCHGSFATASCL--ICKYKVDCEAVRGDIFNQVVPRCPRCPAD---------------EPLAIMKPEIVFFGEN---LPEQFHRAMKYDKDEV   433
hSIRT2/Q8IXJ6  ERIAGLEQEDLVEAHGTFYTSHCVSAS-CRHEYPLSWMKEKIFSEVTPKCEDCQS--------------LVKPDIVFFGES---LPARFFSCMQSDFLKV   254
hSIRT3/Q9NTG7  ERVSGIPASKLVEAHGTFASAITC--VQCQFPRGVLDRADVMADRVPRCPVCTG--------------VVKDIVFFGEP---LPQRFLLHVV-DFPMA   312
hSIRT4/Q9Y6E7  HTKAGSRR---LLELHGCMDRVLCL--DCGEQTPRGVLQERFQVLNPTWSAEAHGLAPD--------GDVFLSEEQVRSFQVPTCVQCGGHLKPDVVFFGDT---VNPDKVDFVHKRVKEA   253
hSIRT5/Q9NXA8  HRKAGTKN--LLEIIGSLPKTRCTGVVAENYKSPICPALSGKGAPEPGTQDASIP----------------VEKLPRCEEAGGCLLRPHVVWFGEN---LDPAILEEVDRELAHC   242
hSIRT6/Q8N6T7  HVRSGFPRDKLAELHGNMFVEECA--KCKTQYVRDTVVGTMGLKATGRLCTVAKARG---------LRACRGELRDTILDWEDS---LPDRDLALADEASRNA   207
hSIRT7/Q9NRC8  HLRSGLPRTAISELHGNMYIEVCT--SCVPNREYVRVFDVTERTALHRHQTGRTCHK------CGTQLRDTIVHFGERGTLGQPLNWEAAATEAASRA   261


ySIR2/P06700   DLLICIGTSLKVAP-VSEIVNMVPSHVPQVLINRD-----------------PVKHAEFDLSLLGYCDDI----------------A   517
hSIRT1/Q96EB6  DLLIVIGSSLKVRP-VALIPSSIPHEVPQILINREPLPHLHFDVELLGDCDVIINELCHRLGGEYAKLCCNPVKLSEITEKPPRTQKELAYLSELPPTPLHVSEDSSPERTSPPDSSVI   552
hSIRT2/Q8IXJ6  DLLLVMGTSLQVQP-FASLISKAPLSTPRILINKE-----------------KAGQSDPFIGMIMAGLGGGGMDFDSKYAIRDVAWLGECDQG   328
hSIRT3/Q9NTG7  DLLLILGTSLEVEP-FASLTEAVRSSVPRLLINRD-----------------LVGP-----------------LAWHEP--SRDVAQLGDVVHG--------------V   370
hSIRT4/Q9Y6E7  DSLLVVGSSLQVYSGYRFILTAWEKKLPIAILNIG------------------PTRSDDLACL-----------------   298
hSIRT5/Q9NXA8  DLCLVVGTSSVVYPAAMFAPQVAARGVPVAEFNTET----------------TPATNRFRFHFQG--------------   291
hSIRT6/Q8N6T7  DLSITLGTSLQIRPSGNLPLAATKREGGRLVIVNLQP----------------TKHDRHADLRIHGYVDEVMTRLMKHLGLEIPAWDGPRVLE   283
hSIRT7/Q9NRC8  DTILCLGSSLKVYPRIWCMTKPPSRRPKLYIVN-----------------LQWTPKDDWAALKLHGKCDDVMRLLMAELG------   327


ySIR2/P06700   AMVAQKCGWTIPHKKWNDLKNKNFKCQ-----------------EKDKGVYVTSDEHPKTL---   562
hSIRT1/Q96EB6  VTLLDQAAKSNDDLDVSESKGCMEEKPQEVQTSRNVESIAEQMENPDLKNVGSSTGEKNERTSVAGTVRKCWPNRVAKEQISRRLDGNQYLFLPPNRYIFHGAEVYSDSEDDVLSSSSCG   672
hSIRT2/Q8IXJ6  LALAELLGWKKELEDIVRREHASIDAQSGAGVPNPSTSASPKKSPPPAKDEARTEREKPQ----------   389
hSIRT3/Q9NTG7  ESLVELLGWTEEMRDIVQRETGKLDG-----------------PDK----------------   399
hSIRT4/Q9Y6E7  -KLNSRCGELLPLIDPC-----------------   314
hSIRT5/Q9NXA8  ----PCGTTLPEALACHENETVS-----------------   310
hSIRT6/Q8N6T7  RALPLPRPPTKLKEFPSPTRINGSIPAGPKQEFCAQHNGSEPASPKREPTSPAPHRPKRVKAKAVPS----------   355
hSIRT7/Q9NRC8  LEIPAYSRWQDPIFSLATPLRAGEEGSHSRKSLCRSREEAFPGDRGAPLSSAPILGGWFGRGCTKRTKRKKVT----   400
```

Fig. 2. Alignment of centre core of human sirtuins family in comparison with yeast sir2 (previous page). The figure represent Clustalw alignment from indicated yeast and human sirtuins indicated by their UniProtKB accession numbers. In yeast sir2, the deacetylase sirtuin-type dominium is from 245 to 529 (red arrow) that correspond with the highest homology sequence among the members of the family. The NAD+ binding domains are indicated in green, there are the most conserved domains in the whole family. The active site is determined by a histidine at 364 position of sir2 that acts as a proton acceptor, the key histidines in other members have been determined in silico by homology. Although it has been indicated that these enzymes do not bind zinc, probable cystein residues able to bind zinc are also conserved in some of the members of the family (in blue). Regarding regulation, in sir2 two points of regulation by phosphorylation, phosphorylation at serine 23 and at tyrosine 400 have been determined (violet residues). None of them are conserved residues in human sirt forms. Further, in SIRT1, modifications at cysteines 395 and 398 by s-nitrosylation impede the binding of NAD+ and then, the activity of the enzyme.

	S. cerev. SIR2	S. pombe SIR2	D. melanog. SIR2.1	C. elegans SIR2	D. rerio SIR2	M. musc. SIRT1	H. sapiens SIRT1
S. cerev. SIR2	100	41	43	39	45	40	40

Table 2. Pairwise comparison of the aminoacid sequences among yeast (*Saccharomyces cerevisiae*) SIR2 and higher homologues in model animals: fission yeast (Saccharomyces pombe); fly (*Drosophila melanogaster*); worm (*Caenorhabditis elegans*); zebrafish (*Dario rerio*); mice (*Mus musculus*) and human (*Homo sapiens*). The percentage of identity in comparison with S. *cerevisiae* sir2 protein is indicated. From BLAST analysis of the indicated proteins.

Among the other human sirtuins, SIRT4 and SIRT5 are mitochondrial sirtuins that show predominant ADP-ribosyl-transferase activity and a weak deacetylase activity and are involved in urea cycle regulation (Nakagawa & Guarente, 2009). On the other hand, SIRT6 and SIRT7 are considered as members of another subclass of sirtuins involved in reparation of DNA and the control of ribosomal RNA production through cell cycle (Lombard et al., 2008). Although it has been described that sirtuins does not bind zinc, Sir2, SIRT1, -2 and 3 share four proximal cysteines that can indicate the possibility of binding zinc (figure 2). These four Cys are highly conserved among these sirtuins and just following the catalytic histidine. Then, although this Zinc ion must be not involved in the catalytic activity, their presence can be important for the maintenance of the structure of the sirtuin. In fact, recently Sanders and coworkers (Sanders et al., 2010) have shown that the four-cysteine metal binding site resembles the Zn-ribbon structure of transcription factors such as TF-IIS, TF-IIN and RNA polymerase II subunit RPB9. Further, although the Zinc-binding site is too far from catalytic domain, its presence is important for the activity of the enzyme since the change of any cystein to alanine or addition of zinc chelators inhibits the *in vitro* deacetylase activity of sirtuins (Min et al., 2001).

Regarding post-translational regulatory mechanisms, sirtuins can be regulated by phosphorylation and sumoylation. In fact, in SIRT1 thirteen residues have been found to be phosphorylated *in vivo* (Sasaki et al., 2008) indicating a high ratio of regulation by kinases. Further, dephosphorylation by protein phosphatases *in vitro* results in the decrease of the NAD+-dependent deacetylase activity in SIRT1. On the other hand, sumoylation of SIRT1 at

Lys734 residue has been also reported (Yang et al., 2007). Sumoylation consist in the binding of small ubiquitin-like modifier (SUMO) proteins to lysine residues (Hay, 2001). Binding of SUMO protein to SIRT1 increases its deacetylase activity and mutation of SIRT1 at the Lys734 residue or desumoylation by the nuclear desumoylase SENP1 reduces the activity (Yang et al., 2007).

Finally, another regulatory mechanism also establishes a relationship of sirtuins with metabolism. Glyceraldehyde-3-phosphate dehydrogenase (GAPDH) is physiologically nitrosylated at its Cys150 residue and binds to Siah1. Further, the complex moves to the nucleus since Siah1 show a nuclear localization signal. In the nucleus, Siah1 interacts with SIRT1 and other proteins. By this mechanism, S-nitrosylation of SIRT1 by GAPDH inhibits its deacetylase activity but specifically in the nucleus (Kornberg et al., 2010).

2.1 Enzymatic activity of sirtuin

As it has been above indicated (figure 1) sirtuins bind a Nε-acetyl-lysine of the target protein and deacetylate it by using NAD^+ as substrate and producing NAM and 2'-O-AADPR (Hirsch & Zheng, 2011). However, this mechanism is no completely clear. In the case of SIRT6, although the deacetylation of histone 3 by SIRT6 has been described (Kawahara et al., 2009), other authors indicate that the main activity of this sirtuin is the ADP-ribosylation (Liszt et al., 2005). However, more recent studies indicate that the ADP-ribosyl-transferase of sirtuins could be only some inefficient side reactions of the deacetylase activity without any relevant physiological role (Du et al., 2009).

2.2 Subcellular localization of sirtuins

One of the key facts that determine the main targets of the different members of the sirtuin family is their respective subcellular localization. SIRT1 is found in the cell in both, the cytosol and the nucleus although it seems that nuclear localization is the most prevalent. However, recent research has demonstrated that SIRT1 is mainly sequestered in cytosol in highly glycolitic tumoral cells (Stunkel et al., 2007) indicating a metabolic-dependent localization of this deacetylase. On the other hand, SIRT3 is predominantly found in mitochondrial matrix (Schwer et al., 2002) although some studies have shown nuclear and also cytosolic locations (Sundaresan et al., 2008) whereas other authors have reported an exclusive mitochondrial localization (Cooper & Spelbrink, 2008). In the case of SIRT2, this sirtuin appears to be exclusively cytoplasmic (North & Verdin, 2007). SIRT4 and SIRT5 are located in the inner mitochondrial membrane or matrix (Michishita et al., 2005) and SIRT6 and SIRT7 are located in the nucleus (Schwer & Verdin, 2008).

2.3 Modulation of sirtuin levels

Acting as metabolic sensors, these proteins respond to many processes that affect the energetic balance in the organism including aging, dietary interventions, fasting or exercise. Aging progress is associated with a gradual decline of several physiological processes in the organism. In heart, age-related in SIRT1, decline is accompanied by a higher level of oxidative stress and the decrease in the expression of endogenous antioxidant enzymes and their regulators (Ferrara et al., 2008). In central nervous system, aging results in decreased activity of SIRT1 in cerebellum that leads to the increase in acetylation of protein residues specially affecting motor function (Marton et al., 2010). In cell culture models, cellular senescence induced by ionizing radiation is accompanied by the decrease in the levels of SIRT1 (Hong et al., 2010). On the other hand, the contrary effect of aging has been reported. In rats, an age-related increase in SIRT1 levels has been shown in skeletal muscle (Koltai et al., 2010).

Caloric restriction (CR) is the only dietary modification able to extend median and maximum lifespan in a number of organisms from yeast to mammals (Lomb et al., 2010). The effect of CR on lifespan extension is thought to be dependent on multiple different signaling pathways. CR decreases the activity of pro-aging pathways such as oxidative stress and insulin and growth hormone signaling whereas it stimulates the endogenous capacity of the cells against stress including antioxidant mechanisms (Qiu et al., 2010), DNA repair capacity and autophagy (Morselli et al., 2010). Further, the activity of mitochondria is modified in CR. Under CR, mitochondria show higher efficiency with lower reactive oxygen species production (Lopez-Lluch et al., 2006).

Many of the effects of CR on longevity have been associated to the induction of sirtuin activity in cells (Cohen et al., 2004). Studies performed in mice have demonstrated that SIRT1 protein levels increases during CR in many tissues including brain, white adipose tissue, muscle, liver and kidney (Kanfi et al., 2008). Moreover, loss-of-function and gain-of-function mouse studies have provided genetic evidences that indicate that SIRT1 is a key factor in the physiological response to CR (Imai, 2009). It is also important to highlight that SIRT1 has been related to the central response to low nutritional availability at the hypothalamus level probably playing an important role in the regulation of the whole metabolism in mammals (Satoh et al., 2010). Further, SIRT6 levels are also modulated by nutrient availability in a p53-independent mechanism. SIRT6 modulation is mainly through the stabilization of protein levels but not via increase of SIRT6-gene transcription (Kanfi et al., 2008).

The practice of exercise has been also considered to promote longevity and activate common pathways to CR probably by producing a metabolic stress in the organism (Lanza et al., 2008). Then, as in the case of CR, exercise also modulates the levels of sirtuins. In muscle, SIRT1 levels increases along aging and exercise training further increase the relative activity of this sirtuin (Koltai et al., 2010) indicated by an strong inverse correlation between nuclear activity of SIRT1 and the level of acetylated proteins. On the other hand, age-associated increase in SIRT6 levels is attenuated by exercise (Koltai et al., 2010). Exercise also increases SIRT3 expression in muscle and its activity is associated with a higher activity of AMP-dependent protein kinase (AMPK), cAMP-response element binding (CREB) and Peroxisome proliferator-activated receptor gamma coactivator 1-α (PGC1α) indicating its importance in mitochondrial biogenesis in muscle fibers especially in respiratory type I fibers (Palacios et al., 2009).

In heart, the aging-related decrease in MnSOD and catalase expression accompanied by the increase in oxidative damage levels indicated by TBARS and 4-HNE has been related to the decrease in the expression of SIRT1 (Ferrara et al., 2008). Exercise increases SIRT1 levels in heart reverting aging-related effect on MnSOD and catalase levels and its regulatory transcription factor, FOXO3a levels (Ferrara et al., 2008). Exercise not only modulates sirtuin levels in muscle or heart but also can modulate sirtuin levels and activity in central nervous system. Further, the modulation of SIRT1 by natural polyphenolic flavonoids such as resveratrol or quercetin may exert important beneficial effects in exercise performance (Lappalainen, 2011).

Sirtuin expression is also altered in cancer cells (Ashraf et al., 2006). This fact is important because these cells show a distinctive metabolism and higher growth in comparison with non-transformed cells. The different pattern of sirtuin expression in tumoral cells would confer to these cells higher resistance against exogenous agents and also control a different metabolism.

Other important issue in the regulation of sirtuin levels is the complex and new world of microRNAs (miR)-dependent regulation. Currently, the study of the regulation of sirtuin expression by microRNAs has demonstrated that these proteins are also regulated by this system. MiR-34a is markedly reduced in p53-null PC3 cells and its overexpression inhibits SIRT1 expression at the transcriptional level indicating a p53-dependent regulation of SIRT1 levels (Fujita et al., 2008). On the other hand, in mesenchimal transition processes in breast cancer, the downregulation of miR-200 has been related to the increase in the levels of SIRT1 in these cells contributing to the tumoral phenotype (Eades et al., 2011). On the other hand, the release of proinflammatory mediators in adipocytes in serum-free conditions is regulated by the inhibition of SIRT1 expression mediated by miR-132 (Strum et al., 2009). Further, miR-199a also represses SIRT1 in cardiomyocytes and its downregulation in low oxygen tension conditions derepresses SIRT1 expression at the same time than HIF-1α (Rane et al., 2009). Other interesting miRs, miR-33a and b, are involved in the regulation of fatty acid oxidation including the levels of SIRT6. Increase in the levels of miR-33a and b decrease fatty acid oxidation and also insulin signaling in hepatic cell lines indicating a regulatory role of these miRs in important metabolic pathways in the cell (Davalos et al., 2011). Taken together, it is clear that the, to date, poorly clarified regulatory mechanisms depending on miRs complicate the regulatory mechanisms of sirtuin levels at posttranscriptional level.

During last years, small polyphenol molecules have also demonstrated capacity to increase sirtuin activity. Some years ago, we and others demonstrated that resveratrol, a polyphenol of the family of stilbenes found in grapes, dry fruits and berries, is able to extend lifespan in mice fed under high fat conditions (Baur et al., 2006). In this process, sirtuin activity was considered as an important factor. From them, several other works have demonstrated the importance of resveratrol and related compounds in sirtuin-dependent metabolic modifications. In fact, resveratrol is able to modulate insulin response (Zhang, 2006), and also regulate AMPK activity (Dasgupta & Milbrandt, 2007). In some cases, these effects have been related to sirtuin activity and in others, a sirtuin-independent effect has been suggested. In any case, in our hands resveratrol have shown capacity to increase SIRT1 and SIRT3 levels in cultured cells indicating the capacity to modulate sirtuin expression (Santa-Cruz Calvo et al., unpublished results), accordingly with already published results (Costa Cdos et al., 2011; Kao et al., 2010; Sulaiman et al., 2010).

3. Sirtuin-dependent metabolic regulation

As its can be concluded by the complexity of sirtuin interactions, the different partners and regulatory processes, this family of deacetylases is involved in many different physiological mechanisms in cells. In the following sections we are going to resume the most important findings about the role of these enzymes in metabolic control in relationship with the aging process. Taken into consideration that metabolic processes are involved in all the cellular processes, metabolic control by sirtuins is the most important function of these enzymes (Yu & Auwerx, 2009).

3.1 SIRT1

Looking inside this deacetylase enzymes family, SIRT1 is one of the members that show more interactions and that respond to more factors. Mammalian SIRT1 has multiple targets including histones, transcription factors and other molecules that collectively modulate several processes such as energy metabolism, stress response and cell survival (Tang, 2011). Its activity may decline with aging in many tissues and it has been proposed that its reactivation can produce beneficial effects (Tang, 2011).

One of the most important factors involved in the metabolic control regulated by SIRT1 is PGC1α. SIRT1 functionally interacts with PGC1α and deacetylates it (Nemoto et al., 2005). Deacetylation of PGC1α activates this transcription factor that induce the expression of nuclear respiratory factor 1 (NRF1) and then, mitochondrial biogenesis. In fact, activation of SIRT1 induces deacetylation of PGC1α and FOXO1 that finally control the transcriptional modulation for lipid catabolism (Canto et al., 2010). Further, deletion of SIRT1 alters fatty acid metabolism resulting in hepatic steatosis and inflammation (Purushotham et al., 2009). SIRT1-dependent regulatory mechanisms regulate the switch from carbohydrate to lipid as main energy sources in muscle. Limitation in glucose availability during fasting or exercise induces AMPK activity in muscle that acts as a prime initial sensor that activates SIRT1. PGC1α is acetylated by the acetyltransferase GCN5 that together with SIRT1 control its regulation depending on nutritional status (Dominy et al., 2010). Activity of this GCN5 or inhibition by nicotinamide reduces SIRT1-dependent PGC1α acetylation and decreases the expression of genes involved in mitochondrial biogenesis in muscle (Gerhart-Hines et al., 2007). Further, PGC1β is also acetylated on at least 10 lysine residues by GCN5 repressing its transcriptional activity, SIRT1 activity also deacetylates it and restores transcriptional activity (Kelly et al., 2009).

As a cycle of regulation, SIRT1 also controls the expression levels of PGC1α in skeletal muscle through stimulation of its promoter activity probably with the activity of myogenic factors such as MEF2 and MyoD (Amat et al., 2009). On the other hand, PGC1α is also involved in sirtuin expression since, as mitochondrial biogenesis is activated and some sirtuins are located in mitochondria, the expression of SIRT3 gene is also controlled by PGC1α. This regulation is key in the differentiation of brown adipocytes (Giralt et al., 2011). Besides the high number of evidences demonstrating the relationship of SIRT1 activity and PGC1α-dependent mitochondrial biogenesis, some other works indicate that SIRT1 overexpression reduces mitochondrial biogenesis (Gurd et al., 2009). This last paper is based on the correlation of SIRT1 levels with mitochondrial biogenesis. In this context, a recent paper indicate that there are a direct relationship between mitochondrial biogenesis and activity of PGC1α with nuclear activity of SIRT1 although not with its protein content in skeletal muscle cells (Gurd et al., 2011) indicating that sirtuins levels are not necessarily related to the activity of these enzymes.

Another of the most studied targets of SIRT1 is the tumor suppressor p53. SIRT1 deacetylates K382 of p53/TP53 and inhibits its transcriptional activity impairing then, its ability to induce proapoptotic mechanisms and to modulate cell senescence. Further, it has been also reported that H_2O_2-induced cell senescence is accompanied by accumulation of acetylated p53 by decrease in the function of SIRT1 (Furukawa et al., 2007). Taken into consideration the role of p53 in nuclear and mitochondrial apoptosis (Moll & Zaika, 2001), SIRT1 seems to be a p53-dependent antiapoptotic factor.

On the other hand, modulation of p53 by SIRT1 also produces effects on cell metabolism since p53 seems to regulate mitochondrial respiration and glycolysis (Ma et al., 2007). In fact, p53 regulates the transcription of cytochrome c oxidase 2, an important factor in assembly of the cytochrome c oxidase complex (Fields et al., 2007), and then, an important factor in mitochondrial respiration. Then, high levels of SIRT1 in tumor cells will block p53-

dependent SCO2 transcription and contribute to the Warburg effect found in these cells. However, the relationship between SIRT1 and p53 is more complex at the transcriptional level. Transcription of SIRT1 is repressed by p53 via p53 response elements in its proximal promoter (Naqvi et al., 2010). However, another p53 binding site has been reported in the distal promoter of SIRT1. This binding site is necessary for SIRT1 induction under caloric restriction (Naqvi et al., 2010). In this site, p53 competes with the Hypermethylated-In-Cancer-1 (HIC1) transcriptional repressor and, then, activation of p53 derepresses SIRT1 transcription. Taken together all the information available about p53 and SIRT1 interaction, more research is be necessary to clarify the complex system p53-SIRT1 and regulation of mitochondrial activity.

All these regulations implicate a contradictory role of SIRT1 in modulation of mitochondrial biogenesis and respiratory metabolism. If SIRT1 is activating PGC1α by deacetylation and, then, inducing mitochondrial biogenesis, downregulation of the activity of p53 by the same SIRT1 will reduce the respiratory capacity by affecting SCO2 levels and Complex IV assembly. This contradictory effect would be explained by the different location of SIRT1 and then, modulation of different regulatory processes. Further, some other studies have indicate that despite the role of SIRT1 as deacetylase of p53, SIRT1 has little effect on p53-dependent transcription and does not affect many of the p53-mediated biological activities (Kamel et al., 2006). If these results are confirmed, they would explain how mitochondrial biogenesis and p53-repression can occur at the same time. Future research will clarify this complex regulation of mitochondrial respiratory metabolism.

3.2 SIRT2

The human SIRT2 is predominantly a cytosolic protein known to be a tubulin deacetylase (North et al., 2003). SIRT2 deacetylates lysine-40 of α-tubulin and then, its knockdown results in tubulin hyperacetylation. Levels of SIRT2 increases dramatically during mitosis and is also phosphorylated during the G_2/M transition. Then, SIRT2 is an important factor in the control of mitotic exit in the cell cycle (Dryden et al., 2003). Further, its interaction with the homeobox transcription factor, HOXA10, raises the possibility that SIRT2 also plays a role in mammalian development (Bae et al., 2004). The importance of tubulin activity in neuronal activity probably explains the important role ascribed to SIRT2 in neurodegenerative diseases (Harting & Knoll, 2010).

Regarding metabolism, information about SIRT2 is very limited and seems to indicate a lipid inhibitory role of this sirtuin in contrast with the role found with SIRT1 and 3. Its presence has been inversely correlated with the differentiation of preadipocytes to adipocytes by modulating the activity of FOXO1 (Jing et al., 2007). SIRT2 deacetylates FOXO1 and enhances its repressive interaction with PPAR-γ, an essential factor in adipocyte differentiation (Wang & Tong, 2009). In contrast with the neuroprotection reported in some neurodegenerative processes, a neuroprotective effect of the decrease of sterol biosynthesis through SIRT2 inhibition has been also shown in the case of Huntington's disease (Luthi-Carter et al., 2010). On the other hand, silencing of SIRT2 induces intracellular ATP drop and cell death in neuronal PC12 cells (Nie et al., 2011) indicating metabolic regulatory mechanisms of this sirtuin. Further, the role of SIRT2 in expression of antioxidant systems has been also reported. Induction of MnSOD seems to depend on deacetylation of FOXOa by SIRT2 (F. Wang et al., 2007).

3.3 SIRT3

One of the most important sirtuin in metabolic regulation is SIRT3. Its localization into mitochondria and the diversity of targets including both metabolic and antioxidant components makes it in one of the immediate regulators of mitochondrial activity. In a recent paper, Kendrick and co-authors found that mice fed under high-fat diet develop fatty liver and show high levels of acetylated proteins in parallel with a decrease in SIRT3 activity (Kendrick et al., 2011). Moreover, deletion of SIRT3 further increases acetylation in high-fat fed animals and reduces the activity of respiratory complexes III and IV indicating a key role of this sirtuin in mitochondrial activity control. One of the direct targets of SIRT3 is succinate dehydrogenase (SDH, Complex II). Acetylated SDH show low activity and deacetylation by SIRT3 activates it (Finley et al., 2011). These papers indicate the important role of SIRT3 in the regulation of mitochondrial activities by deacetylation.

The role of SIRT3 in acetate metabolism has been also related to aging (Shimazu et al., 2010). Acetate plays an important role in cell metabolism being an important product of ethanol and fatty acid metabolism especially during fasting or starvation (Seufert et al., 1974). Acetate can be converted into acetyl-CoA by the activity of acetyl-CoA synthase enzymes in cytosol (AceCS1) or mitochondria (AceCS2). These enzymes are activated by deacetylation by both SIRT1 in cytosol and SIRT3 in mitochondria (Shimazu et al., 2010). An important role for SIRT3 in acetate metabolism has been suggested since both, SIRT3 KO and AceCS2 KO mice show overlapping phenotypes. However, to date no clear data about the role of acetate in aging process have been shown although AceCS-mediated synthesis in yeast has been associated with higher longevity (Falcon et al.).

The importance of SIRT3 in the protection against oxidative stress is also important since the protective effect of CR on oxidative stress is diminished in mice lacking SIRT3. This sirtuin is involved in the reduction of cellular ROS levels depending on the manganese-dependent mitochondrial superoxide dismutase 2 (MnSOD or SOD2) (Qiu et al., 2010). SIRT3 adjusts MnSOD activity to the mitochondrial nutrients availability and then, the production of mitochondrial ROS (Ozden et al., 2011). In this regulation, SIRT3 deacetylates two important lysine residues on SOD2 promoting its antioxidant activity.

Levels of SIRT3 are also regulated by the energetic status of the organism. SIRT3 levels increase by caloric restriction of exposure to low temperatures in brown adipocytes. Forced expression of SIRT3 and activity in a cell line for brown adipocytes enhances the expression of PGC1α, UCP1 and another mitochondria-related genes whereas mutation of SIRT3 inhibits PGC1α-dependent UCP1 expression (Shi et al., 2005). Diet and exercise signals also regulate SIRT3 and activate the AMPK and PGC1α in skeletal muscle cells (Palacios et al., 2009) whereas this activation is much lower in SIRT3 KO animals. On the other hand, PGC1α strongly stimulates mouse SIRT3 gene expression in muscle cells and hepatocytes (Kong et al., 2010) through binding to an oestrogen-related receptor binding element (ERRE) in its promoter region. Induction of SIRT3 is also essential for mitochondrial biogenesis and the expression of several of mitochondrial components including antioxidant systems (Kong et al., 2010).

Taken together, SIRT3 seems to be a key sirtuin that senses metabolic status through $NAD^+/NADH$ levels at the mitochondria and then, integrates respiratory metabolism and antioxidant systems.

3.4 SIRT4 and SIRT5

SIRT4 and SIRT5 are also mitochondrial sirtuins involved in the regulation of other metabolic processes essentially related with the urea cycle (Li & Kazgan, 2011). One of the main activities of SIRT4 in mitochondria is the downregulation of insulin secretion by beta cells by repressing the activity of glutamate dehydrogenase in response to aminoacids (Argmann & Auwerx, 2006; Haigis et al., 2006). Depletion of SIRT4 have also shown that this sirtuin seems to exert an opposite role than SIRT1 and SIRT3 since its depletion increases gene expression of fatty acid metabolism enzymes in hepatocytes (Nasrin et al., 2010). However, this effect is indirect due to compensatory mechanisms involving higher expression of SIRT1. However, putative contrary effects of this sirtuin in relationship with other members of the family cannot be discarded and further research involving specific inhibitors instead of gene depletion is needed.

In the case of SIRT5, this sirtuin, mainly located at the matrix of the mitochondria, is also involved in the regulation of the urea cycle (Nakagawa & Guarente, 2009). SIRT5 deacetylates the mitochondrial carbamoyl phosphate synthetase 1. This enzyme is the first and rate-limiting step of the urea cycle (Nakagawa & Guarente, 2009). Furthermore, deacetylation of cytochrome c has been also reported although the effect of this deacetylation in both, respiration or apoptosis, is not clear (Schlicker et al., 2008). To date, no other substrates of SIRT5 have been reported.

3.5 SIRT6 and SIRT7

SIRT6 is predominantly a nuclear protein broadly expressed in tissues showing the highest levels in muscle, brain and heart (Liszt et al., 2005). In any case, SIRT6 is mainly involved in DNA damage repair and is located in the nucleus. It is recruited to the sites of DNA double-strand breaks (DSBs) and stimulates DSB repair through both, nonhomologous end joining and homologous recombination by stimulating PARP-1 poly-ADP-rybosilase activity (Mao et al., 2011).

It seems that this sirtuins is closely involved in neural degeneration related to aging. In fact, a mice model lacking SIRT6 develops a degenerative disorder that mimics models of accelerated aging (Lombard et al., 2008, Mostoslavsky et al., 2006). This effect depends on a higher instability through the DNA base excision repair pathway, then, the accumulation of mutations in the genome leads to aging-associated degenerative phenotypes (Mostoslavsky et al., 2006). Furthermore, SIRT6-deficient mice show deficiency in growth and show severe metabolic defects indicating that the higher DNA-damage found in these animals is linked to a systemic metabolic deregulation that leads to age-related processes and death.

Neural SIRT6 has been also recently related to metabolic homeostasis in mammals. Neural-specific deletion of SIRT6 in mice produces postnatal growth retardation since these animals show low growth hormone (GH) and also insulin-like growth factor 1 (IGF1) levels (Schwer et al., 2010). However, unlike SIRT6-KO animals that die by hypoglycaemia and other severe metabolic defects (Mostoslavsky et al., 2006), neural-SIRT6 KO animals, reach normal size and even become obese. It seems that at the central nervous system, SIRT6 acts as a central regulator of somatic growth and metabolism by modulating neuroendocrine system. It seems that the main mechanism of action of SIRT6 in the regulation of gene expression and the control or systemic metabolism and aging is through the deacetylation of lysine 9 in histone H3. Recently it has been shown that hyperacetylation of H3K9 found in SIRT6-deficient cells leads to a higher NF-κB-dependent modulation of gene expression, proinflammatory processes, apoptosis and cellular senescence (Kawahara et al., 2009). Then, the control of gene expression by histone modulation seems to be a key factor in SIRT6-dependent prolongevity effect.

Furthermore, the activity of SIRT6 in liver has been also reported. Rosiglitazone (RGZ) is used to protect liver against steatosis. This compound increases the levels of SIRT6 in liver at the same time that ameliorates hepatic liver accumulation affecting PGC1α and FOXO1 (Yang et al., 2011). However, in SIRT6-deficient mice, RGZ was unable to decrease fat accumulation in hepatocytes and to affect PGC1α and FOXO1 activity indicating an important role of this sirtuin in fat storage in liver. In this mechanism, SIRT1 could be also involved since it forms a complex with FOXO3 and NRF1 and activates the expression of SIRT6 (Kim et al., 2010). In this case SIRT6 would be the sirtuin that negatively regulates glycolysis, triglyceride synthesis and fat metabolism by deacetylating H3K9 and then, modifying the activity of the promoters of many genes involved in metabolic processes. In fact, the specific deletion of SIRT6 in liver causes profound alterations in gene expression that produce the contrary effects in glycolysis, triglyceride synthesis and fat metabolism.

The last member of sirtuins, SIRT7, is widely expressed in nucleolus and has been associated with active rRNA genes interacting with RNA polymerase I and with histones (Ford et al., 2006). This sirtuin is controlled by CDK1-cyclin B-dependent phosphorylation and dephosphorylation indicating that its activity is required to resume rDNA transcription in late telophase (Grob et al., 2009). In the case of SIRT7, studies performed by using murine cells lacking or overexpressing this sirtuin demonstrate that it is related with the tumorigenic potential and may enable cells to sustain critical metabolic functions because it inhibits cell growth under severe stress conditions (Vakhrusheva, Braeuer et al., 2008). These studies have also demonstrated the important role of this sirtuin in lifespan. In fact, mice lacking SIRT7 undergo reduction in mean and maximum lifespans and develop heart hypertrophy and inflammatory cardiopathy (Vakhrusheva, Smolka et al., 2008) probably by the impossibility to deacetylate p53 and regulate p53-dependent apoptosis.

4. Sirtuins: Antioxidant mechanisms and autophagy

Other of the important roles of sirtuins related to metabolism and aging is based on their activity to maintain cellular antioxidant mechanisms and autophagy systems. A great body of evidence has accumulated indicating that at the same time that sirtuins are modulating metabolism, they also regulate, in a coordinated mechanism, antioxidant systems and recycling systems in cells.

Altered ROS levels are observed in several age-related illnesses including carcinogenesis, neurodegenerative, fatty liver, insulin resistance, cardiac resistance, etc. In mitochondria MnSOD is the primary ROS scavenging enzyme to converts superoxide to hydrogen peroxide that is further converted to water by catalase and other peroxidases. In this mechanism SIRT3 exert a key role since changes in lysine acetylation modifies MnSOD activity in mitochondria (Ozden et al., 2011). Further, CR effect depends, at least in part, on sirtuin regulation but at the same time oxidative stress is reduced in CR by activation of antioxidant systems such as SOD2 in mitochondria by SIRT3 (Qiu et al., 2010).

In heart, the aging-related decrease in MnSOD and catalase expression accompanied by increase in the levels of oxidative damage indicated by TBARS and 4-HNE has been related to the decrease in the expression of SIRT1 (Ferrara et al., 2008).

Regarding recycling mechanism a correct balance between biogenesis and recycling of damaged structures is essential to maintain a correct homeostasis in the cell. Caloric restriction induces autophagy through induction of SIRT1. Transgenic expression of SIRT1 in human cells and in *C. elegans* induces autophagy whereas knockout of SIRT1 in the same cells and organisms prevents autophagy induced by resveratrol or nutrient deprivation (Morselli et al., 2010). Autophagy induction has been also related to the extension of lifespan by some agents such as spermidine and resveratrol in organism such as yeast, nematodes and flies (Morselli et al., 2009). In this process, deacetylation of FOXO3 by SIRT1 seems to be essential to the induction of the expression of genes involved in autophagy in caloric restriction (Kume et al., 2010). FOXO is an essential factor in the induction of autophagy and, as it has been above commented, in the antitumoral role of sirtuins (Zhao et al., 2010). All these works and some other more indicate that sirtuins not only control metabolism regulating essentially mitochondrial respiration and fatty acid oxidation but also regulate in a coordinated way the expression and activity of endogenous antioxidant systems and autophagy processes to eliminate damaged structures including mitochondria.

5. Sirtuins, prolongevity or healthspan effect?

The prolongevity effect of sirtuins was initially determined in yeast (Kaeberlein et al., 1999) and lower metazoan such as *C. elegans* (Tissenbaum & Guarente, 2001) and in *D. melanogaster* (Rogina & Helfand, 2004). However, very recently, these results have been revised and the prolongevity effect of sir2 in these animals has been related to transgene-linked genetic effects other than overexpression or sir2.1 in *C. elegans* or dSir2 in *D. melanogaster* (Burnett et al., 2011; Viswanathan & Guarente, 2011). Further, along last year, a considerable body of evidences indicates the controversial aspect of sirtuins in longevity studies. Calorie restriction clearly exerts a prolongevity effect on many organisms. In this effect, sirtuins were described as important factors in yeasts (Lin et al., 2000), *C. elegans* (Y. Wang & Tissenbaum, 2006) and *D. melanogaster* (Rogina & Helfand, 2004). However, other studies in yeast and *C. elegans* have argued about the role of sirtuins in caloric restriction-dependent longevity (Kaeberlein, 2010; Kenyon, 2010). Further, in mammals, overexpression of SIRT1 in mice does not increase lifespan (Herranz & Serrano, 2010).

These new concerns about the promising role of sirtuins in longevity do not affect other important functions of sirtuins in cell physiology. There is also an overwhelming body of evidences indicating that sirtuins play a crucial role in metabolic homeostasis. As, the activity of sirtuins depends strictly on the levels of NAD^+ which acts as co-substrate in the deacetylation activity catalyzed by sirtuins, changes in NAD^+ levels, reflecting modifications in the metabolic status of the cells, would modulate sirtuin activity. NAD^+-dependence for sirtuin activity in cells confers to sirtuins the integrative role of metabolic sensors that modulates cell changes depending on the metabolic status of the cells. Furthermore, the broad group of targets of sirtuins activity in cells confers to these proteins the capacity to modulate executive proteins and also to influence transcription factors and histone proteins to change not only protein activity but also gene expression profile in cells accordingly to changes in metabolism (Canto & Auwerx, 2009).

In mammals, SIRT1 mediates the metabolic and transcriptional adaptations after nutrient deprivation or energy stress changes. These adaptations are centered in a higher respiratory activity of mitochondria. Calorie restriction induces the expression of sirtuins in many tissues and likely this regulation is related to the changes in metabolism found under dietary restrictions (Bamps et al., 2009; Imai, 2009). Overexpression of SIRT1 in mice protects animals against metabolic damage caused by a fat-rich diet (Herranz et al., 2010). On the other hand, mice lacking SIRT1 show deficiencies in metabolism and are unable to increase lifespan in calorie restriction conditions (Herranz & Serrano, 2010). Further, resveratrol, a polyphenol considered as activator of sirtuins protects against metabolic and age-related diseases (Lagouge et al., 2006) and also increase lifespan in animals fed with fat-rich diets (Baur et al., 2006). However, accordingly with the above indicated recent studies that indicate that sirtuins do not affect longevity, in normal diet conditions, resveratrol is unable to increase lifespan in mice although delays age-related deterioration (Pearson et al., 2008). Many researchers have also demonstrated the role of sirtuins in protection of cell and tissues against different forms of injury through activation of FoxO and intracellular antioxidant systems (Hsu et al., 2010).

It has been recently proposed that energy metabolism can be importantly involved in the accumulation of high levels of advanced-glycosylation end (AGES)-products into cells and, then, in the impairment of cell and tissue activity (Hipkiss, 2008). In this process, $NAD^+/NADH$ ratio is importantly involved. Decrease of NAD^+ availability in ad libitum conditions decreases metabolism of triose phosphate glycolytic intermediates such as glyceraldehydes-3-phosphate and dihydroxyacetone-phosphate. These compounds can spontaneously decompose into methylglyoxal (MG), a highly toxic glycating aging that produces AGES. AGES and MG can be involved in mitochondrial dysfunction, the increase in ROS production and also affect gene expression and intracellular signaling. However, under CR or exercise NADH is oxidized to NAD^+ and also NAD^+ synthesis is activated. NAD^+ not only activate sirtuins but also reduces the levels of MG and then, reduces the deleterious effects of this compound (Hipkiss, 2008). This hypothesis directly links metabolism and its regulation to cell damage and then to aging indicating that sirtuins are directly involved in a more balanced metabolism and then, are important factors to be considered in aging, longevity and healthspan. Taken together, it seems clear that sirtuins are key factors in metabolic homeostasis and can increase healthspan and also show prolongevity effects in conditions of metabolic stress such as western food rich in unsaturated fat.

6. Conclusion

In the present chapter we have resumed the complex system regulated by sirtuins and involved in metabolic aspects that affect aging. Aging is a process that courses with the accumulation of damage into cells and organs. Most of the energy spent by cells is used to maintain the biological structures and the order into cells and tissues. When energy is deficient or the injury increases, damage in cells accumulates in structures that cannot be eliminated and that disturb their correct physiologic mechanisms. Accumulation of aberrant structures ends in the incapacity of cells to function properly and then, produce the decline in functionality found in aging. Sirtuins are key factors in this process. These deacetylases link energetic status of the cell with regulation of aerobic metabolism, reparation activities and antioxidant systems preventing the accumulation of damaged structures. Although the right role of these sirtuins in longevity is currently questioned, their activity as core of several regulatory processes make them important regulators in, at least, the correct physiology of the organism until death.

7. Acknowledgment

The research group is financed by the Andalusian Government as the BIO177 group and P08-CTS-03988 project financed by FEDER funds (European Commission). Research has been also financed by the Spanish Government grants DEP2009-12019 (Spanish Ministry of Science and Innovation), and PI080500 (FIS, Carlos III institute). Authors also thank to the National Institutes of Health grant 1R01AG028125-01A1. Authors are also members of the Centro de Investigación Biomédica en Red de Enfermedades Raras (CIBERER), Instituto Carlos III.

8. References

Amat, R., Planavila, A., Chen, S. L., Iglesias, R., Giralt, M., & Villarroya, F. (2009). SIRT1 controls the transcription of the peroxisome proliferator-activated receptor-gamma Co-activator-1alpha (PGC-1alpha) gene in skeletal muscle through the PGC-1alpha autoregulatory loop and interaction with MyoD. *J Biol Chem*, Vol.284, No.33, (Aug 14), pp, 21872-21880, ISSN 0021-9258

Argmann, C., & Auwerx, J. (2006). Insulin secretion: SIRT4 gets in on the act. *Cell*, Vol.126, No.5, (Sep 8), pp, 837-839, ISSN 0092-8674

Asha Devi, S. (2009). Aging brain: prevention of oxidative stress by vitamin E and exercise. *ScientificWorldJournal*, Vol.9, pp, 366-372, ISSN 1537-744X

Ashraf, N., Zino, S., Macintyre, A., Kingsmore, D., Payne, A. P., George, W. D., & Shiels, P. G. (2006). Altered sirtuin expression is associated with node-positive breast cancer. *Br J Cancer*, Vol.95, No.8, (Oct 23), pp, 1056-1061, ISSN 0007-0920

Bae, N. S., Swanson, M. J., Vassilev, A., & Howard, B. H. (2004). Human histone deacetylase SIRT2 interacts with the homeobox transcription factor HOXA10. *J Biochem*, Vol.135, No.6, (Jun), pp, 695-700, ISSN 0021-924X

Balcerczyk, A., & Pirola, L. (2010). Therapeutic potential of activators and inhibitors of sirtuins. *Biofactors*, Vol.36, No.5, (Sep), pp, 383-393, ISSN 1872-8081

Bamps, S., Wirtz, J., Savory, F. R., Lake, D., & Hope, I. A. (2009). The Caenorhabditis elegans sirtuin gene, sir-2.1, is widely expressed and induced upon caloric restriction. *Mech Ageing Dev*, Vol.130, No.11-12, (Nov-Dec), pp, 762-770, ISSN 1872-6216

Baur, J. A., Pearson, K. J., Price, N. L., Jamieson, H. A., Lerin, C., Kalra, A., Prabhu, V. V., Allard, J. S., Lopez-Lluch, G., Lewis, K., Pistell, P. J., Poosala, S., Becker, K. G., Boss, O., Gwinn, D., Wang, M., Ramaswamy, S., Fishbein, K. W., Spencer, R. G., Lakatta, E. G., Le Couteur, D., Shaw, R. J., Navas, P., Puigserver, P., Ingram, D. K., de Cabo, R., & Sinclair, D. A. (2006). Resveratrol improves health and survival of mice on a high-calorie diet. *Nature*, Vol.444, No.7117, (Nov 16), pp, 337-342, ISSN 1476-4687

Blander, G., & Guarente, L. (2004). The Sir2 family of protein deacetylases. *Annu Rev Biochem*, Vol.73, pp, 417-435, ISSN 0066-4154

Burnett, C., Valentini, S., Cabreiro, F., Goss, M., Somogyvari, M., Piper, M. D., Hoddinott, M., Sutphin, G. L., Leko, V., McElwee, J. J., Vazquez-Manrique, R. P., Orfila, A. M., Ackerman, D., Au, C., Vinti, G., Riesen, M., Howard, K., Neri, C., Bedalov, A., Kaeberlein, M., Soti, C., Partridge, L., & Gems, D. (2011). Absence of effects of Sir2 overexpression on lifespan in C. elegans and Drosophila. *Nature*, Vol.477, No.7365, pp, 482-485, ISSN 1476-4687

Canto, C., & Auwerx, J. (2009). Caloric restriction, SIRT1 and longevity. *Trends Endocrinol Metab*, Vol.20, No.7, (Sep), pp, 325-331, ISSN 1879-3061

Canto, C., Gerhart-Hines, Z., Feige, J. N., Lagouge, M., Noriega, L., Milne, J. C., Elliott, P. J., Puigserver, P., & Auwerx, J. (2009). AMPK regulates energy expenditure by modulating NAD+ metabolism and SIRT1 activity. *Nature*, Vol.458, No.7241, (Apr 23), pp, 1056-1060, ISSN 1476-4687

Canto, C., Jiang, L. Q., Deshmukh, A. S., Mataki, C., Coste, A., Lagouge, M., Zierath, J. R., & Auwerx, J. (2010). Interdependence of AMPK and SIRT1 for metabolic adaptation to fasting and exercise in skeletal muscle. *Cell Metab*, Vol.11, No.3, (Mar 3), pp, 213-219, ISSN 1932-7420

Cohen, H. Y., Miller, C., Bitterman, K. J., Wall, N. R., Hekking, B., Kessler, B., Howitz, K. T., Gorospe, M., de Cabo, R., & Sinclair, D. A. (2004). Calorie restriction promotes mammalian cell survival by inducing the SIRT1 deacetylase. *Science*, Vol.305, No.5682, (Jul 16), pp, 390-392, ISSN 1095-9203

Cooper, H. M., & Spelbrink, J. N. (2008). The human SIRT3 protein deacetylase is exclusively mitochondrial. *Biochem J*, Vol.411, No.2, (Apr 15), pp, 279-285, ISSN 1470-8728

Costa Cdos, S., Rohden, F., Hammes, T. O., Margis, R., Bortolotto, J. W., Padoin, A. V., Mottin, C. C., & Guaragna, R. M. (2011). Resveratrol upregulated SIRT1, FOXO1, and adiponectin and downregulated PPARgamma1-3 mRNA expression in human visceral adipocytes. *Obes Surg*, Vol.21, No.3, (Mar), pp, 356-361, ISSN 1708-0428

Dali-Youcef, N., Lagouge, M., Froelich, S., Koehl, C., Schoonjans, K., & Auwerx, J. (2007). Sirtuins: the 'magnificent seven', function, metabolism and longevity. *Ann Med*, Vol.39, No.5, pp, 335-345, ISSN 0785-3890

Dasgupta, B., & Milbrandt, J. (2007). Resveratrol stimulates AMP kinase activity in neurons. *Proc Natl Acad Sci U S A*, Vol.104, No.17, (Apr 24), pp, 7217-7222, ISSN 0027-8424

Davalos, A., Goedeke, L., Smibert, P., Ramirez, C. M., Warrier, N. P., Andreo, U., Cirera-Salinas, D., Rayner, K., Suresh, U., Pastor-Pareja, J. C., Esplugues, E., Fisher, E. A., Penalva, L. O., Moore, K. J., Suarez, Y., Lai, E. C., & Fernandez-Hernando, C. (2011). miR-33a/b contribute to the regulation of fatty acid metabolism and insulin signaling. *Proc Natl Acad Sci U S A*, Vol.108, No.22, (May 31), pp, 9232-9237, ISSN 1091-6490

Dominy, J. E., Jr., Lee, Y., Gerhart-Hines, Z., & Puigserver, P. (2010). Nutrient-dependent regulation of PGC-1alpha's acetylation state and metabolic function through the enzymatic activities of Sirt1/GCN5. *Biochim Biophys Acta*, Vol.1804, No.8, (Aug), pp, 1676-1683, ISSN 0006-3002

Dryden, S. C., Nahhas, F. A., Nowak, J. E., Goustin, A. S., & Tainsky, M. A. (2003). Role for human SIRT2 NAD-dependent deacetylase activity in control of mitotic exit in the cell cycle. *Mol Cell Biol*, Vol.23, No.9, (May), pp, 3173-3185, ISSN 0270-7306

Du, J., Jiang, H., & Lin, H. (2009). Investigating the ADP-ribosyltransferase activity of sirtuins with NAD analogues and 32P-NAD. *Biochemistry*, Vol.48, No.13, (Apr 7), pp, 2878-2890, ISSN 1520-4995

Eades, G., Yao, Y., Yang, M., Zhang, Y., Chumsri, S., & Zhou, Q. (2011). miR-200a regulates SIRT1 expression and epithelial to mesenchymal transition (EMT)-like transformation in mammary epithelial cells. *J Biol Chem*, Vol.286, No.29, (Jul 22), pp, 25992-26002, ISSN 1083-351X

Elliott, P. J., & Jirousek, M. (2008). Sirtuins: novel targets for metabolic disease. *Curr Opin Investig Drugs*, Vol.9, No.4, (Apr), pp, 371-378, ISSN 1472-4472

Falcon, A. A., Chen, S., Wood, M. S., & Aris, J. P. Acetyl-coenzyme A synthetase 2 is a nuclear protein required for replicative longevity in Saccharomyces cerevisiae. *Mol Cell Biochem*, Vol.333, No.1-2, (Jan), pp, 99-108, ISSN 1573-4919

Ferrara, N., Rinaldi, B., Corbi, G., Conti, V., Stiuso, P., Boccuti, S., Rengo, G., Rossi, F., & Filippelli, A. (2008). Exercise training promotes SIRT1 activity in aged rats. *Rejuvenation Res*, Vol.11, No.1, (Feb), pp, 139-150, ISSN 1549-1684

Fields, J., Hanisch, J. J., Choi, J. W., & Hwang, P. M. (2007). How does p53 regulate mitochondrial respiration? *IUBMB Life*, Vol.59, No.10, (Oct), pp, 682-684, ISSN 1521-6543

Finley, L. W., Haas, W., Desquiret-Dumas, V., Wallace, D. C., Procaccio, V., Gygi, S. P., & Haigis, M. C. (2011). Succinate dehydrogenase is a direct target of sirtuin 3 deacetylase activity. *PLoS One*, Vol.6, No.8, pp, e23295, ISSN 1932-6203

Fleming, J. E., & Bensch, K. G. (1991). Oxidative stress as a causal factor in differentiation and aging: a unifying hypothesis. *Exp Gerontol*, Vol.26, No.5, pp, 511-517, ISSN 0531-5565

Ford, E., Voit, R., Liszt, G., Magin, C., Grummt, I., & Guarente, L. (2006). Mammalian Sir2 homolog SIRT7 is an activator of RNA polymerase I transcription. *Genes Dev*, Vol.20, No.9, (May 1), pp, 1075-1080, ISSN 0890-9369

Frye, R. A. (2000). Phylogenetic classification of prokaryotic and eukaryotic Sir2-like proteins. *Biochem Biophys Res Commun*, Vol.273, No.2, (Jul 5), pp, 793-798, ISSN 0006-291X

Fujita, Y., Kojima, K., Hamada, N., Ohhashi, R., Akao, Y., Nozawa, Y., Deguchi, T., & Ito, M. (2008). Effects of miR-34a on cell growth and chemoresistance in prostate cancer PC3 cells. *Biochem Biophys Res Commun*, Vol.377, No.1, (Dec 5), pp, 114-119, ISSN 1090-2104

Furukawa, A., Tada-Oikawa, S., Kawanishi, S., & Oikawa, S. (2007). H2O2 accelerates cellular senescence by accumulation of acetylated p53 via decrease in the function of SIRT1 by NAD+ depletion. *Cell Physiol Biochem*, Vol.20, No.1-4, pp, 45-54, ISSN 1015-8987

Gambini, J., Gomez-Cabrera, M. C., Borras, C., Valles, S. L., Lopez-Grueso, R., Martinez-Bello, V. E., Herranz, D., Pallardo, F. V., Tresguerres, J. A., Serrano, M., & Vina, J. (2011). Free [NADH]/[NAD(+)] regulates sirtuin expression. *Arch Biochem Biophys*, Vol.512, No.1, (Aug 1), pp, 24-29, ISSN 1096-0384

Gerhart-Hines, Z., Rodgers, J. T., Bare, O., Lerin, C., Kim, S. H., Mostoslavsky, R., Alt, F. W., Wu, Z., & Puigserver, P. (2007). Metabolic control of muscle mitochondrial function and fatty acid oxidation through SIRT1/PGC-1alpha. *Embo J*, Vol.26, No.7, (Apr 4), pp, 1913-1923, ISSN 0261-4189

Giralt, A., Hondares, E., Villena, J. A., Ribas, F., Diaz-Delfin, J., Giralt, M., Iglesias, R., & Villarroya, F. (2011). Peroxisome proliferator-activated receptor-gamma coactivator-1alpha controls transcription of the Sirt3 gene, an essential component of the thermogenic brown adipocyte phenotype. *J Biol Chem*, Vol.286, No.19, (May 13), pp, 16958-16966, ISSN 1083-351X

Grob, A., Roussel, P., Wright, J. E., McStay, B., Hernandez-Verdun, D., & Sirri, V. (2009). Involvement of SIRT7 in resumption of rDNA transcription at the exit from mitosis. *J Cell Sci*, Vol.122, No.Pt 4, (Feb 15), pp, 489-498, ISSN 0021-9533

Guarente, L. (2000). Sir2 links chromatin silencing, metabolism, and aging. *Genes Dev*, Vol.14, No.9, (May 1), pp, 1021-1026, ISSN 0890-9369

Guarente, L. (2007). Sirtuins in aging and disease. *Cold Spring Harb Symp Quant Biol*, Vol.72, pp, 483-488, ISSN 0091-7451

Guarente, L. (2008). Mitochondria--a nexus for aging, calorie restriction, and sirtuins? *Cell*, Vol.132, No.2, (Jan 25), pp, 171-176, ISSN 1097-4172

Gurd, B. J., Yoshida, Y., Lally, J., Holloway, G. P., & Bonen, A. (2009). The deacetylase enzyme SIRT1 is not associated with oxidative capacity in rat heart and skeletal muscle and its overexpression reduces mitochondrial biogenesis. *J Physiol*, Vol.587, No.Pt 8, (Apr 15), pp, 1817-1828, ISSN 1469-7793

Gurd, B. J., Yoshida, Y., McFarlan, J. T., Holloway, G. P., Moyes, C. D., Heigenhauser, G. J., Spriet, L., & Bonen, A. (2011). Nuclear SIRT1 activity, but not protein content, regulates mitochondrial biogenesis in rat and human skeletal muscle. *Am J Physiol Regul Integr Comp Physiol*, Vol.301, No.1, (Jul), pp, R67-75, ISSN 1522-1490

Haigis, M. C., Mostoslavsky, R., Haigis, K. M., Fahie, K., Christodoulou, D. C., Murphy, A. J., Valenzuela, D. M., Yancopoulos, G. D., Karow, M., Blander, G., Wolberger, C., Prolla, T. A., Weindruch, R., Alt, F. W., & Guarente, L. (2006). SIRT4 inhibits glutamate dehydrogenase and opposes the effects of calorie restriction in pancreatic beta cells. *Cell*, Vol.126, No.5, (Sep 8), pp, 941-954, ISSN 0092-8674

Harting, K., & Knoll, B. (2010). SIRT2-mediated protein deacetylation: An emerging key regulator in brain physiology and pathology. *Eur J Cell Biol*, Vol.89, No.2-3, (Feb-Mar), pp, 262-269, ISSN 1618-1298

Hay, R. T. (2001). Protein modification by SUMO. *Trends Biochem Sci*, Vol.26, No.5, (May), pp, 332-333, ISSN 0968-0004

Herranz, D., Munoz-Martin, M., Canamero, M., Mulero, F., Martinez-Pastor, B., Fernandez-Capetillo, O., & Serrano, M. (2010). Sirt1 improves healthy ageing and protects from metabolic syndrome-associated cancer. *Nat Commun*, Vol.1, pp, 3, ISSN 2041-1723

Herranz, D., & Serrano, M. (2010). SIRT1: recent lessons from mouse models. *Nat Rev Cancer*, Vol.10, No.12, (Dec), pp, 819-823, ISSN 1474-1768

Hipkiss, A. R. (2008). Energy metabolism, altered proteins, sirtuins and ageing: converging mechanisms? *Biogerontology*, Vol.9, No.1, (Feb), pp, 49-55, ISSN 1389-5729

Hirsch, B. M., & Zheng, W. (2011). Sirtuin mechanism and inhibition: explored with N(epsilon)-acetyl-lysine analogs. *Mol Biosyst*, Vol.7, No.1, (Jan), pp, 16-28, ISSN 1742-2051

Hong, E. H., Lee, S. J., Kim, J. S., Lee, K. H., Um, H. D., Kim, J. H., Kim, S. J., Kim, J. I., & Hwang, S. G. (2010). Ionizing radiation induces cellular senescence of articular chondrocytes via negative regulation of SIRT1 by p38 kinase. *J Biol Chem*, Vol.285, No.2, (Jan 8), pp, 1283-1295, ISSN 1083-351X

Hsu, C. P., Zhai, P., Yamamoto, T., Maejima, Y., Matsushima, S., Hariharan, N., Shao, D., Takagi, H., Oka, S., & Sadoshima, J. (2010). Silent information regulator 1 protects the heart from ischemia/reperfusion. *Circulation*, Vol.122, No.21, (Nov 23), pp, 2170-2182, ISSN 1524-4539

Imai, S. (2009). SIRT1 and caloric restriction: an insight into possible trade-offs between robustness and frailty. *Curr Opin Clin Nutr Metab Care*, Vol.12, No.4, (Jul), pp, 350-356, ISSN 1535-3885

Jin, K. (2010). Modern Biological Theories of Aging. *Aging Dis*, Vol.1, No.2, (Oct 1), pp, 72-74, ISSN 2152-5250

Jing, E., Gesta, S., & Kahn, C. R. (2007). SIRT2 regulates adipocyte differentiation through FoxO1 acetylation/deacetylation. *Cell Metab*, Vol.6, No.2, (Aug), pp, 105-114, ISSN 1550-4131

Kaeberlein, M. (2010). Lessons on longevity from budding yeast. *Nature*, Vol.464, No.7288, (Mar 25), pp, 513-519, ISSN 1476-4687

Kaeberlein, M., McVey, M., & Guarente, L. (1999). The SIR2/3/4 complex and SIR2 alone promote longevity in Saccharomyces cerevisiae by two different mechanisms. *Genes Dev*, Vol.13, No.19, (Oct 1), pp, 2570-2580, ISSN 0890-9369

Kamel, C., Abrol, M., Jardine, K., He, X., & McBurney, M. W. (2006). SirT1 fails to affect p53-mediated biological functions. *Aging Cell*, Vol.5, No.1, (Feb), pp, 81-88, ISSN 1474-9718

Kanfi, Y., Peshti, V., Gozlan, Y. M., Rathaus, M., Gil, R., & Cohen, H. Y. (2008). Regulation of SIRT1 protein levels by nutrient availability. *FEBS Lett*, Vol.582, No.16, (Jul 9), pp, 2417-2423, ISSN 0014-5793

Kanfi, Y., Shalman, R., Peshti, V., Pilosof, S. N., Gozlan, Y. M., Pearson, K. J., Lerrer, B., Moazed, D., Marine, J. C., de Cabo, R., & Cohen, H. Y. (2008). Regulation of SIRT6 protein levels by nutrient availability. *FEBS Lett*, Vol.582, No.5, (Mar 5), pp, 543-548, ISSN 0014-5793

Kao, C. L., Chen, L. K., Chang, Y. L., Yung, M. C., Hsu, C. C., Chen, Y. C., Lo, W. L., Chen, S. J., Ku, H. H., & Hwang, S. J. (2010). Resveratrol protects human endothelium from H(2)O(2)-induced oxidative stress and senescence via SirT1 activation. *J Atheroscler Thromb*, Vol.17, No.9, (Sep 30), pp, 970-979, ISSN 1880-3873

Kawahara, T. L., Michishita, E., Adler, A. S., Damian, M., Berber, E., Lin, M., McCord, R. A., Ongaigui, K. C., Boxer, L. D., Chang, H. Y., & Chua, K. F. (2009). SIRT6 links histone H3 lysine 9 deacetylation to NF-kappaB-dependent gene expression and organismal life span. *Cell*, Vol.136, No.1, (Jan 9), pp, 62-74, ISSN 1097-4172

Kelly, T. J., Lerin, C., Haas, W., Gygi, S. P., & Puigserver, P. (2009). GCN5-mediated transcriptional control of the metabolic coactivator PGC-1beta through lysine acetylation. *J Biol Chem*, Vol.284, No.30, (Jul 24), pp, 19945-19952, ISSN 0021-9258

Kendrick, A. A., Choudhury, M., Rahman, S. M., McCurdy, C. E., Friederich, M., Van Hove, J. L., Watson, P. A., Birdsey, N., Bao, J., Gius, D., Sack, M. N., Jing, E., Kahn, C. R., Friedman, J. E., & Jonscher, K. R. (2011). Fatty liver is associated with reduced SIRT3 activity and mitochondrial protein hyperacetylation. *Biochem J*, Vol.433, No.3, (Jan 14), pp, 505-514, ISSN 1470-8728

Kenyon, C. J. (2010). The genetics of ageing. *Nature*, Vol.464, No.7288, (Mar 25), pp, 504-512, ISSN 1476-4687

Kim, H. S., Xiao, C., Wang, R. H., Lahusen, T., Xu, X., Vassilopoulos, A., Vazquez-Ortiz, G., Jeong, W. I., Park, O., Ki, S. H., Gao, B., & Deng, C. X. (2010). Hepatic-specific disruption of SIRT6 in mice results in fatty liver formation due to enhanced glycolysis and triglyceride synthesis. *Cell Metab*, Vol.12, No.3, (Sep 8), pp, 224-236, ISSN 1932-7420

Koltai, E., Szabo, Z., Atalay, M., Boldogh, I., Naito, H., Goto, S., Nyakas, C., & Radak, Z. (2010). Exercise alters SIRT1, SIRT6, NAD and NAMPT levels in skeletal muscle of aged rats. *Mech Ageing Dev*, Vol.131, No.1, (Jan), pp, 21-28, ISSN 1872-6216

Kong, X., Wang, R., Xue, Y., Liu, X., Zhang, H., Chen, Y., Fang, F., & Chang, Y. (2010). Sirtuin 3, a new target of PGC-1alpha, plays an important role in the suppression of ROS and mitochondrial biogenesis. *PLoS One*, Vol.5, No.7, pp, e11707, ISSN 1932-6203

Kornberg, M. D., Sen, N., Hara, M. R., Juluri, K. R., Nguyen, J. V., Snowman, A. M., Law, L., Hester, L. D., & Snyder, S. H. (2010). GAPDH mediates nitrosylation of nuclear proteins. *Nat Cell Biol*, Vol.12, No.11, (Nov), pp, 1094-1100, ISSN 1476-4679

Kume, S., Uzu, T., Horiike, K., Chin-Kanasaki, M., Isshiki, K., Araki, S., Sugimoto, T., Haneda, M., Kashiwagi, A., & Koya, D. (2010). Calorie restriction enhances cell adaptation to hypoxia through Sirt1-dependent mitochondrial autophagy in mouse aged kidney. *J Clin Invest*, Vol.120, No.4, (Apr 1), pp, 1043-1055, ISSN 1558-8238

Kuzmichev, A., & Reinberg, D. (2001). Role of histone deacetylase complexes in the regulation of chromatin metabolism. *Curr Top Microbiol Immunol*, Vol.254, pp, 35-58, ISSN 0070-217X

Lagouge, M., Argmann, C., Gerhart-Hines, Z., Meziane, H., Lerin, C., Daussin, F., Messadeq, N., Milne, J., Lambert, P., Elliott, P., Geny, B., Laakso, M., Puigserver, P., & Auwerx, J. (2006). Resveratrol improves mitochondrial function and protects against metabolic disease by activating SIRT1 and PGC-1alpha. *Cell*, Vol.127, No.6, (Dec 15), pp, 1109-1122, ISSN 0092-8674

Lanza, I. R., Short, D. K., Short, K. R., Raghavakaimal, S., Basu, R., Joyner, M. J., McConnell, J. P., & Nair, K. S. (2008). Endurance exercise as a countermeasure for aging. *Diabetes*, Vol.57, No.11, (Nov), pp, 2933-2942, ISSN 1939-327X

Lappalainen, Z. (2011). Sirtuins: a family of proteins with implications for human performance and exercise physiology. *Res Sports Med*, Vol.19, No.1, (Jan), pp, 53-65, ISSN 1543-8635

Li, X., & Kazgan, N. (2011). Mammalian sirtuins and energy metabolism. *Int J Biol Sci*, Vol.7, No.5, pp, 575-587, ISSN 1449-2288

Lin, S. J., Defossez, P. A., & Guarente, L. (2000). Requirement of NAD and SIR2 for life-span extension by calorie restriction in Saccharomyces cerevisiae. *Science*, Vol.289, No.5487, (Sep 22), pp, 2126-2128, ISSN 0036-8075

Liszt, G., Ford, E., Kurtev, M., & Guarente, L. (2005). Mouse Sir2 homolog SIRT6 is a nuclear ADP-ribosyltransferase. *J Biol Chem*, Vol.280, No.22, (Jun 3), pp, 21313-21320, ISSN 0021-9258

Lomb, D. J., Laurent, G., & Haigis, M. C. (2010). Sirtuins regulate key aspects of lipid metabolism. *Biochim Biophys Acta*, Vol.1804, No.8, (Aug), pp, 1652-1657, ISSN 0006-3002

Lombard, D. B., Schwer, B., Alt, F. W., & Mostoslavsky, R. (2008). SIRT6 in DNA repair, metabolism and ageing. *J Intern Med*, Vol.263, No.2, (Feb), pp, 128-141, ISSN 1365-2796

Lopez-Lluch, G., Hunt, N., Jones, B., Zhu, M., Jamieson, H., Hilmer, S., Cascajo, M. V., Allard, J., Ingram, D. K., Navas, P., & de Cabo, R. (2006). Calorie restriction induces mitochondrial biogenesis and bioenergetic efficiency. *Proc Natl Acad Sci U S A*, Vol.103, No.6, (Feb 7), pp, 1768-1773, ISSN 0027-8424

Luthi-Carter, R., Taylor, D. M., Pallos, J., Lambert, E., Amore, A., Parker, A., Moffitt, H., Smith, D. L., Runne, H., Gokce, O., Kuhn, A., Xiang, Z., Maxwell, M. M., Reeves, S. A., Bates, G. P., Neri, C., Thompson, L. M., Marsh, J. L., & Kazantsev, A. G. (2010). SIRT2 inhibition achieves neuroprotection by decreasing sterol biosynthesis. *Proc Natl Acad Sci U S A*, Vol.107, No.17, (Apr 27), pp, 7927-7932, ISSN 1091-6490

Ma, W., Sung, H. J., Park, J. Y., Matoba, S., & Hwang, P. M. (2007). A pivotal role for p53: balancing aerobic respiration and glycolysis. *J Bioenerg Biomembr*, Vol.39, No.3, (Jun), pp, 243-246, ISSN 0145-479X

Mao, Z., Hine, C., Tian, X., Van Meter, M., Au, M., Vaidya, A., Seluanov, A., & Gorbunova, V. (2011). SIRT6 promotes DNA repair under stress by activating PARP1. *Science*, Vol.332, No.6036, (Jun 17), pp, 1443-1446, ISSN 1095-9203

Marton, O., Koltai, E., Nyakas, C., Bakonyi, T., Zenteno-Savin, T., Kumagai, S., Goto, S., & Radak, Z. (2010). Aging and exercise affect the level of protein acetylation and SIRT1 activity in cerebellum of male rats. *Biogerontology*, Vol.11, No.6, (Dec), pp, 679-686, ISSN 1389-5729

Maynard, S., Schurman, S. H., Harboe, C., de Souza-Pinto, N. C., & Bohr, V. A. (2009). Base excision repair of oxidative DNA damage and association with cancer and aging. *Carcinogenesis*, Vol.30, No.1, (Jan), pp, 2-10, ISSN 1460-2180

Michishita, E., Park, J. Y., Burneskis, J. M., Barrett, J. C., & Horikawa, I. (2005). Evolutionarily conserved and nonconserved cellular localizations and functions of human SIRT proteins. *Mol Biol Cell*, Vol.16, No.10, (Oct), pp, 4623-4635, ISSN 1059-1524

Min, J., Landry, J., Sternglanz, R., & Xu, R. M. (2001). Crystal structure of a SIR2 homolog-NAD complex. *Cell*, Vol.105, No.2, (Apr 20), pp, 269-279, ISSN 0092-8674

Moll, U. M., & Zaika, A. (2001). Nuclear and mitochondrial apoptotic pathways of p53. *FEBS Lett*, Vol.493, No.2-3, (Mar 30), pp, 65-69, ISSN 0014-5793

Morselli, E., Galluzzi, L., Kepp, O., Criollo, A., Maiuri, M. C., Tavernarakis, N., Madeo, F., & Kroemer, G. (2009). Autophagy mediates pharmacological lifespan extension by spermidine and resveratrol. *Aging (Albany NY)*, Vol.1, No.12, (Dec), pp, 961-970, ISSN 1945-4589

Morselli, E., Maiuri, M. C., Markaki, M., Megalou, E., Pasparaki, A., Palikaras, K., Criollo, A., Galluzzi, L., Malik, S. A., Vitale, I., Michaud, M., Madeo, F., Tavernarakis, N., & Kroemer, G. (2010). Caloric restriction and resveratrol promote longevity through the Sirtuin-1-dependent induction of autophagy. *Cell Death Dis*, Vol.1, pp, e10, ISSN 2041-4889

Mostoslavsky, R., Chua, K. F., Lombard, D. B., Pang, W. W., Fischer, M. R., Gellon, L., Liu, P., Mostoslavsky, G., Franco, S., Murphy, M. M., Mills, K. D., Patel, P., Hsu, J. T., Hong, A. L., Ford, E., Cheng, H. L., Kennedy, C., Nunez, N., Bronson, R., Frendewey, D., Auerbach, W., Valenzuela, D., Karow, M., Hottiger, M. O., Hursting, S., Barrett, J. C., Guarente, L., Mulligan, R., Demple, B., Yancopoulos, G. D., & Alt, F. W. (2006). Genomic instability and aging-like phenotype in the absence of mammalian SIRT6. *Cell*, Vol.124, No.2, (Jan 27), pp, 315-329, ISSN 0092-8674

Nakagawa, T., & Guarente, L. (2009). Urea cycle regulation by mitochondrial sirtuin, SIRT5. *Aging (Albany NY)*, Vol.1, No.6, (Jun), pp, 578-581, ISSN 1945-4589

Naqvi, A., Hoffman, T. A., DeRicco, J., Kumar, A., Kim, C. S., Jung, S. B., Yamamori, T., Kim, Y. R., Mehdi, F., Kumar, S., Rankinen, T., Ravussin, E., & Irani, K. (2010). A single-nucleotide variation in a p53-binding site affects nutrient-sensitive human SIRT1 expression. *Hum Mol Genet,* Vol.19, No.21, (Nov 1), pp, 4123-4133, ISSN 1460-2083

Nasrin, N., Wu, X., Fortier, E., Feng, Y., Bare, O. C., Chen, S., Ren, X., Wu, Z., Streeper, R. S., & Bordone, L. (2010). SIRT4 regulates fatty acid oxidation and mitochondrial gene expression in liver and muscle cells. *J Biol Chem,* Vol.285, No.42, (Oct 15), pp, 31995-32002, ISSN 1083-351X

Nemoto, S., Fergusson, M. M., & Finkel, T. (2005). SIRT1 functionally interacts with the metabolic regulator and transcriptional coactivator PGC-1{alpha}. *J Biol Chem,* Vol.280, No.16, (Apr 22), pp, 16456-16460, ISSN 0021-9258

Nie, H., Chen, H., Han, J., Hong, Y., Ma, Y., Xia, W., & Ying, W. (2011). Silencing of SIRT2 induces cell death and a decrease in the intracellular ATP level of PC12 cells. *Int J Physiol Pathophysiol Pharmacol,* Vol.3, No.1, pp, 65-70, ISSN 1944-8171

North, B. J., Marshall, B. L., Borra, M. T., Denu, J. M., & Verdin, E. (2003). The human Sir2 ortholog, SIRT2, is an NAD+-dependent tubulin deacetylase. *Mol Cell,* Vol.11, No.2, (Feb), pp, 437-444, ISSN 1097-2765

North, B. J., & Verdin, E. (2007). Interphase nucleo-cytoplasmic shuttling and localization of SIRT2 during mitosis. *PLoS One,* Vol.2, No.8, pp, e784, ISSN 1932-6203

Ozden, O., Park, S. H., Kim, H. S., Jiang, H., Coleman, M. C., Spitz, D. R., & Gius, D. (2011). Acetylation of MnSOD directs enzymatic activity responding to cellular nutrient status or oxidative stress. *Aging (Albany NY),* Vol.3, No.2, (Feb), pp, 102-107, ISSN 1945-4589

Palacios, O. M., Carmona, J. J., Michan, S., Chen, K. Y., Manabe, Y., Ward, J. L., 3rd, Goodyear, L. J., & Tong, Q. (2009). Diet and exercise signals regulate SIRT3 and activate AMPK and PGC-1alpha in skeletal muscle. *Aging (Albany NY),* Vol.1, No.9, (Sep), pp, 771-783, ISSN 1945-4589

Pallas, M., Verdaguer, E., Tajes, M., Gutierrez-Cuesta, J., & Camins, A. (2008). Modulation of sirtuins: new targets for antiageing. *Recent Pat CNS Drug Discov,* Vol.3, No.1, (Jan), pp, 61-69, ISSN 1574-8898

Pearson, K. J., Baur, J. A., Lewis, K. N., Peshkin, L., Price, N. L., Labinskyy, N., Swindell, W. R., Kamara, D., Minor, R. K., Perez, E., Jamieson, H. A., Zhang, Y., Dunn, S. R., Sharma, K., Pleshko, N., Woollett, L. A., Csiszar, A., Ikeno, Y., Le Couteur, D., Elliott, P. J., Becker, K. G., Navas, P., Ingram, D. K., Wolf, N. S., Ungvari, Z., Sinclair, D. A., & de Cabo, R. (2008). Resveratrol delays age-related deterioration and mimics transcriptional aspects of dietary restriction without extending life span. *Cell Metab,* Vol.8, No.2, (Aug), pp, 157-168, ISSN 1932-7420

Perez, V. I., Bokov, A., Van Remmen, H., Mele, J., Ran, Q., Ikeno, Y., & Richardson, A. (2009). Is the oxidative stress theory of aging dead? *Biochim Biophys Acta,* Vol.1790, No.10, (Oct), pp, 1005-1014, ISSN 0006-3002

Purushotham, A., Schug, T. T., Xu, Q., Surapureddi, S., Guo, X., & Li, X. (2009). Hepatocyte-specific deletion of SIRT1 alters fatty acid metabolism and results in hepatic steatosis and inflammation. *Cell Metab,* Vol.9, No.4, (Apr), pp, 327-338, ISSN 1932-7420

Qiu, X., Brown, K., Hirschey, M. D., Verdin, E., & Chen, D. (2010). Calorie restriction reduces oxidative stress by SIRT3-mediated SOD2 activation. *Cell Metab,* Vol.12, No.6, (Dec 1), pp, 662-667, ISSN 1932-7420

Rane, S., He, M., Sayed, D., Vashistha, H., Malhotra, A., Sadoshima, J., Vatner, D. E., Vatner, S. F., & Abdellatif, M. (2009). Downregulation of miR-199a derepresses hypoxia-inducible factor-1alpha and Sirtuin 1 and recapitulates hypoxia preconditioning in cardiac myocytes. *Circ Res,* Vol.104, No.7, (Apr 10), pp, 879-886, ISSN 1524-4571

Rogina, B., & Helfand, S. L. (2004). Sir2 mediates longevity in the fly through a pathway related to calorie restriction. *Proc Natl Acad Sci U S A,* Vol.101, No.45, (Nov 9), pp, 15998-16003, ISSN 0027-8424

Ruderman, N. B., Xu, X. J., Nelson, L., Cacicedo, J. M., Saha, A. K., Lan, F., & Ido, Y. (2010). AMPK and SIRT1: a long-standing partnership? *Am J Physiol Endocrinol Metab,* Vol.298, No.4, (Apr), pp, E751-760, ISSN 1522-1555

Sanders, B. D., Jackson, B., & Marmorstein, R. (2010). Structural basis for sirtuin function: what we know and what we don't. *Biochim Biophys Acta,* Vol.1804, No.8, (Aug), pp, 1604-1616, ISSN 0006-3002

Sasaki, T., Maier, B., Koclega, K. D., Chruszcz, M., Gluba, W., Stukenberg, P. T., Minor, W., & Scrable, H. (2008). Phosphorylation regulates SIRT1 function. *PLoS One,* Vol.3, No.12, pp, e4020, ISSN 1932-6203

Satoh, A., Brace, C. S., Ben-Josef, G., West, T., Wozniak, D. F., Holtzman, D. M., Herzog, E. D., & Imai, S. (2010). SIRT1 promotes the central adaptive response to diet restriction through activation of the dorsomedial and lateral nuclei of the hypothalamus. *J Neurosci,* Vol.30, No.30, (Jul 28), pp, 10220-10232, ISSN 1529-2401

Schlicker, C., Gertz, M., Papatheodorou, P., Kachholz, B., Becker, C. F., & Steegborn, C. (2008). Substrates and regulation mechanisms for the human mitochondrial sirtuins Sirt3 and Sirt5. *J Mol Biol,* Vol.382, No.3, (Oct 10), pp, 790-801, ISSN 1089-8638

Schwer, B., North, B. J., Frye, R. A., Ott, M., & Verdin, E. (2002). The human silent information regulator (Sir)2 homologue hSIRT3 is a mitochondrial nicotinamide adenine dinucleotide-dependent deacetylase. *J Cell Biol,* Vol.158, No.4, (Aug 19), pp, 647-657, ISSN 0021-9525

Schwer, B., Schumacher, B., Lombard, D. B., Xiao, C., Kurtev, M. V., Gao, J., Schneider, J. I., Chai, H., Bronson, R. T., Tsai, L. H., Deng, C. X., & Alt, F. W. (2010). Neural sirtuin 6 (Sirt6) ablation attenuates somatic growth and causes obesity. *Proc Natl Acad Sci U S A,* Vol.107, No.50, (Dec 14), pp, 21790-21794, ISSN 1091-6490

Schwer, B., & Verdin, E. (2008). Conserved metabolic regulatory functions of sirtuins. *Cell Metab,* Vol.7, No.2, (Feb), pp, 104-112, ISSN 1550-4131

Seufert, C. D., Graf, M., Janson, G., Kuhn, A., & Soling, H. D. (1974). Formation of free acetate by isolated perfused livers from normal, starved and diabetic rats. *Biochem Biophys Res Commun,* Vol.57, No.3, (Apr 8), pp, 901-909, ISSN 0006-291X

Shi, T., Wang, F., Stieren, E., & Tong, Q. (2005). SIRT3, a mitochondrial sirtuin deacetylase, regulates mitochondrial function and thermogenesis in brown adipocytes. *J Biol Chem,* Vol.280, No.14, (Apr 8), pp, 13560-13567, ISSN 0021-9258

Shimazu, T., Hirschey, M. D., Huang, J. Y., Ho, L. T., & Verdin, E. (2010). Acetate metabolism and aging: An emerging connection. *Mech Ageing Dev,* Vol.131, No.7-8, (Jul-Aug), pp, 511-516, ISSN 1872-6216

Sohal, R. S., Ku, H. H., Agarwal, S., Forster, M. J., & Lal, H. (1994). Oxidative damage, mitochondrial oxidant generation and antioxidant defenses during aging and in response to food restriction in the mouse. *Mech Ageing Dev,* Vol.74, No.1-2, (May), pp, 121-133, ISSN 0047-6374

Strum, J. C., Johnson, J. H., Ward, J., Xie, H., Feild, J., Hester, A., Alford, A., & Waters, K. M. (2009). MicroRNA 132 regulates nutritional stress-induced chemokine production through repression of SirT1. *Mol Endocrinol*, Vol.23, No.11, (Nov), pp, 1876-1884, ISSN 1944-9917

Stunkel, W., Peh, B. K., Tan, Y. C., Nayagam, V. M., Wang, X., Salto-Tellez, M., Ni, B., Entzeroth, M., & Wood, J. (2007). Function of the SIRT1 protein deacetylase in cancer. *Biotechnol J*, Vol.2, No.11, (Nov), pp, 1360-1368, ISSN 1860-7314

Sulaiman, M., Matta, M. J., Sunderesan, N. R., Gupta, M. P., Periasamy, M., & Gupta, M. (2010). Resveratrol, an activator of SIRT1, upregulates sarcoplasmic calcium ATPase and improves cardiac function in diabetic cardiomyopathy. *Am J Physiol Heart Circ Physiol*, Vol.298, No.3, (Mar), pp, H833-843, ISSN 1522-1539

Sundaresan, N. R., Samant, S. A., Pillai, V. B., Rajamohan, S. B., & Gupta, M. P. (2008). SIRT3 is a stress-responsive deacetylase in cardiomyocytes that protects cells from stress-mediated cell death by deacetylation of Ku70. *Mol Cell Biol*, Vol.28, No.20, (Oct), pp, 6384-6401, ISSN 1098-5549

Tang, B. L. (2011). Sirt1's systemic protective roles and its promise as a target in antiaging medicine. *Transl Res*, Vol.157, No.5, (May), pp, 276-284, ISSN 1878-1810

Tissenbaum, H. A., & Guarente, L. (2001). Increased dosage of a sir-2 gene extends lifespan in Caenorhabditis elegans. *Nature*, Vol.410, No.6825, (Mar 8), pp, 227-230, ISSN 0028-0836

Vakhrusheva, O., Braeuer, D., Liu, Z., Braun, T., & Bober, E. (2008). Sirt7-dependent inhibition of cell growth and proliferation might be instrumental to mediate tissue integrity during aging. *J Physiol Pharmacol*, Vol.59 Suppl 9, (Dec), pp, 201-212, ISSN 1899-1505

Vakhrusheva, O., Smolka, C., Gajawada, P., Kostin, S., Boettger, T., Kubin, T., Braun, T., & Bober, E. (2008). Sirt7 increases stress resistance of cardiomyocytes and prevents apoptosis and inflammatory cardiomyopathy in mice. *Circ Res*, Vol.102, No.6, (Mar 28), pp, 703-710, ISSN 1524-4571

Viswanathan, M., & Guarente, L. (2011). Regulation of Caenorhabditis elegans lifespan by sir-2.1 transgenes. *Nature*, Vol.477, No.7365, pp, E1-2, ISSN 1476-4687

Wang, F., Nguyen, M., Qin, F. X., & Tong, Q. (2007). SIRT2 deacetylates FOXO3a in response to oxidative stress and caloric restriction. *Aging Cell*, Vol.6, No.4, (Aug), pp, 505-514, ISSN 1474-9718

Wang, F., & Tong, Q. (2009). SIRT2 suppresses adipocyte differentiation by deacetylating FOXO1 and enhancing FOXO1's repressive interaction with PPARgamma. *Mol Biol Cell*, Vol.20, No.3, (Feb), pp, 801-808, ISSN 1939-4586

Wang, Y., & Tissenbaum, H. A. (2006). Overlapping and distinct functions for a Caenorhabditis elegans SIR2 and DAF-16/FOXO. *Mech Ageing Dev*, Vol.127, No.1, (Jan), pp, 48-56, ISSN 0047-6374

Wolberger, C. (2007). Identification of a new nicotinamide binding site in a sirtuin: a reassessment. *Mol Cell*, Vol.28, No.6, (Dec 28), pp, 1102-1103, ISSN 1097-2765

Yang, S. J., Choi, J. M., Chae, S. W., Kim, W. J., Park, S. E., Rhee, E. J., Lee, W. Y., Oh, K. W., Park, S. W., Kim, S. W., & Park, C. Y. (2011). Activation of peroxisome proliferator-activated receptor gamma by rosiglitazone increases sirt6 expression and ameliorates hepatic steatosis in rats. *PLoS One*, Vol.6, No.2, pp, e17057, ISSN 1932-6203

Yang, Y., Fu, W., Chen, J., Olashaw, N., Zhang, X., Nicosia, S. V., Bhalla, K., & Bai, W. (2007). SIRT1 sumoylation regulates its deacetylase activity and cellular response to genotoxic stress. *Nat Cell Biol*, Vol.9, No.11, (Nov), pp, 1253-1262, ISSN 1465-7392

Yu, J., & Auwerx, J. (2009). The role of sirtuins in the control of metabolic homeostasis. *Ann N Y Acad Sci*, Vol.1173 Suppl 1, (Sep), pp, E10-19, ISSN 1749-6632

Zhang, J. (2006). Resveratrol inhibits insulin responses in a SirT1-independent pathway. *Biochem J*, Vol.397, No.3, (Aug 1), pp, 519-527, ISSN 1470-8728

Zhao, Y., Yang, J., Liao, W., Liu, X., Zhang, H., Wang, S., Wang, D., Feng, J., Yu, L., & Zhu, W. G. (2010). Cytosolic FoxO1 is essential for the induction of autophagy and tumour suppressor activity. *Nat Cell Biol*, Vol.12, No.7, (Jul), pp, 665-675, ISSN 1476-4679

5

Role of Inorganic Polyphosphate in the Energy Metabolism of Ticks

Eldo Campos[1,4], Arnoldo R. Façanha[2,4],
Jorge Moraes[1,4] and Carlos Logullo[3,4]
[1]Universidade Federal do Rio de Janeiro - Macaé
[2]Universidade Estadual do Norte Fluminense
[3]Universidade Estadual do Norte Fluminense
[4]Instituto Nacional de Ciência e Tecnologia - Entomologia Molecular
Brazil

1. Introduction

Inorganic polyphosphates are long chains of a few to several hundred phosphate residues linked by phosphoanhydride bonds (Figure 1). Polyphosphates have been found in all cell types examined to date and have been demonstrated to play diverse roles depending on the cell type and circumstances (Kornberg et al., 1999; Kulaev & Kulakovskaya, 2000). The biological roles played by polyphosphates have been most extensively studied in prokaryotes and unicellular eukaryotes, where they have been shown to regulate many biochemical processes including the metabolism and transport of inorganic phosphate, cation sequestration and storage (Kornberg et al., 1999), and membrane channel formation (Reusch, 1989; Jones et al., 2003), and they have also been found to be involved in cell envelope formation and bacterial pathogenesis (Rashid et al., 2000; Kim et al., 2002), the regulation of gene and enzyme activities (McInerney et al., 2006), the activation of Lon proteases (Kuroda et al., 2001), and KcsA channel regulation (Negoda et al., 2009).

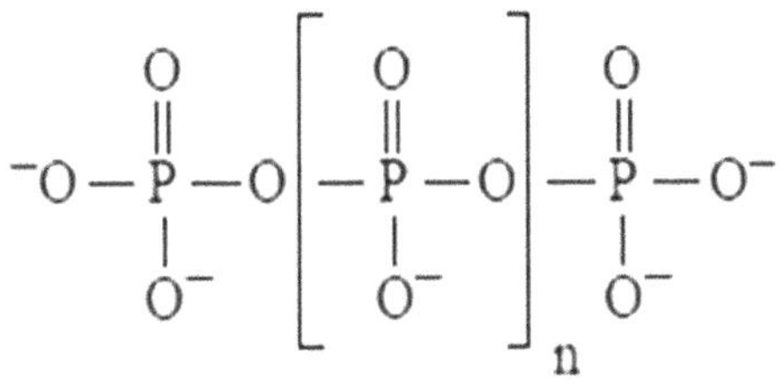

Fig. 1. Inorganic Polyphosphate

Conversely, polyphosphate functions have not been extensively investigated in higher eukaryotes; however, there is a good deal of interest in polyphosphates in mitochondria regarding two circumstances: polyphosphate as a macroenergetic compound with the same energy hydrolysis of the phosphoanhydride bond as an ATP and, according to the endosymbiotic theory, mitochondria originated from ancient prokaryotic cells (Clements et al., 2009; Kulakovskaya et al., 2010), thus, it would be intriguing to discover whether or not mitochondria have preserved polyphosphate functions such as the regulation of energy metabolism and the participation in transport channel formation.

2. Polyphosphate mobilization during *Rhipicephalus (Boophilus) microplus* embryogenesis

The tick *Rhipicephalus microplus* is a one-host tick that causes major losses to bovine herds, especially in tropical regions. In this scenario, major efforts have been made to develop immunoprophylatic tick control tools (Guerrero et al., 2006). Ticks are also vectors of parasites that cause hemoparasitic diseases, which are endemic in many cattle production areas (Sonenshine et al., 2006). *Rhipicephalus microplus* only has one host throughout all three life stages, usually a bovine animal, and a long feeding period (approximately 21 days). The adult female, after becoming completely engorged, drops off of the host and initiates oviposition approximately three days later. Being an oviparous creature, embryogenesis occurs in the absence of exogenous nutrients, and maternal nutrients are packaged in oocytes and mostly stored as yolk granules. Hatching occurs around 21 days after oviposition, and the emerging larvae can survive for several weeks before finding a host, using the remaining yolk as the only source of energy (Fagotto, 1990).

Early *R. microplus* embryonic stages are similar to those of *D. melanogaster* and mosquitoes (Bate & Arias, 1991; Monnerat et al., 2002). Tick embryogenesis is characterized by the formation of a non-cellular syncytium up to day 4 (Campos et al., 2006). After this, the embryo becomes a multicellular organism and starts organogenesis (Campos et al., 2006).

The function of polyphosphate as a phosphate reserve is well known in prokaryotes and also in eukaryote microorganisms (Kulaev & Vagabov, 1983; Kornberg, 1995; Kulaev, 2004). The cells of higher eukaryotes also carry polyphosphate, but in smaller amounts than found in microorganisms. Therefore, as well as being a source of phosphate, these biopolymers probably participate in regulatory processes (Kornberg et al., 1999). Total polyphosphate levels were quantified throughout *R. microplus* embryogenesis and the levels were found to be higher during embryo cellularization and segmentation, from the fifth to the seventh day of development, and declined after that until a plateau was reached. The free phosphate content rapidly decreased during syncytial blastoderm formation on the third day of development, and remained low until the twelfth day of embryogenesis, when it rapidly increased thereafter (Figure 2A). Exopolyphosphatase splits phosphate off from the end of a polyphosphate chain and represents one of the main enzyme types responsible for polyphosphate hydrolysis (Kulaev et al., 2004). The activity of exopolyphosphatase was analyzed during embryogenesis and its activity was in agreement with total polyphosphate mobilization (Figure 2B).

It is interesting to note that in *R. microplus* the decline in total polyphosphate content after the seventh day of embryogenesis did not reflect the increase in the free phosphate content, since this only occurs after the twelfth day, suggesting that polyphosphate also plays roles other than being a phosphate reserve for embryo development. In this case, an alternative source of phosphate could be derived from the dephosphorylation of vitellin, a major yolk protein that is gradually dephosphorylated throughout embryogenesis (Silveira et al., 2006). This source could mainly be used until segmentation of the embryo, on the seventh day of development, because there is no total polyphosphate mobilization during this period.

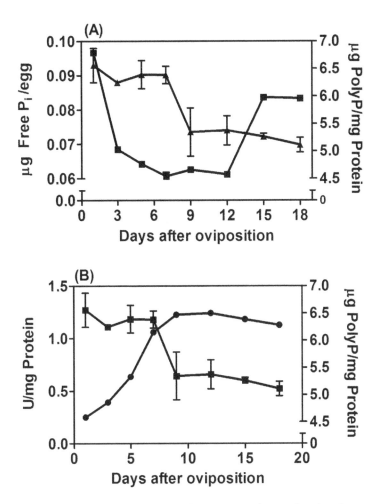

Fig. 2. Characterization of the total polyphosphate content during R. microplus embryogenesis. A) Total polyphosphate (▲) was extracted and quantified and free phosphate (■) was quantified in an egg homogenate on different days after oviposition. B) Total polyphosphate (▲) was extracted and quantified and exopolyphosphatase activity (●) was analyzed in an egg homogenate on different days after oviposition. Activity is expressed as units per milligram of total protein. The results represent the mean ± SD of three independent experiments, in triplicate.

Quantification of the major energy sources in the egg over the course of R. microplus embryogenesis suggests that lipids and carbohydrates are the main energy source used during early development of the embryo. The total lipid contents remained stable until the fifth day, dropped on the seventh day, and remained roughly unchanged until hatching (Figure 3A). The total sugar contents exhibited a similar pattern, although slightly delayed: the values remained stable until the seventh day, dropped on the ninth day and remained constant until hatching (Figure 3B). This pattern suggests the utilization of lipids during the course of cellularization, a maternally driven process (Bate & Arias, 1993). On the other hand, carbohydrates would be the major energy source for the quick segmentation of the embryo, of zygotic nature (Nusslein-Volhard & Roth, 1989; Bate & Arias, 1993).

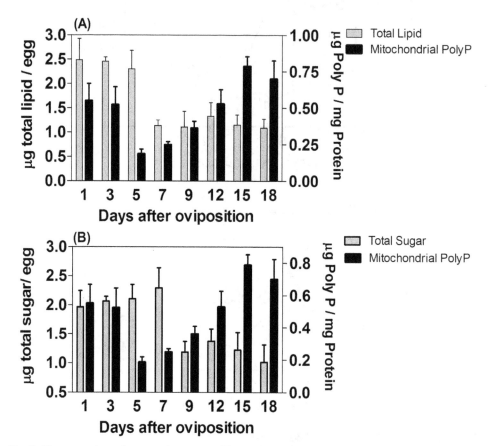

Fig. 3. Consumption of energetic sources. The major egg storage components of R. *microplus* were quantified on different days after oviposition. A) Lipid quantification, determined via the gravimetric method (Bligh & Dyer, 1959); B) total sugar concentration, measured using the method of Dubois (Dubois et al., 1956). The results represent the mean ± SD of three independent experiments, in triplicate.

Interestingly, mitochondrial polyphosphate utilization occurred during blastoderm formation and segmentation of the embryo, between the fifth and seventh days of development, and higher total polyphosphate utilization occurred after blastoderm formation and segmentation of the embryo, after the seventh day of development (Figure 4). Thus, mitochondrial polyphosphate levels seemed to correlate with the energy demand of the embryo during these developmental stages, during which the embryo utilized a large part of its reserve lipids and sugars.

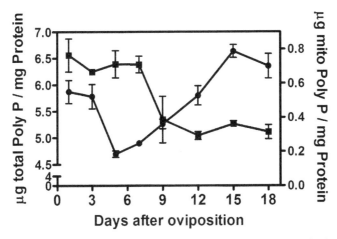

Fig. 4. Polyphosphate metabolism during *R. microplus* embryogenesis. Polyphosphate levels during embryogenesis in the mitochondrial fraction (●) and total polyphosphate (■) during embryogenesis. The results represent the mean ± SD of three independent experiments, in triplicate.

2.1 Inorganic polyphosphate metabolism in tick mitochondria

Mitochondria from tick embryos in the segmentation stage (ninth day after oviposition) were isolated and respiration was measured using pyruvate as the substrate. The rate of oxygen consumption was 30 nmol/min/mg protein, and the respiratory control ratio (RCR) was 6.5. The process was KCN- and oligomycin-sensitive, his fraction exhibited an ATP hydrolyses azide sensitivity, a mitochondrial marker higher than 80%, and no activity of glucose-6-phosphate dehydrogenase, a cytosol marker, was detected (Table 1).

	State 3	State 4	RCR	% F – ATPase azide sensitive	G6PDH (U/mg protein)
Homogenate	---	---	---	49.50 %	2.9 ± 0.4
Mitochondrial fraction	30.2 ± 3.2	4.6 ± 0.7	6.5 ± 0.4	83.45 %	---

Table 1. Mitochondrial characterization

Once the mitochondria were characterized, mitochondria in eggs in the segmentation stage (ninth day after ovoposition) were isolated and exopolyphosphatase activity was measured in order to evaluate the regulation of its activity. The influence of NADH, phosphate, and ADP was investigated in concentrations ranging from 0.1 to 2.0 mM. The activity of exopolyphosphatase was stimulated by a factor of two by NADH, whereas its activity was completely inhibited by 2 mM phosphate and slightly stimulated by ADP (Figure 5A). The activity of exopolyphosphatase was also measured during mitochondrial respiration using pyruvate as the substrate and polyphosphate as the only phosphate source. During this assay, the addition of a small amounts of ADP (0.2 mM) induced state 3 (phosphorylating respiratory rate) followed by state 4 (non-phosphorylating respiratory rate), when all of the ADP was converted to ATP.

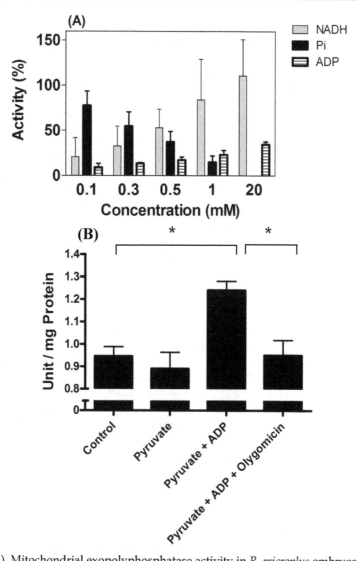

Fig. 5. In (A), Mitochondrial exopolyphosphatase activity in *R. microplus* embryos. Mitochondria from eggs on the ninth day of embryogenesis were isolated and exoolyphosphatase activity was determined using polyphosphate₃ in the presence of 0.1–2 mM NADH, ADP and Pi. The results represent the mean ±SD of three independent experiments, in triplicate. B) Exopolyphosphatase activity was measured in the mitochondria of eggs on the ninth day of development during mitochondrial respiration with pyruvate as the oxidative substrate, polyphosphate₃ as the exopolyphosphatase substrate and olygomicin as ATP synthase. The activity is expressed as units per milligram of total protein and the results represent the mean ± SD of three independent experiments, in triplicate. The asterisk (*) denotes the difference between the populations and the significance was determined by a two-way ANOVA test (Kruskal-Wallis).

Thus, during state 3, a balance existed between the release of phosphate by exopolyphosphatase and ATP synthesis, since exopolyphosphatase activity was measured by the amount of phosphate present. The exopolyphosphatase activity increased during mitochondrial respiration when pyruvate and ADP were added. This increase did not occur without the addition of ADP, indicating that exopolyphosphatase is stimulated during state 3 and that the rate of phosphate release is higher than the rate of ATP synthesis. Indeed, the stimulatory effect was antagonized by olygomicin, an ATP synthase inhibitor (Figure 5B). These data suggest that mitochondrial exopolyphosphatase activity is regulated by phosphate and the energy demand.

Furthermore, it was possible to measure ADP-dependent mitochondrial oxygen consumption in the presence of polyphosphate and in the absence of any other phosphate source. This oxygen consumption was observed using polyphosphate$_3$ and polyphosphate$_{15}$; however, the consumption was higher with polyphosphaste$_3$. On the other hand, heparin, an exopolyphosphatase inhibitor, blocked oxygen consumption, which was recovered when 5 mM phosphate was added and was again interrupted by the addition of oligomycin, an ATP-synthase inhibitor (Figure 6). These results suggest that polyphosphate was used as a phosphate donor for ATP synthesis due to the mitochondrial coupling observed when mitochondrial respiration was interrupted by oligomycin and the existence of membrane exopolyphosphatase in this process, due to the inhibition by heparin, which cannot cross the mitochondrial membrane and has its active site oriented toward the external face of the membrane. In fact, after mitochondrial subfractionation, the main exopolyphosphatase activity was recovered in the membrane fraction, supporting this hypothesis (Table 2).

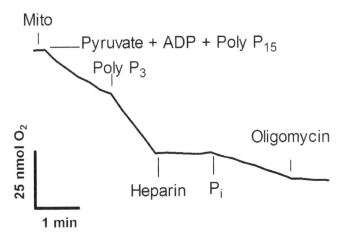

Fig. 6. Polyphosphate as a source for ATP synthesis. Oxygen consumption was monitored using a reaction buffer in the absence of a phosphate source in the eggs on the ninth day of development. The addition of 1 mM ADP, 5 mM pyruvate, 0.5 µM polyphosphate$_3$ and $_{15}$, 20 µg/mL heparin, 5 mM phosphate and 0.5 µM oligomycin is represented in the figure. This experiment was repeated at least three times with different preparations, and this figure shows a representative experiment.

	Exopolyphosphatase activity (U / mg protein)	Heparin (% inhibition)
Mitochondria	0.60 ± 0.19	98
Soluble fraction (intermembrane space and matrix)	0.35 ± 0.06	98
Membrane fraction (mixture of inner and outer membranes)	1.11 ± 0.16	98

Table 2. Exopolyphosphatase activity in mitochondrial preparations. Exopolyphosphatase activity was measured using eggs on the ninth day of development using polyphosphate$_3$ as the substrate. The activity is expressed as units per milligram of total protein and the results represent the mean ± SD of three independent experiments, in triplicate.

2.1.1 A mitochondrial membrane exopolyphosphatase

Exopolyphosphatases have been found in prokaryotes and eukaryotes and, although in bacteria these enzymes mostly hydrolyze high molecular weight polyphosphates (Kumble & Kornberg, 1996), at least some of the enzymes from *Saccharomyces cerevisiae* and *Leishmania major* are more active in hydrolyzing short chain polyphosphates, such as polyphosphate$_3$ (Kumble & Kornberg, 1996; Rodrigues et al., 2002). Exopolyphosphatase from *Escherichia coli* requires divalent cations and K$^+$ for maximum activity, while exopolyphosphatase from yeast only requires divalent cations (Lichko et al., 2003). Membrane mitochondrial exopolyphosphatase activity from the hard tick *R. microplus* was found to be stimulated by Mg^{2+} and was insensitive to K$^+$. Only a few compounds that inhibit exopolyphosphatase have been identified (Kornberg et al., 1999): treatment with molybdate (a common phosphohydrolase inhibitor) and fluoride (a pyrophosphatase inhibitor) showed that exopolyphosphatase present in the mitochondrial membrane fractions was insensitive to these compounds. However, heparin, a good inhibitor of other well-characterized exopolyphosphatases (Lichko et al., 2003), was effective in almost 100% (Figure 7). In order to obtain an insight into membrane exopolyphosphatase kinetics, the apparent Km was measured using polyphosphate$_3$ and polyphosphate$_{15}$ as substrates and the results were expressed as the average of three independent experiments. The membrane exopolyphosphatase affinity for polyphosphate$_3$ was 10 times stronger than for polyphosphate$_{15}$ (Table 3). These results are in contrast with those found in a mitochondrial membrane-bound exopolyphosphatase of *Saccharomyces cerevisiae*, in which case the affinity was stronger for long-chain polyphosphates (Lichko et al., 1998). However, the data demonstrated that membrane exopolyphosphatase kinetics were in agreement with the oxygen consumption rate, which was much higher for polyphosphate$_3$ than polyphosphate$_{15}$. These results reinforce the theory of coupling between the activity of this enzyme and mitochondrial ADP phosphorylation (Figure 8).

Substrates	Km (µM)	Vmax (µmol·min^{-1}·mg protein^{-1})
PolyP$_3$	0.2	2.4
PolyP$_{15}$	2.2	1.1

Table 3. Kinetics characterization of exopolyphosphatase activity in membrane preparations of mitochondria from *R. microplus* embryos on the ninth day of embryogenesis.

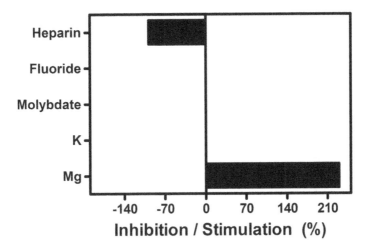

Fig. 7. The effect of some reagents on membrane exopolyphosphatase activity. Mitochondrial membrane fractions of *R. microplus* embryos in eggs on the ninth day of embryogenesis were isolated and the membrane exopolyphosphatase activity was determined using polyphosphate$_3$ as the substrate in the presence of 2.5 mM Mg^{2+}, 50–200 mM K$^+$, 10–100 µM molybdate, 1–10 mM NaF and 20µg/mL heparin.

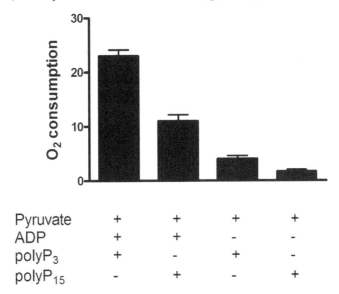

Pyruvate	+	+	+	+
ADP	+	+	-	-
polyP$_3$	+	-	+	-
polyP$_{15}$	-	+	-	+

Fig. 8. Involvement of membrane exopolyphosphatase in mitochondrial respiration. Oxygen consumption was monitored using a reaction buffer in the absence of a phosphate source in the eggs on the ninth day of development in the presence of 1 mM ADP, 5 mM pyruvate, and 0.5 µM polyphosphate$_3$ and $_{15}$. The results represent the mean ± SD of three independent experiments, in triplicate.

To further investigate the regulation of membrane exopolyphosphatase during mitochondrial respiration, the activity was measured using pyruvate as the substrate and polyphosphte as the only source of phosphate. Membrane exopolyphosphatase activity increased during mitochondrial respiration when pyruvate and ADP were added and the stimulatory effect was antagonized by **potassium cyanide** addition (decreased electron flux) and increased by protonophore carbonyl cyanide-p-trifluoromethoxyphenylhydrazone (increased electron flux), suggesting that membrane exopolyphosphatase could be modulated by the electron flux (Figure 9). These findings are consistent with those of Pavlov et al., 2010, who demonstrated that the production and consumption of mitochondrial polyphosphate depends on the activity of the oxidative phosphorylation machinery in mammalian cells. Furthermore, heparin completely inhibited exopolyphosphatase activity, reinforcing the role of membrane exopolyphosphatase during mitochondrial respiration, and the respiration activation by membrane exopolyphosphatase activity indicated that exopolyphosphatase could be close to the site of ATP production.

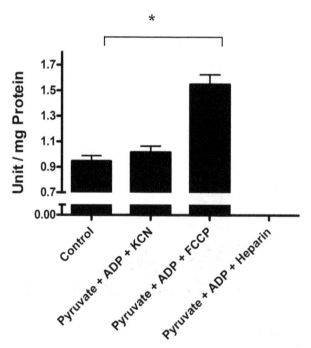

Fig. 9. Regulation of mitochondrial exopolyphosphatase activity during mitochondrial respiration. The activity of exopolyphosphatase was measured in the mitochondria of the eggs on the ninth day of development during mitochondrial respiration, using pyruvate as the oxidative substrate, polyphosphate$_3$ as the exopolyphosphatase substrate, KCN as the respiratory chain inhibitor, FCCP as the un-coupler and heparin as the exopolyphosphatase inhibitor. The activity was expressed as units per milligram of total protein and the results represent the mean ± SD of three independent experiments, in triplicate. The asterisk (*) denotes the difference between the populations and the significance was determined by a two-way ANOVA test (Kruskal-Wallis).

Despite the regulation of membrane exopolyphosphatase by an increased or decreased electron flux, the sensitivity of this enzyme according to the redox state using polyphosphate$_3$ as the substrate was evaluated. The influence of 1.0 mM dithiothreitol (DTT) and 1.0 mM hydrogen peroxide (H_2O_2) was investigated at different times and the exopolyphosphatase activity was stimulated and inhibited by 50% of both, suggesting that exopolyphosphatase is tightly regulated by the redox state (Figure 10).

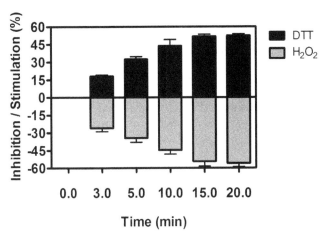

Fig. 10. The redox regulation of mitochondrial membrane exopolyphosphatase. Exopolyphosphatase activity was measured in the mitochondria of the eggs on the ninth day of development using polyphosphate$_3$ as the substrate. The mitochondria were treated with 1 mM DTT and 1 mM H_2O_2 for 0–20 min. The results represent the mean ± SD of three independent experiments, in triplicate.

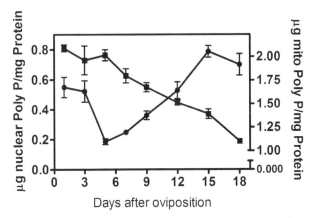

Fig. 11. Polyphosphate quantification in the nuclear and mitochondrial fractions. Polyphosphate levels during embryogenesis in the nuclear fraction (■) and mitochondrial fraction (•) during embryogenesis. The results represent the mean ± SD of three independent experiments, in triplicate.

Additionally, mitochondrial polyphosphate can form polyphosphate/Ca^{2+}/PHB complexes (Reusch, 1989) with ion-conducting properties similar to those of the native mitochondrial permeability transition pore (Pavlov et al., 2005). Polyphosphatases localized in the membrane can not only degrade, but they can also synthesize polyphosphate inside these complexes (Lichko et al., 1998). During the embryogenesis of R. microplus, the synthesis of polyphosphate occurs in mitochondria but not in the nuclei (Figure 11). As polyphosphate kinases have only been found in prokaryotes, the observation that polyphosphate synthesis in ticks only occurs in the mitochondrial fraction supports the possibility that such synthesis probably occurs via the action of these complexes, as already suggested for other organisms (Reusch and Sadoff, 1988; Lichko et al., 1998; Reusch et al., 1998; Abramov et al., 2007).

3. Conclusion

The ubiquity of polyphosphate and the variation in its chain length, location and metabolism indicate the relevant functions of this polymer, including those in animal systems. The present study showed that electron flux and the redox state may exert some influence on and be influenced by the activity of membrane exopolyphosphatase, and its describes a role for polyphosphate in the energy supply and ATP synthesis during embryogenesis of the hard tick R. microplus. In this sense, a more comprehensive understanding of polyphosphate biochemistry during tick embryo development may unravel additional targets that could be effective in the control of this ectoparasite and shed new light on polyphosphate metabolism.

4. Acknowledgment

This work was supported by grants from Fundação de Amparo à Pesquisa do Estado do Rio de Janeiro–FAPERJ, Conselho Nacional de Desenvolvimento Científico e Tecnológico–CNPq, Programa de Núcleos de Excelência–PRONEX, Programa Nacional de Cooperação Acadêmica–PROCAD-CAPES, Instituto Nacional de Ciência e Tecnologia/Entomologia Molecular–INCT/EM and FUNEMAC.

5. References

Abramov, A. Y., Fraley, C., Diao, C. T., Winkfein, R., Colicos, M. A., Duchen, M. R., French, R. J. & Pavlov, E. (2007). Targeted polyphosphatase expression alters mitochondrial metabolism and inhibits calcium-dependent cell death. *Proceedings of the National Academy of Sciences*. Vol.104, (November, 2007), pp. 18091–18096, ISSN 0027-8424

Bate, M. & Arias, A, M. (1991). The embryonic origin of imaginal discs in Drosophila. *Development*, Vol.112, (July, 1991), pp. 755-761, ISSN 0950-1991

Bligh, E. G. & Dyer, W. J. (1959). A rapid method of total lipid extraction and purification. *Canadian Journal of Biochemistry and Physiology*, Vol.37, (August, 1959), pp. 911-917, ISSN 0576-5544

Campos, E., Moraes, J., Facanha, A, R., Moreira, E., Valle, D., Abreu, L., Manso, P, P., Nascimento, A., Pelajo-Machado, M., Lenzi, H., Masuda, A., Vaz Ida S Jr. & Logullo, C. (2006). Kinetics of energy source utilization in *Boophilus microplus* (Canestrini, 1887) (Acari: Ixodidae) embryonic development. *Veterinary Parasitology*, Vol.138, (February, 2006), pp. 349-357, ISSN 0304-4017

Clements, A., Bursa, D., Gatsos, X., Perry, A, J., Civciristov, S., Celik, N.,Likic, V, A., Poggio, S., Jacobs-Wagner, C., Strugnell, R, A. & Trevor Lithgow, T. (2009). The reducible complexity of a mitochondrial molecular machine. *Proceedings of the National Academy of Sciences*, Vol.106, (September 2009), pp. 15791-15795, ISSN 0027-8424

Dubois, M., Gilles, K. A., Hamilton, J. K., Rebers, P. A. & Smith, F. (1956). Calorimetricmethod for determination of sugar and related substances. *Analytical Chemistry*, Vol.25, (March, 1956), pp. 350-356, ISSN 0003-2700

Fagotto, F. (1990). Yolk degradation in tick eggs: I. Occurrence of a cathepsin L-like acid proteinase in yolk spheres. *Archives of Insect Biochemistry and Physiology*, Vol.14, (February, 1990), pp. 217-235, ISSN 0739-4462

Guerrero, F, D., Nene, V, M., George, J, E., Barker, S, C. & Willadsen, P. (2006). Sequencing a new target genome: the *Boophilus microplus* (Acari: Ixodidae) genome project. *Journal of Medical Entomology*, Vol.43, (January, 2006), pp. 9-16, ISSN 0022-2585

Jones, H, E., Holland, I, B., Jacq, I, B., Wall, T. & Campbell, A.K. (2003). Escherichia coli lacking the AcrAB multidrug efflux pump also lacks nonproteinaceous, PHB–polyphosphate Ca^{2+} channels in the membrane. *Biochimica et Biophysica Acta Biomembranes*, Vol. 1612, pp. 90-97, ISSN 0005-2736

Kim, K, S., Rao, N, N., Fraley, C, D. & Kornberg, A. (2002). Inorganic polyphosphate is essential for long-term survival and virulence factors in Shigella and Salmonella spp. *Proceedings of the National Academy of Sciences*, Vol.99, (May 2002), pp. 7675-7680, ISSN 0027-8424

Kornberg, A. (1995). Inorganic polyphosphate: toward making a forgotten polymer unforgettable. *Journal of Bacteriology*, Vol.177, (February, 1995), pp. 491-496, ISSN 0021-9193

Kornberg, A., Rao, N, N. & ult-Riche, D. (1999). Inorganic polyphosphate: a molecule of many functions. *Annual Review of Biochemistry*, Vol. 68, (July 1999), pp. 89-125, ISSN 0066-4154

Kulaev, I, S. & Vagabov, V, M. (1983). Polyphosphate metabolism in micro-organisms. *Advances in Microbiol Physiology*, Vol.24, pp. 83-171, ISSN 0065-2911

Kulaev, I. & Kulakovskaya, T. (2000). Polyphosphate and phosphate pump. *Annual Review of Microbiology*, Vol. 54, (October 2000), pp. 709-734, ISSN 0066-4227

Kulaev, I, S.; Vagabov, V, M. & Kulakovskaya, T, V. (2004). *The Biochemistry of Inorganic Polyphosphate*, Wiley, ISBN 0 470 85810 9, Chichester, England

Kulakovskaya, T, V., Lichko, L, P., Vagabov, V, M., & Kulaev, I, S. (2010). Inorganic Polyphosphates in Mitochondria. *Biochemistry (Moscow)*, Vol.75, (July 2010), pp. 825-831, ISSN 0006-2979

Kumble, K. D. & Kornberg, A. (1996). Endopolyphosphatases for long chain inorganic polyphosphate in yeast and mammals. *Journal of Biological Chemistry*, Vol.271, (October, 1996), pp. 27146–27151, ISSN 0021-9258

Kuroda, A., Nomura, K., Ohtomo, R., Kato, J., Ikeda, T., Takiguchi, N., Ohtake, H. & Kornberg, A. (2001). Role of inorganic polyphosphate in promoting ribosomal protein degradation by the Lon protease in E. coli. *Science*, Vol. 293, (July 2001), pp. 705-708, ISSN 0036-8075

Lichko, L. P., Andreeva, N. A., Kulakovskaya, T. V. & Kulaev, I. S. (2003). Exopolyphosphatases of the yeast Saccharomyces cerevisiae. *FEMS Yeast Research*, Vol.3, (January, 2003), pp. 233–238, ISSN 1567-1356

Lichko, L. P., Kulakovskaya, T. V. & Kulaev, I. S. (1998). Membrane-bound and soluble polyphosphatases of mitochondria of Saccharomyces cerevisiae: identification and comparative characterization. *Biochimica et Biophysica Acta – Biomembranes*, Vol.1372, (January, 1998), pp. 153–162. ISSN 0005-2736

McInerney, P., Mizutani, T. & Shiba, T. (2006). Inorganic polyphosphate interacts with ribosomes and promotes translation fidelity in vitro and in vivo. *Molecular Microbiology*, Vol.60, (April 2006), pp. 438-447, ISSN 0950-382X

Monnerat, A, T., Machado, M, P., Vale, B, S., Soares, M, J., Lima, J, B., Lenzi, H, L. & Valle, D. (2002). *Anopheles albitarsis* embryogenesis: morphological identification of major events. *Memorias do Instiuto Oswaldo Cruz*, Vol.97. (June, 2002), pp. 589-596, ISSN 0074-0276

Negoda, A., Negoda, E., Xian, M. & Reusch, R.N. (2009). Role of polyphosphate in regulation of the Streptomyces lividans KcsA channel. *Biochimica et Biophysica Acta*, Vol.1788, (March 2009) pp. 608-614, ISSN 0006-3002

Nusslein-Volhard, C. & Roth, S. (1989). Axis determination in insect embryos. Ciba *Foundation Symposium*, Vol.144, pp. 37-55, ISBN: 9780471923060

Pavlov, E., Aschar-Sobbi, R., Campanella, M., Turner, R. J., Gómes-García, M. & Abramov, A. Y. (2010). Inorganic polyphosphate and energy metabolism in mammalian cells. *Journal of Biological Chemistry*. Vol.285, (March, 2010), pp. 9420–9428, ISSN 0021-9258

Pavlov, E., Zakharian, E., Bladen, C., Diao, C. T. M., Grimbly, C., Reusch, R. N. & French, R. J. (2005). A large, voltage-dependent channel, isolated from mitochondria by water-free chloroform extraction. *Biophysical Journal*. Vol.88, (April, 2005), pp. 2614–2625, ISSN 0006-3495

Rashid, M, H., Rumbaugh, K., Passador, L., Davies, D, G., Hamood, A, N., Iglewski, B, H. & Kornberg, A. (2000). Polyphosphate kinase is essential for biofilm development, quorum sensing, and virulence of Pseudomonas aeruginosa. *Proceedings of the National Academy of Sciences*, Vol.97, (August 2000), pp. 9636-9641, ISSN 0027-8424

Reusch, N. M. & Sadoff, H. L. (1988). Putative structure and functions of a poly-beta-hydroxybutyrate/ calcium polyphosphate channel in bacterial plasma membranes. *Proceedings of the National Academy of Sciences*. Vol.85, (June, 1988), pp. 4176–4180, ISSN 0027-8424

Reusch, R, N. (1989). Poly-beta-hydroxybutyrate/calcium polyphosphate complexes in eukaryotic membranes. *Proceedings of the Society for Experimental Biology and Medicine*, Vol. 191, (September 1989), pp. 377-381, ISSN 1525-1373

Rodrigues, C. O., Ruiz, F. A., Vieira, M., Hill, J. E. & Docampo, R. (2002). An acidocalcisomal exopolyphosphatase from Leishmania major with high affinity for short chain polyphosphate. *Journal of Biological Chemistry*, Vol.277, (October, 2002), pp. 50899-50906, ISSN 0021-9258

Silveira, A, B., Castro-Santos, J., Senna, R., Logullo, C., Fialho, E. & Silva-Neto, M, A. (2006). Tick vitellin is dephosphorylated by a protein tyrosine phosphatase during egg development: effect of dephosphorylation on VT proteolysis. *Insect Biochemistry and Molecular Biology*, Vol.36, (March, 2006), pp. 200-209, ISSN 0965-1748

Sonenshine, D, E., Kocan, K, M. & de la, F, J. (2006). Tick control: further thoughts on a research agenda. *Trends in Parasitology*, Vol.22, (September, 2006), pp. 550-551, ISSN 1471-4922

6

Energy Metabolism in Children and Adolescents

Valentin Son'kin and Ritta Tambovtseva
Institute for Developmental Physiology
Russian Academy of Education, Moscow
Russian Federation

1. Introduction

Energy metabolism is the most integral body function, and, as any functional activity, it has an effect on energy expenditure. Body energy expenditures are composed of three unequal parts: basal metabolism, energy supply of functional activity and energy expenditure on growth, development and the adaptive processes. The proportion of these expenditures is determined by the stage of individual development and specific life conditions.

Basal metabolism (the minimum level of energy production in conditions of complete rest), in its turn, is composed of three primary energy expenditure types: minimum level of vital physiological functions; intracellular futile cycles of biochemical processes; and reparative processes, including growth and development expenditure.

With age, basal metabolism expenditure and growth and development expenditure is considerably reduced, while functional expenditure can increase (for instance, muscle energy expenditure of an adult can be sometimes more than that of a child), but in any case they undergo important qualitative changes.

2. Age changes of basal metabolism

Methodological requirements for basal metabolism measuring are hard to be implemented outside a clinic, therefore metabolism in a state of rest is most commonly measured (lying down, comfortable temperature, 2-4 hours after food intake, without any stress factors), which is approximately 10-20% more than the level of basal metabolism. Present-day children have resting metabolism values even lower than standard norms of basal metabolism (Kornienko, 1979), proposed by Harris and Benedict (1919), which might be the result of acceleration of growth and development, observed up to the 1980s (Godina, 2009). With age the rate of resting metabolism (per body mass unit) is reduced – from infancy to the adulthood– by 1.5 – 2 times. The reasons for this reduction have been discussed for the last 150 years.

Since Max Rubner's time (1883) it has been known that as mammals gain body mass, heat production per mass unit is reduced, while the metabolism rate, relative to the surface space, is practically constant ("the rule of surface"). These metabolism changes were primarily explained by thermoregulation expenditure, but it turned out that in a thermoneutral conditions, without any extra heat production, this alignment persists. Moreover, this relation between metabolism rate and body size is observed in invertebrates (Schmidt – Nielsen, 1987; Ivanov, 1990).

For a long time the increased metabolic rate in infants has been attributed to metabolic expenditure on growth (Karlberg, 1952). But this hypothesis was not corroborated by facts. An infant's growth is most intense in the first 6 months after birth. The growth coefficient during this period is 4.0 (Schmal'hausen, 1935). At one year after birth, the coefficient is sharply reduced, by more than 10 times - to 0.3. Basal metabolism rate is at its peak at the age of one. Special calculations (Kornienko & Gohblit, 1983) proved that true expenditure on synthesis, associated with growth processes, even in the first 3 postnatal months, when the infant's growth velocity is at its peak, is no more than 20 kcal /day, which is 7−8% of the total expenditure. According to King et al. (1994), total energy expenditure of a woman body during pregnancy is on the average 325 MJ (77621 kcal). It is approximately a 20% increase in metabolism rate, compared to basal energy expenditure of a female. Evidently most of this energy expenditure is the expenditure on extra functions of maternal body systems, including the ones associated with the required adaptation to the increased physical load: during the second half of the pregnancy period the condition of the mother is bearing an extra load weighing from 2-3 kg to 10-12 kg (that includes the weight of the foetus, placenta, amniotic fluid, grown uterus, etc.). In fact, the growth processes takes a small part of the volume of energy expenditure. The energy expenditure on proliferative processes of kids older than one year is even less (under 1%), when the growth velocity becomes 12-15 times slower compared to intrauterine period.

Empiric formulae are used to express the relation between body size and metabolic rate. Kleiber (Kleiber, 1961) proposed the following formula for mammals, including humans:

$$M = 67.7 \bullet P^{0.75} \text{kcal} / \text{day} \tag{1}$$

Where M is the heat production of the whole body, and P is body mass.

But age changes in basal metabolism cannot be calculated using this equation. During the first year after birth, heat production is not reduced as required by the Kleiber equation, but stays on the same level or even increases, while the body mass during this period is tripled. Only one year after birth is the metabolism rate of 55 kcal/kg per day reached, "proposed" by the Kleiber equation for the body with the mass of 10 kg.

Only after 3 years does the basal metabolic rate starts to gradually reduce, and reaches the level of an adult person (25 kcal/kg per day) only during puberty.

Increase in the basal metabolic rate within the first year of an infant's life is correlated by some authors with a decrease in volume of intracellular space in most tissues. According to Brück (1970), if the oxygen consumption rate per mass unit of newborn infants in rest is 5.0 ml O2/kg/min, and in one-year infants − 8.2 ml O2/kg/min, then recalculated per an active cellular mass unit, it turns out that a newborn consumes 9.0 ml O2/kg/min, and a one year-old child− 10.9 ml O2/kg/min.

Rate of basal metabolism starts reducing from the age of three (Fig. 1). The first place among the probable reasons for this phenomenon is change in body composition correlated with age – increase in relative mass of tissue with a small rest metabolism rate (bone tissue, fat tissue, skeletal muscles, etc.). M. Holliday (1971) has already proven that a gradual decrease in the metabolism rate of children can be easily explained by the uneven growth of organs, presuming that the metabolic rate of tissue growth in the process of postnatal development is constant. For instance, it is known that mass of the brain (which greatly contributes to the level of basal metabolism) for newborns is 12% of their body mass, while in adults it's only 2%. Internal organs (liver, kidneys, etc.) also grow unevenly, and have a high level of energy metabolism even during rest— 300 kcal/kg/day. At the same time, the muscle tissue whose relative quantity is almost doubled in the period of postnatal development, is characterized by a very slow resting metabolism rate — 18 kcal/kg/day.

It should be noted, that the dynamics of age changes in resting metabolism is not just a simple decrease in metabolism rate. As it is given in Fig. 1, periods characterized by a rapid decrease in metabolism rate, are replaced by age intervals where resting metabolism values are stabilized (Kornienko, 1979; Kornienko & Gohblit 1983; Kornienko et al, 2000). Taking this into consideration, a close correlation of changes in metabolic rate and growth velocity is found. Columns in Fig. 1 show relative annual increase in body mass. Turns out that, excluding the first year after birth, the higher the relative growth velocity, the higher the rate of resting metabolism lowering during this period. Inhibition of growth processes at the age of 1.5-2 coincides with the highest values of resting metabolism, and the increase in growth velocity by ages 6 through 7 is accompanied by a considerable decrease in metabolism. After this there is the next inhibition of growth, during which the level of metabolism is stabilized, and the next value decrease coincides with a new acceleration of growth processes. The last peak of resting metabolism is observed at about the age of 14 years, before the puberty growth spurt, and soon after that the energy metabolism rate is stabilized on the level typical for adults. According to longitudinal observations, all these changes are typical both for boys and girls, but in girls they are usually observed 0.5-1 years earlier (Kornienko & Gohblit, 1983).

Rate value of basal metabolism is especially important for diagnosing and treating several endocrinological diseases, as well as obesity. Because of that there are ongoing discussions in scientific literature about methods of calculating basal metabolism values using various formulae – Harris & Benedict (1919), the WHO committee and others (White & Seymour, 2005; Frankenfield et al., 2005; Garrel et al., 1996; Hayter & Henry, 1994; Tverskaya et al., 1998, etc.). Most contemporary authors consider the volume of cellular mass or the value of lean body mass the most important factor, as well as age, sex, constitution, race and ethnicity (Bosy-Westphal et al., 2009; McDuffie et al., 2004; St-Onge & Gallagher, 2010; Vermorel et al., 2005).

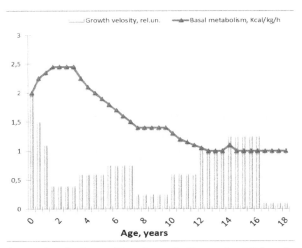

Fig. 1. Dynamics of growth velocity and basal metabolic rate in children from birth to maturity (after: Kornienko, 1979; modified)

Another factor might play an important role – the change in metabolic activity of tissues in a growing organism, that occurs with age (Conrad & Miller, 1956; Nagorny et al., 1963). To test this assumption, our laboratory researched age changes in the mitochondrial apparatus of various tissues (Demin, 1983; Kornienko, 1979). Using Chance's differential spectrophotometer we measured cytochrome **a** concentration, which is a terminal ferment of the oxidative chain of mitochondria, in tissue homogenates of Wistar rats during ontogeny (Fig.2). The higher cytochrome **a** concentration, the higher oxidation activity is developed by a given tissue under the influence of an appropriate stimulus, provided it is adequately supplied with substrates and oxygen. This data allows to compare not just the potential metabolic activity of various tissues, but also to observe its changes, including changes occurring with age.

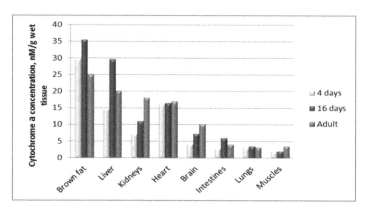

Fig. 2. Cytochrome **a** concentration in tissue homogenates of Wistar rat during ontogeny (after: Kornienko, 1979; modified).

Brown adipose tissue has the highest potential metabolic activity, both in young and adult rats; the liver is second in potential. Both tissues are characterized by the fast rise of cytochrome a concentration at the age of 16-20 days, which can be explained by the fact that at that time young rats leave their nest and start their separate life, which requires the activation of thermoregulatory processes. The food type changes at the same age, which has an effect on cytochrome **a** concentration in the intestine tissue. When adulthood is reached, cytochrome content in all these tissues is considerably reduced.

Content of oxidation ferments in cardiac and lung tissues is the most stable – it stays almost unchanged with age. But oxidation ferments in tissues of rat kidneys and brain increase approximately by a factor of 2.5 by adulthood. If the increase in tissue mass is considered, it turns out that the metabolic potential of brown adipose tissue during postnatal ontogenesis grows 10 times, the brain's - 11.4 times, the liver's –38.5 times, the kidney's –57 times, the skeletal muscles' –87 times. That directly affects the level and structure of energy metabolism.

It has been proven using this method (based on post-mortem materials from a trauma clinic) that cytochrome concentration is increased considerably in some grey matter areas in the brain cortex (4, 6, 10th and 17th fields according to Broadman), in the subcortical structures, and in the homogenates of children's whole brain at the age of 1–1.5, compared to the first months of life (Kornienko, 1979). Since at this age the human brain accounts for at least 50% of basal metabolism, it can be assumed that a more active oxidation processes of this tissue will have an effect basal metabolism of the whole body. Important qualitative changes in nervous and mental activity occur in children age 1-2, due to differentiation processes in neural tissue (Farber & Machinskaya, 2009; Tsekhmistrenko et al., 2009). Meanwhile it has been proven 40 years ago that tissue differentiation in ontogenesis always starts with mitochondria development and a more active oxidation metabolism (Macler et al., 1971; Makhin'ko & Nikitin 1975).

Calculations by Kornienko (1979) have demonstrated that for humans the contribution of various organs to the basal metabolism is changes with age. The adult human brain accounts for 24% of basal metabolism, the liver for 20%, the heart for 10.2% and the skeletal muscles for 28%. A one-year old child's brain accounts for 53% of basal metabolism, while the skeletal muscles account only for 8%. The liver's contribution is about 18% (Fig.3).

3. Energy supply for functions during ontogenesis

3.1 Daily energy expenditure structure

Unlike basal metabolism, which is the minimum level of body energy expenditure, average daily metabolism includes the sum of all expenditures associated with the realization of various body functions. Food processing and digestion, thermoregulation and muscle activity are the most power-consuming functions. Unfortunately, there is almost no data in literature on the energy value of mental activity (not taking into consideration the indirect calculations by Holliday, 1971).

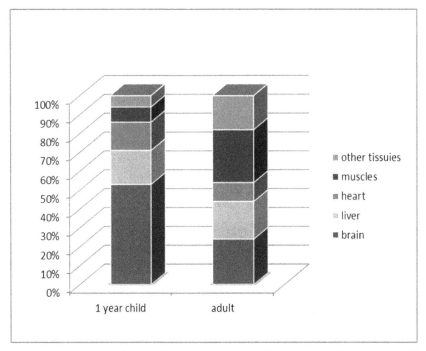

Fig. 3. Age changes in contribution of various organs and tissues into the structure of human basal metabolism (After: Holliday, 1971; Kornienko, 1979; modified)

3.2 Functional range

Various body tissues can change their metabolic activity to a different degree, ranging from rest to maximum functional activity. It depends on the organization of metabolic paths in cells that form the corresponding tissue. Based on content and activity data of vital energy ferments in body tissues, Demin (1983) calculated the hypothetic values of minimum and maximum metabolic activity for the liver, brain and muscles of a young man (Fig. 4). Characteristics of the functional range in skeletal muscles obtained using this method are close to the actual measured maximum energy expenditure (Son'kin, 1990; Kornienko, et al, 2000). As seen in fig. 3, muscles have both the highest potential for metabolic activity and the widest functional range. It should be noted that the maximum level of functional activity is carried out through the anaerobic metabolic processes, not limited by the possibilities of mitochondrial oxidation. The functional range value of various tissues can be affected differently depending on age. According to Demin (1983), this value for the brain is at its peak in childhood; for the liver it stays more or less the same at all periods of postnatal ontogenesis; for skeletal muscles it considerably increases from birth to the end of puberty.

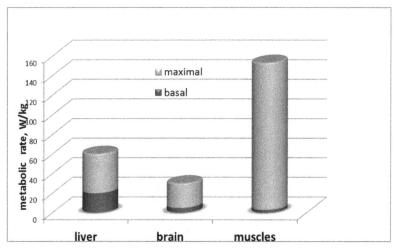

Fig. 4. Comparative characteristic of functional range in various tissues of human body (calculation for a young man) (After: Demin, 1985; modified)

The intensity of body functions in a child is much higher than of an adult. The rate of children's basal metabolism is 1.5-2 times higher, but the maximum activity level is considerably lower than that of an adult. This results in a smaller functional range, and that makes the child body's existence more stressful.

The high intensity of energy metabolism in children becomes particularly obvious when you look at the fact that the child's body reacts with higher intensity to an impact equal in power, demonstrating a higher lability of autonomic systems and metabolic processes. These differences are well known, when talking about muscle activity (Åstrand, 1952; Kornienko et al, 2000). The fact that similar differences are evident in metabolic reactions to other functional loads, in particular food load, is less known.

3.3 Specific – Dynamic (thermogenic) effect of food

The rate of heat production is considerably increased after food consumption, despite of lack of muscle activity, and it remains elevated for 2-3 hours (depending on the structure of food and other factors). Though the thermogenic effect of food is a known phenomenon and has been studied since the end of the 19th century, there is still no single opinion about its reasons and occurrence mechanisms. The simplest explanation, which states that extra energy production is required to activate a motor function of the gastrointestinal tract, has not been borne out by experiments: the thermogenic effect of glucose cannot be recovered in patients with diabetes and lab animals, even though carbohydrate is absorbed to the bloodstream and is extracted with urine (Lusk, 1919).

Today the most probable reason considered for the thermogenic effect of food is the effect of enterohormones produced by the duodenum epithelium. For example, it has been proven that the lack of these hormones in blood results in a lower body temperature, meaning heat production (Ugolev et al, 1976). But even in this case it is unclear which particular tissue accounts for extra heat production. Recently brown adipose tissue has been considered as the reason (Himms-Hagen, 1989; Nedergaard & Cannon, 2010), which, according to recent data, is preserved in adults (Nedergaard et al., 2007) and maintains substrate homeostasis (Son'kin et al., 2010).

A unique systematic research of age changes in the thermogenic effect of food substance (glucose) in school children was made in the laboratory of I.A.Kornienko (Kornienko et al, 1984). A standard test was used to evaluate glucose tolerance during this study: glucose was taken orally on an empty stomach in quantity proportional to mass (1g/kg). Content of glucose in blood in such a probe is usually increased during 30 minutes after intake and gradually normalizes within 2-3 hours. As glucose increases, with some delay (about 0.5 hours) oxidation processes start in the body, and in 3 hours the oxygen consumption levels return to the primary level. Total intensification of energy production with age for children of 7-8 up to 15-17 years is considerably reduced, especially for boys (Table 1).

The given data proves that reactivity of oxygen metabolism decreases with age, meaning efficiency of mechanisms providing homeostasis increases. The difference in age dynamics of thermogenic glucose effect for boys and girls is the most interesting phenomenon. It is known that adult women, on average, have a better tolerance to glucose than men (Korkushko & Orlov, 1974). Possibly, the given data reflects the formation of such sex differences.

Age, years	Boys	Girls
7 – 8	2.125 + 0.16 (n=18)	1.825 + 0.14 (n=23)
11 – 12	1.255 + 0.10 (n=22)	1.365 + 0.11 (n=21)
15 – 17	0.585 + 0.06 (n=16)	1.060 + 0.08 (n=14)

Table 1. Total thermogenic effect of glucose (per oral 1 g/kg) for school children for 3 hours of observation (kcal/kg, M±m) (After: Kornienko et al., 1984)

In the same research (Son'kin et al., 1975) it was proved that glucose put into the body depends considerably on their body constitution: children of 11-12 years with a low fat content in the body of no less 1/3 introduced glucose is oxidized in process of a thermogenic response to its putting into the body, while children with a high fat content a thermogenic response to input of glucose is considerably less. Similar results for adult persons are described in press of the last years (Nedergaard & Cannon, 2010).

Results obtained in such studies as well as other data about thermogenic effect of food cast doubt on the validity of widely used calculations of caloric food value. All such calculations do not take into account energy expenditure on digestion of food substances which are known to take from 1/5 up to 1/3 caloric value of the taken food substance. The problem is complicated by the fact that fats and proteins have a greater thermogenic effect than carbohydrates (Kassirsky, 1934), while mixed products have a smaller thermogenic effect than the total thermogenic effects of food substances they contain (Forbes & Swift, 1944). It is proven that liquid food, with similar calorie value has a less specifically dynamic effect than solids (Habas & Macdonald, 1998). We think that mechanical calculations of food caloricity based on the caloric equivalent of proteins, fats and carbohydrates in it which is widely used in clinic and health-improving systems, including paediatric practice (Morgan, 1980; Young et al. 1991; Schmelzle et al., 2004) need to be corrected.

3.4 Thermoregulation development in ontogenesis

Thermoregulation, support of constant temperature in the body core is determined by two basic processes: heat production and heat dissipation. Heat production (thermogenesis) depends primarily on the rate of metabolic processes, while heat dissipation is defined by heat insulation provided by cutaneous coverings, vascular reactions, active outer respiration and perspiration. Because of this, thermogenesis is considered a mechanism of chemical thermoregulation, and heat dissipation regulation – a mechanism of physical thermoregulation. Both these processes change with age, as well as their role in providing a constant body temperature.

As a result of laws of physics, increase in mass and body absolute dimensions reduces the contribution of chemical thermoregulation. Thus, the value of thermoregulation heat production for newborn children makes about 0.5 kcal/kg • hour • °C, and for adults – 0.15 kcal/kg• hour•°C.

A newborn child, if temperature of the environment lowers, can enlarge heat production to adult levels - to 4 kcal/kg•hour. But because of lower heat insulation (0.15 °C•m²•hour/kcal) the chemical thermoregulation range of a newborn is small—no more than 5°.

At that it should be accounted that the critical temperature level (Th), switching thermogenesis for a healthy newborn is 33°C, by the adult period it falls down to 27−23°C. But in clothes with heat insulation usually making 2.5 CLO, or 0.45 °C •m²•hour/kcal, Th value falls down to 20°C, therefore a child in his usual clothes at room temperature is in a thermoneutral environment, meaning that in these conditions a child requires no extra expenditure to support body temperature.

If the temperature falls down below threshold values (for instance, during the change of a child's clothes), mechanisms of extra heat production switch on. For a child they are mainly, "nonshivering thermogenesis", localized in metabolically active tissues – liver and brown adipose tissue (Brück, 1970; Kornienko, 1979). Researches of the latest years have revealed that an acute short-term cooling of adults also results in activation of nonshivering thermogenesis in brown adipose tissue (Nedergaard et al., 2007; Son'kin et al., 2010), which is proved to be preserved for most adults residing in a moderate climatic zone (Nedergaard & Cannon, 2010). Another mechanism of thermogenesis is a cold-induced muscle tremor which is usually observed in adults when the cooling effect is strengthened or prolonged. For children this physiological mechanism turns out inefficient due to particular features of the child's body constitution, therefore it is activated in the last turn, if temperature of the body core falls down despite the processes (Kornienko, 1979).

High activity of special mechanisms of thermogenesis in infants is connected not only with small size and large relative surface increasing heat insulation, not only with low heat insulation of cutaneous coverings, but with a relatively low level of basal metabolism, which has been noted before in this paper. Within the first year of life all these parameters are changing and the chemical thermoregulation activity is reduced. For a child of 5–6 months the importance of physical thermoregulation is considerably increased; it makes the temperature threshold and latent period of an interscapular brown adipose tissue activation almost double compared with the same parameters for infants 1-2 months old (Gohblit et al, 1975; Kornienko, 1979).

Under usual conditions the child older than 3 years old has a high value of heat flow in relation to the body surface unit, and heat insulation of cutaneous coverings is low, therefore children's skin is practically always warm. Even at the age of 4.5–5 years for girls and 5.5–6 years for boys the body heat insulation is very low: 0.226 ± 0.003 °C • m^2 • hour/kcal, not changed a lot compared to infants. Their mechanisms of physical thermoregulation are poorly developed. Therefore, if such a child is in conditions of room temperature (+ 20°C) in underwear and T-shirt, in 80 cases of 100 his thermoregulatory heat production is activated (Kornienko, 1979).

Intensification of growth processes at the age of 5-7 years results in accelerating the length and surface area of extremities, providing a regulated heat exchange of the body with the environment. It is, in turn, results in the fact that from the age of 5.5– 6 years (it is especially visible for girls) the thermoregulation function is considerably changed. The body heat insulation is increased, and the chemical thermoregulation activity is substantially reduced. This method of body temperature regulation is more efficient and it becomes predominant in further development with age. In girls this transformation of thermoregulation happens, on average, one year earlier than boys.

At the age of 10 years for girls and 11 years for boys quicker growth processes and considerably lower rate of basal metabolism which are typical for them are observed again. According to thermoregulation conditions, this age can be marked out as a crucial period: physical thermoregulation is activated again, with chemical thermoregulation becoming less important. For boys these changes are distinctly expressed at the age of 12 years.

The next stage of thermoregulation development is during pubescence, becoming apparent in the frustration of the forming functional system. For 11–12-year old girls and 13-year old boys, despite the continuous decrease in the resting metabolism rate, there is no corresponding adjustment of vascular regulation. Worse heat insulation facilities of covering tissues result in the fact that, notwithstanding the age tendency, the critical temperature shifts to higher values with the temporarily growing role of chemical thermoregulation – most teenagers (up to 80%) enlarge their heat production even under slight cooling conditions effects.

Distinct sex differences in dynamics of thermoregulation development are seen during pubescence (Kornienko & Gohblit, 1988). Parallel to decrease in basal metabolism, for girls after the second stage of pubescence (according to Tanner), heat insulation properties are rapidly increased, and the function of physical regulation is restored in full. By the age of 16 this process is usually over, and all thermoregulation parameters reach the values typical for adults. The same tendencies exist in boys, but by the age of 16 years, processes forming mechanisms of physical thermoregulation are incomplete. Only in youths after pubescence do thermoregulation facilities reach their final level. Increase in tissue heat insulation to the level of 1.1 CLO allows to function without activating the chemical thermoregulation (meaning extra heat production) even when the environment temperature falls down by 10-15 degrees below thermoneutral. Such body reaction is naturally more economical and efficient.

The given data prove that in the process of postnatal ontogenesis the primary line of the system development providing temperature homeostasis is indirect (Falk, 1998). At each stage of individual organism development there is a complex dependence of thermoregulation active mechanisms on growth and development, the rate of metabolic processes and conditions of some autonomic functions. It is this dependence that determines a primary activity of physical or chemical thermoregulation mechanisms, providing temperature homeostasis at the corresponding stage of development.

4. Muscle function energy supply development

Muscle activity is the most energy-intensive function: even for a person engaged in mental work about half of daily energy expenditure is used to provide a contracting activity of somatic muscles. One of the first works researching ontogenesis of the muscle function energy supply was made by Robinson (1938), who discovered age changes in maximum oxygen consumption in children, teenagers and adults. The research of P.-O. Astrand (1952), still a classic, presents data of maximal aerobic capacity of people from 6 up to 60 years old.

As compared with other tissues, skeletal muscles have the greatest functional range (Fig.4) – metabolic process can change its velocity in muscles by a factor of dozens. Such amplitude of metabolic activity change is impossible to be explained through the work of mitochondrial apparatus; therefore muscles can get the energy required for contraction even from the glycolysis process in cytoplasm and macroergs reserves accumulated in cells in the form of ATP and creatine phosphate. It forms a specific character of muscle tissue energetics. These specifics were expressed by Margaria (1963) in his conception of three energy sources for the muscle activity: aerobic (oxidative, mitochondrial); anaerobic glycolytic (lactacide); anaerobic phosphagenic (alactacide). In combination with the anaerobic threshold conception (Mader & Heck, 1986; Skinner, 1993), these presentations are now the theoretic base of muscle bioenergetics (Volkov, 2010).

Ample data prove an uneven development of aerobic and both anaerobic sources with age, like, for example, heterochrony determines a qualitative peculiarity of skeletal muscles energetics at separate stages of ontogenesis (Guminskiy et al., 1985; Demin, 1983; Kornienko, 1979; Kornienko et al., 2000, 2005; Son'kin, 1988; 2007; Tambovtseva, 2003; Van Praagh, Dore, 2002).

4.1 Aerobic (oxidative) source

Facilities of aerobic energy supply in skeletal muscles are considerably changed in the course of individual development. It is provided by both the change in content of most important oxidizing ferments in the somatic muscle tissue (Kornienko, 1979), and structural change in the oxidative chain of mitochondria (Demin, 1983, 1985; Demin et al., 1987; Son'kin & Tambovtseva, 2011). Naturally, the most important factor is absolute and relative age increase in mass of somatic muscles. Generally, the maximum oxygen consumption

(MOC) rises proportionally to the muscle mass growth (Kornienko et al., 2000), but it lacks information about qualitative changes in muscle energy supply of children and teenagers (Son'kin & Tambovtseva, 2011).

Age, years	Cytochrome **a** concentration, nM/g raw mass of muscle tissue	Muscle mass, kg (after: Holliday, 1971)	MOC, l/min (after: Åstrand, 1952)	MOC, ml/kg muscle mass
Newborn	0,9	0,6	-	-
5 - 7	4,6	6,7	1,01	151
9	5,2	10,5	1,8	171
11	6,6	11,6	2,1	181
14	4,8	21,2	3,5	165
20	4,5	25,0	4,1	164
36-40	3,7	28,3	3,9	138

Table 2. Age changes in cytochrome **a** content in thigh muscle and human aerobic capacity (After: Kornienko, 1979)

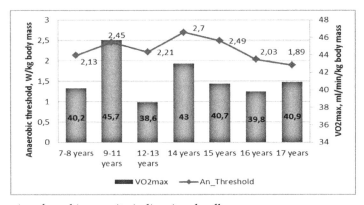

Fig. 5. Dynamics of aerobic capacity indices in schoolboys

Age dynamics of cytochrome **a** content – a terminal site of the oxidative chain – in human skeletal muscles (Kornienko, 1979; Demin et al., 1987) is given in Table 2. Calculations of the estimated value for specific MOC (per 1 kg of somatic muscle mass) are shown. As obvious from the given data, the highest cytochrome **a** concentration is registered in skeletal muscles for boys 9-11 years old. It is also proved by data of electron microscopic researches (Kornienko, 1979; Kornienko et al, 1987): the number of mitochondria in relation to the area of myofibrils for 11-year old boy is considerably more than in an adult man (Table 3). It is remarkable, that, according to data of morphologists, a capillary network in the muscles of extremities turns out to be more developed in children of 9-11 years (Topol'sky, 1951), which is the age when there is the highest content of oxidizing ferments in the muscle tissue. Thus, an age development of the energy production aerobic source in skeletal muscles does not happen monotonously, but gets the expressed maximum during prepubescence (Fig.5). These conditions have a considerable effect on the functioning of the muscle energy supply system.

Index	11-year old boy	35-year old man	Difference, %
Electron micro photos of somatic muscle lengthwise cuts (m. Quadriceps Femori)			
Mean diameter of mitochondria, micron	236	175	-35
Mean thickness of myofibrils, micron	505	590	+14
Ratio of mitochondria area to myofibril area	0,034	0,016	-113
Ratio of mitochondria total area to myofibril total area	0,153	0,097	-58

Table 3. Morphometric indices of skeletal muscles mitochondria in 11-year boy and adult man according to electron microscopy (After Kornienko, 1979; modified)

Such special energetic structure of skeletal muscles for children in prepubescence, as we see it, is caused by the fact that this ontogenesis period is the preparation for the radical reconstructing structural and functional characteristics of somatic muscles, occurring during pubescence under the influence of sex hormones. We used special histochemical tests to prove this (Tambovtseva, 2003).

4.2 Morpho-functional changes of skeletal muscles during postnatal ontogenesis

Fiber structure of mixed skeletal muscles is usually considered to be determined genetically and not dependent on age and training (Van Praagh & Dore, 2002; Yazvikov et al., 1978). But according to results of histochemical investigations, the ratio of various fiber types in the structure of skeletal muscles is not constant in ontogenesis (Kornienko et al., 2005; Son'kin & Tambovtseva, 2011; Tambovtseva & Kornienko, 1986a,b, 1987).

Research primarily made on laboratory animals – Wistar rats and Guinea pigs – made possible a conclusion that at an early age the most part of mammals is non-differentiated fibers which further acquires features of red oxidative fibers. The share of quick fibers is rapidly increased during pubescence, which become predominant after pubescence (Tambovtseva & Kornienko, 1986a; 1987).

These studies continued on post mortem material of males within the age bracket from birth to adulthood (Tambovtseva & Kornienko, 1986b). On Fig. 6 there are results characterizing the structure of large skeletal muscles in human extremities, achieved by the method of histochemical revelation of ATP myosin activity.

It turns out that all large human skeletal muscles are characterized by the same age tendency: undifferentiated embryo fibers are changed by fibers with a slow actomyosin ATP-ase, by 4-7 years an "aerobic profile" of mixed muscles is formed, which prevails up to 11-12 years. Then, with start of pubescence there is a two-phase transformation of the muscle structural-functional composition, which results in considerable reduction in a share of red oxidizing fibers (I type), some increase in a share of intermediate fibers (IIA type), and a considerable increase in presented glycolytic fibers (IIB type).

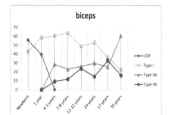
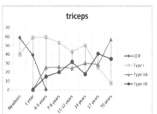
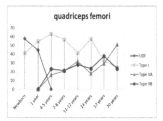

Fig. 6. Age changes of muscle fiber composition in men (% of each fiber type). By X-line – age from 4 months intrauterine development up to 70 years. By Y-line –fiber share (%%): MB undifferentiated, MB I type, MB IIA type, MB IIB type. UDF – undifferentiated fibers.

Only by 17-18 years is a definite picture formed, which is characterized by predominance of anaerobic – glycolytic fibers in all large muscles. Such muscle structure is likely to be preserved up to the start of involutive processes at the old age, which might be connected with decreased activity of genital glands.

It should be noted that in the world literature there is no definite view on the age development of somatic muscle structural components. The relatively scarce researches of age features in human skeletal muscles provide conflicting results (Blimkie & Sale, 1998; Van Praagh, 2000). According to some authors, there is a relatively large share of undifferentiated fibers at birth (10-20%). Amount of I-type fibers grows rapidly after birth, and II-type fibers reduces. By the age of one year the structure which is similar to adults is formed (Bell et al., 1980; Colling-Saltin, 1980; Elder & Kakulas, 1993). The ratio of IIA-type and IIB-type fibers is also disputable (Colling-Saltin, 1980; du Plessis et al., 1985; Jansson, 1996). But, according to other authors, children until pubescence are more characterized by I type than adults (Eriksson & Saltin, 1974; Lexell et al., 1992; Lundberg et al., 1979). According to Jansson (1996), development of muscles from birth up to 35 years old for men corresponds to ∩-model: from birth up to 9 years old a substantial increase in percent of I-type fibers is observed, at the age of about 9 there is a maximum, whereupon their share is reduced considerably by 19 years old. It is evident that this model is very close to results produced by the I.A.Kornienko laboratory.

4.3 Age changes of anaerobic metabolism ferments activity in muscles

In 1971 Swedish scientists demonstrated, by means of needle biopsy, that untrained boys at the age of 12 have a sharp (twofold) increase in the activity of phosphofructokinase (Eriksson et al., 1971). That was the first work where age changes in possible human anaerobic – glycolytic source at the tissue level were discovered. Next by means of biochemical (Eriksson, 1980; Ferretti et al., 1994; Kornienko et al, 1980; Ratel et al, 2002), histochemical (Tambovtseva, 2003; Kornienko et al, 2000) and physiological (Kornienko et al, 2000; Pyarnat & Viru, 1975; Son'kin, 1988;) methods numerous confirmations of an abrupt activation of anaerobic – glycolytic energy production in the process of pubescence reconstructions were obtained, especially for boys (Boisseau & Delamarche, 2000; Van Praagh & Dore, 2002).

In ontogenesis of rats an activity of glycolysis key ferment – lactate dehydrogenase (LDG) was traced in detail (Musaeva, 1986; Demin et al., 1987). LDG molecule consists of 4 monomers and each of them can have one of two following isoforms: "H" – subunits which are typical for LDG from a cardiac muscle, "M" – subunits which are mainly in skeletal muscles of adult mammals (Lehninger, 1965). These isoferment forms differently participate in a cycle of glycolysis reactions, therefore the ratio of "H" and "M" activities – LDG subunits can be used as a sufficiently informative activity ratio index for aerobic and anaerobic – glycolytic sources.

In these studies it was proven that an age increase in "aerobic" ferment activity in muscles of male rates occurs generally parallel to increasing facilities of the oxidizing source and is complete by the start of pubescence, while total LDG activity rapidly grows in pubescence and even after it. Therefore, in process of pubescence qualitatively reconstructed is organization of energy metabolism in somatic muscle cells: an abrupt extension of facilities for anaerobic – glycolytic energy production in terms of stabilization and even some decrease in a relative capacity of aerobic energy production.

The given facts have proposed an important role of sex hormones in regulation of muscle energetics. Direct evidence of this hypothesis was obtained by Musaeva (1986) in tests on male rates with orchotomy at the age of 3 weeks or artificial androgenization: androgenization accelerates and orchotomy inhibits the formation of ferment systems, which are responsible for anaerobic mechanisms producing energy in somatic muscles, and practically does not have an effect on conditions of mitochondria (aerobic) energy production. Under the influence of exogenous testosterone, the fraction of muscle fibers with a high activity of ATP-ase in the structure of extremities considerably increased, meaning those, which are mainly characterized by anaerobic energy supply. Orchotomy has the opposite results (Musaeva, 1986; Son'kin & Tambovtseva, 2011).

Probably, male sex hormones play a role that is not less significant in formation of a morphological–functional status of human skeletal muscles (Boisseau & Delamarche, 2000; Ferretti et al., 1994; Jansson, 1996; Round et al., 1999; Tambovtseva & Kornienko, 1986; Van Praagh & Dore, 2002). It is remarkable that for girls the same effects of pubescence processes on the structure and function of their skeletal muscles are not revealed (Petersen et al., 1999; Tambovtseva, 2003; Treuth et al., 2001), which can be explained by various structural – metabolic consequences of androgen and estrogen effects. For boys testosterone content is increased 4 times at primary stages of pubescence and more than 20 times – at its last stages. For girls the testosterone is only 4 times increased from the primary to the last stages of pubescence (Blimkie & Sale, 1998).

In literature there are no data on creatine phosphokinase (CrK) activity in human muscles with age. In recent years, with development of magnetic resonance research methods (Ross et al., 1992), data on creatine phosphate (CrP) content in muscles in rest as well as under physical load and recreation have been obtained (Zanconato et al., 1993; Ferretti et al., 1994;

Boisseau & Delamarche, 2000; Van Praagh & Dore, 2002). Data available in press are contradictory.

At that same time, dynamics of these indices for various tissues in ontogenesis of rats was studied by Demin in detail (1983, 1985). According to these results, CrK activity in muscles of animal hind extremities in a nest life period makes 2.5 – 2.7 μmol/min/g and it is practically unchanged in the first 2 weeks of life. At the same age CrK activity in cardiac tissue is somehow less than in leg muscles while in neck muscles, which perform the most thermoregulation function – it is 2 times more, than in leg muscles. With start of an active independent motion activity for young rats (3 weeks), CrK in leg muscles grows intensely reaching by pubescence (60 days) the level of 39 μmol/min/g. This is 4.5 times more than in neck and cardiac muscles, 12 times more than in brain tissue and 50 times more than in the liver.

For this period CrP content in muscles is approximately increased by 3 times - from 5.4 up to 15.9 μmol/g (in a 60-day rat's heart CrP content – 7.1 μmol/g; in brain – 5.7 μmol/g; in liver – 2.43 μmol/g).

CrP content and CrK activity are increased in skeletal muscles asynchronously with age and it provides age changes in potential duration of CrP expenditure at maximum activity. The most substantial increase in CrK activity is observed at the last stages of pubescence, which provides a considerable acceleration of ATP forming velocity in a creatine kinase reaction, meaning the capacity of an alactacide energy system. As a result, according to Demin (1985), ATP formation velocity in process of a creatine kinase reaction in muscles of rat extremities is increased from 20 μmol/g/min at the age of 12 days up to 80 μmol/g/min at the age of 40-45 days (an active phase of pubescence), and by the end of pubescence changes it grows up to 160 μmol/g/min.

The given facts prove that the pubescence period is a "divide" between two qualitative conditions in energetics of somatic muscles. Prior to pubescence changes in muscles, like in other mammal tissues, the predominant role in energy supply is played by mitochondria oxidation. After pubescence changes, muscles acquire that colossal functional range and those specific features of organizing energy metabolism, which differentiates them from other tissues of an adult body, and the role of anaerobic energy sources is rapidly increased (Kornienko et al., 2000). Such reconstruction of energetics in skeletal muscles allows after some time the increase of the realized capacity of outer mechanical production, considerably extending the functional range, as well as promoting repeated growth of efficiency and reliable body function under strenuous muscle activity (Kornienko & Son'kin 1999). However, it should be noted, that the data of Demin, like most other similar results, were obtained for male rats. Sex differences in dynamics of energy facilities for skeletal muscles are studied insufficiently. According to the results of research made by means of up-to-date methods (Boisseau & Delamarche, 2000; Petersen et al., 1999; Treuth et al., 2001; Van Praagh, Dore, 2002), for girls at pre- and post-pubescent age such considerable differences in energy metabolism structure under muscular load were not revealed.

5. Conclusion

Energy metabolism, presenting the most integral body function, demonstrates logical age changes reflecting qualitative and quantitative redevelopments of a child's organism. The principle of functional economy is likely to be the most vivid of these changes with age development. This principle is implemented in age-related reduction of basal metabolism, in slower rate of thermoregulatory reactions, decrease of food thermogenic effect with age, in change of daily energy expenditure structure with age. This is the structure of daily expenditure, where there are most vivid qualitative changes reflecting heterochronic development of most energy-intensive functions. If at an early age the energy metabolism priority is the brain and neural processes associated with it, with growth of the muscular system and formation of its functional facilities energy expenditure on kinesis starts taking a greater share in the daily energy balance.

This, together with the general tendency to fall in relative heat production, corresponding to the views of progressive functional economy in rising ontogenesis, with growth and development of skeletal muscles the maximum energy production is considerably increased, which is provided by activation of the least economic anaerobic – glycolytic source of energy production. In other words, by the example of age changes in energy metabolism the most important principle of development can be distinctly demonstrated – a principle of biological suitability, which is sometimes implemented due to the breach of other principles with lower value, to implement biosocial objectives of the corresponding ontogenesis stage.

For many decades the researchers of age changes in energy metabolism have been paying attention to reduction in the rate of exchange processes in rest with age. It was attempted to be explained by smaller relative surface of the body (Rubner, 1883), growth of relative muscular mass value (Arshavskii, 1967), lower relative mass of internal organs with a high rate of oxidizing metabolism (Holliday, 1971; Javed et al., 2010; Kornienko, 1979; Wang et al., 2010). But now we think that the most important age redevelopment is the combined reduction of basal and increased maximum energy expenditure (including expenditure pursuant to anaerobic ways to transform energy with realization of the intense muscle activity), which results in the considerable development of a functional range. That is the biological development objectives of energy production mechanisms, as a vast functional range provides implementation of a wide spectrum of social and biological problems facing the adult organism (Son'kin & Tambovtseva, 2011).

Theoretical views of the laws valid for age changes in energy metabolism can considerably effect the formation and implementation of practical methods and means, firstly in such directions as conditioning to the cold, organization of proper nutrition and rational physical activity of children and teenagers.

6. Acknowledgments

The authors express their sincere gratitude to the colleagues with whom they have for years been investigating age-related changes of energy of the body: Dr. I. Gohblit, Dr. V. Demin, Dr. G. Maslowa and Dr. Z. Musaeva.

We would like to honour the memory of our teacher Professor I.A. Kornienko, whose ideas and experiments formed the basis of this paper.

7. References

Arshavskii, I. (1967). *Ocherki po vozrastnoi fiziologii [Essays on the physiology of age]*, Medicina, Moscow, USSR (rus.)

Astrand, P.-O. (1952). *Experimental studies of physical working capacity in relation to sex and age*, Munksgaard, Copenhagen, Danmark

Bell, RD, MacDougall, JD, Billeter, R, et al. (1980) Muscle fiber types and morphometric analysis of skeletal muscle in six-year-old children. *Med Sci Sports Exerc, Vol.*12, pp. 28-31

Blimkie, C. & Sale, D. (1998). Strength development and trainability during childhood. In: *Pediatric anaerobic performance,* Ed. Van Praagh, E., pp. 193-224, Human Kinetics, Champaign (IL), USA

Boisseau, N, Delamarche, P. (2000) Metabolic and hormonal responses to exercise in children and adolescents. *Sports Med., Vol.* 30, No.6, pp. 405-22.

Bosy-Westphal, A., Kossel, E., Goele, K., Later, W., Hitze, B., Settler, U., Heller, M., Glüer, C.-Ch., Heymsfield, S.B., & Müller, M. J (2009) Contribution of individual organ mass loss to weight loss–associated decline in resting energy expenditure. *Am J Clin Nutr, Vol.*90, pp.993–1001, ISSN 1938-3207

Bruck, K. (1970). Heat production and temperature regulation. In: *Physiology of perinatal period,* pp. 493–557, N. Y., USA

Colling-Saltin, A-S. (1980). Skeletal muscle development in the human fetus and during childhood. In: *Children and exercise,* Berg, K., Eriksson, B., editors, pp. 193-207, University Park Press, Baltimore (MD), USA

Conrad, M. & Miller, A. (1956) Age changes in body size, body composition and basal metabolism. *Amer.J.Physiol.,* Vol. 186, pp.207-210, ISSN 0002-9513

Demin, V. (1983). [Indicators of the mitochondrial respiratory chain, anaerobic glycolysis and creatine kinase system of skeletal muscles in ontogenesis]. In: *Osobennosti razvitiya fiziologicheskih sistem shkol'nika [On the evolution of physiological systems of schoolchildren],* pp. 77-82, APN SSSR, Moscow, USSR (rus.)

Demin, V.I. (1985) [Age-related changes of creatine kinase system]. In: *Novye issledovaniya po vozrastnoi fiziologii [New research on the physiology of age],* No.1(24), pp. 39-43, Pedagogika, Moscow, USSR (rus.)

Demin, V., Kornienko, I., Maslova, G. et al. (1987). [Peculiarities of Energetic Metabolism Organization in Different Organs]. In: *Molekulyarnye mehanizmy i regulyaciya energeticheskogo obmena [Molecular Mechanisms and Regulation of Energy Metabolism],* pp. 174-183, Pushino, USSR, (rus)

du Plessis, M., Smit, P., du Plessis, L., et al. (1985). The composition of muscle fibers in a group of adolescents. In: *Children and exercise XI,* Binkhorst, R., Kemper, H., & Saris, W., editors, pp. 323-328, Human Kinetics Publishers, Champaign (IL), USA

Elder, G.C. & Kakulas, B.A. (1993) Histochemical and contractile property changes during human development. *Muscle Nerve, Vol.* 16, No.11, pp. 1246-1253, ISSN 0148-639X

Eriksson, B.O., Karlsson, J. & Saltin, B. (1971) Muscle metabolites during exercise in pubertal boys. *Acta Paediatr Scand Suppl;* Vol. 217, pp. 154-157, ISSN 0300-8843

Eriksson, B.O. & Saltin, B. (1974) Muscle metabolism during exercise in boys aged 11 to 16 compared to adults. *Acta Paediatr Belg,* Vol. 28 Suppl., pp. 257-65, ISSN 0001-6535

Eriksson, B.O. (1980) Muscle metabolism in children--a review. *Acta Paediatr Scand Suppl.,* Vol.283, pp.20-28, ISSN 0300-8843

Falk, B. (1998) Effects of thermal stress during rest and exercise in the paediatric population. *Sports Med., Vol.* 25, No.4, pp. 221-40, ISSN 0112-1642

Farber, D. & Machinskaya, R. (2009). [Functional organization of the brain in ontogenesis and its reflection in the electroencephalogram of peace]. In: *Razvitie mozga i formirovanie poznavatel'noi deyatel'nosti rebenka. [The development of brain and cognitive development of children]*, Ed. Farber, D.A. & Bezrukih, M.M., pp. 76-118, Izdatel'stvo moskovskogo psihologo-social'nogo instituta, Moscow, Russia, ISBN 978-5-9770-0361-2 (rus.)

Ferretti, G., Narici, M.V., Binzoni, T., Gariod, L., Le Bas, J.F., Reutenauer, H. & Cerretelli, P. (1994) Determinants of peak muscle power: effects of age and physical conditioning. *Eur J Appl Physiol Occup Physiol.; Vol.* 68, No. 2, pp.111-115, ISSN 0301-5548

Forbes, E. & Swift, R. (1944) Associative dynamic effects of proteins, carbohydrate and fat. *Science*, Vol.99, pp.476-478, ISSN 0036-8075

Frankenfield, D., Roth-Yousey, L. & Compher, C. (2005) Comparison of predictive equations for resting metabolic rate in healthy nonobese and obese adults: a systematic review. *J Am Diet Assoc. Vol.*105, No. 5, pp.775-789. ISSN 0002-8223

Garrel, D.R., Jobin, N. & de Jonge, L.H. (1996) Should we still use the Harris and Benedict equations? *Nutr Clin Pract.* Vol. 11, No. 3, pp. 99-103, ISSN 0884-5336

Godina, E.Z. (2009) The secular trend: history and prospects. Human Physiology, Vol. 35, No. 6, pp. 770-776, ISSN 0362-1197

Gohblit, I.I., Bogachev, V.N. & Kornienko, I.A. (1975) [Thermoregulatory responses in children during the first months of life]. *Fiziologiya cheloveka,* Vol. 1, No.4, pp.541-548. (rus.) ISSN 0131-1646

Guminskii, A.A., Tupitsina, L.P. & Feoktistova, S.V. (1985) [Age characteristics of energy metabolism in girls during puberty]. *Fiziologiya cheloveka,* Vol. 11, No.2, pp. 286-292. (rus.), ISSN 0131-1646

Habas, M.E. & Macdonald, I.A. (1998) Metabolic and cardiovascular responses to liquid and solid test meals. *Br J Nutr,* Vol. 79, No. 3, pp. 241-247, ISSN 0007-1145

Harris, JA.; Benedict, FG. (1919). *A biometric study of basal metabolism in man,* Carnegie Institute of Washington, Washington, DC, USA

Hayter, J.E. & Henry, C.J. (1994) A re-examination of basal metabolic rate predictive equations: the importance of geographic origin of subjects in sample selection. *Eur J Clin Nutr.* Vol. 48, No. 10, pp. 702-707, ISSN 0954-3007

Himms-Hagen, J. (1989) Role of thermogenesis in the regulation of energy balance in relation to obesity. *Can J Physiol Pharmacol,* Vol. 67, No. 4, pp. 394-401, ISSN 0008-4212

Holliday, M. (1971) Metabolic rate and organ size during growth from infancy to maturity and during late gestation and early infancy. *Pediatrics,* Vol. 47, Pt. 2. pp. 169–179.ISSN 0031-4005

Ivanov, K. (1990). *Osnovy energetiki organizma. Teoreticheskie i prakticheskie aspekty. [Basis of Organism Energetic. Theoretical and applied aspects] Vol.1. Obshaya energetika, teploobmen i termoregulyaciya [General Energetic, Thermal Turnover and Thermoregulation].* Nauka, Leningrad, USSR, ISBN: 978-5-02-026169-3 (rus)

Jansson, E. (1996). Age-related fiber type changes in human skeletal muscle. In: *Biochemistry of exercise IX.* Maughan, RJ, Shirreffs, SM, editors, pp. 297-307, Human Kinetics, Champaign (IL), USA

Javed, F., He, Q., Davidson, L.E., Thornton, J.C., Albu, J., Boxt, L., Krasnow, N., Elia, M., Kang, P., Heshka, S., & Gallagher, D. (2010) Brain and high metabolic rate organ mass: contributions to resting energy expenditure beyond fat-free mass. *Am J Clin Nutr,* Vol. 91, pp. 907–912, ISSN 1938-3207

Karlberg, P. (1952) Determination of standard energy metabolism (basal metabolism) in normal infant. *Acta pediat.* (Uppsala), Vol. 41 (suppl.83), pp. 3-151, ISSN 0803-5253

Kassirskii, I. (1934). *Osnovnoi obmen i ego klinicheskoe znachenie. [Basal Metabolism and it's Clinical Significance].* Gosizdat, Sredneaziat. Otdelenie, Moskva-Tashkent, USSR (rus)

King, J.C., Butte, N.F., Bronstein, M.N., Kopp, L.E. & Lindquist, S.A. (1994) Energy metabolism during pregnancy: influence of maternal energy status. *Am J Clin Nutr.,* Vol. 59 (2 Suppl), pp. 439-445, ISSN 1938-3207

Kleiber, M. (1961). *The Fire of Life.* John Wiley and Sons, Inc., New York, London, USA, GB

Korkushko, O.V. & Orlov, P.A. (1974) [Calorygenic action of Glucose in Humans of Different Age]. *Voprosy pitaniya,* №1, pp. 54-58. (rus)

Kornienko, I.A. (1979). *Vozrastnye izmeneniya energeticheskogo obmena i termoregulyacii. [Age Development of Energy Metabolism and Thermoregulation].* Nauka, Moscow, USSR (rus)

Kornienko, I.A. & Gohblit, I.I. (1983) [Age-related conversion of energy metabolism]. In: *Fiziologiya razvitiya rebenka [Physiology of Child Development],* pp. 89-114, Pedagogika, Moscow, USSR (rus.)

Kornienko, I.A., Son'kin, V.D. & Urakov T.U. (1984). Calorigenic action of glucose in schoolchildren. *Hum Physiol.* Vol. 10, No. 4, pp. 276-82 ISSN 0362-1197

Kornienko, I.A., Demin, V.I., Maslova, G.M. & Son'kin, V.D. (1987) [Development of skeletal muscle energetics]. In: *[Proceedings of the XV All-Union Congress of Physiologists],* Vol. 2, P.166, Leningrad, USSR, June 1987 (rus.)

Kornienko, I.A., Gohblit, I.I. & Son'kin, V.D. (1988) [Characterization of energy metabolism]- In.: *Fiziologiya podrostka [Physiology of teenager. Ed. Farber, D.],* pp.71-93, Pedagogika, Moscow, USSR (rus.)

Kornienko, I.A. & Son'kin, V.D. (1999) "Biological Reliability," Ontogeny, and Age-Related Dynamics of Muscular Efficiency. *Human Physiology,* Vol.25, No.1, pp.83-92, ISSN 0362-1197

Kornienko, I.A., Son'kin, V.D., Tambovtseva, R.V., Bukreeva, D.P. & Vasil'eva, R.M. (2000) [Age-related development of skeletal muscle and exercise performance] In: *Fiziologiya razvitiya rebenka: teoreticheskie i prikladnye aspekty [Physiology of Child Development: Theoretical and Applied Aspects],* pp.209-238, Obrazovanie ot A do Ya, Moscow, Russia (rus.)

Kornienko, I.A., Son'kin, V.D. & Tambovtseva, R.V. (2005) Development of the Energetics of Muscular Exercise with Age: Summary of a 30-Year Study: I. Structural and Functional Rearrangements. *Human Physiology,* Vol. 31, No.4, pp. . 402-406, ISSN 0362-1197

Lehninger, A.Z. (1965) *Bioenergetics. The molecular basis of biological energy transformations.* Benjamin, New-York, USA, Amsterdam, Holand

Lexell, J., Sjöström, M. & Nordlund, A-S. (1992) Growth and development of human muscle: a quantitative morphological study of whole vastus lateralis from childhood to adult age. *Muscle Nerve,* Vol. 15, pp. 404-409, ISSN 0148-639X

Lundberg, A., Eriksson, B.O. & Mellgren, G. (1979) Metabolic substrates, muscle fibre composition and fibre size in late walking and normal children. *Eur J Pediatr,* Vol. 130, pp. 79-92, ISSN 1432-1076

Lusk, G. (1919) Calorigenic cosporet de l'ingestion de viande decide lacticue et datanine chez l'anisal. *Compt. Rend. Acad. D.sc.*, Paris, Vol. XVIII, pp.1012-1015.

Macler, B., Grace, R. & Duncan, H. (1971) Studies of mitochondrial development during embryogenesis in the rat. *Arch. Biochem. and Biophysics*, Vol. 144, pp.603-610, ISSN 0003-9861.

Mader, A. & Heck, H. (1986) A theory of the metabolic origin of "Anaerobic threshold". *Int. J. Sports Med*, Vol. 7, Suppl., pp. 45-65, ISSN 1439-3964

Mahin'ko, V.I. & Nikitin, V.N. (1975) Obmen veshestv i energii v ontogeneze. [Substrate and energy metabolism in ontogeny] In: *Rukovodstvo po fiziologii. Vozrastnaya fiziologiya [Guide to Physiology. Developmental physiology]*, pp. 249-266, Nauka, Moscow, URSS (rus.)

Margaria, R. (1963) Biochemistry of muscular contraction and recovery. *J.Sports Med .and Physical Fitness*, Vol. 168, No. 3, pp. 145-156, ISSN 0022-4707

McDuffie, J.R., Adler-Wailes, D.C., Elberg, J., Steinberg, E.N., Fallon, E.M., Tershakovec, A.M., Arslanian, S.A., Delany, J.P., Bray, G.A., & Yanovski, J.A. (2004) Prediction equations for resting energy expenditure in overweight and normal-weight black and white children. *Am J Clin Nutr*. Vol. 80, No. 2, pp.365–373, ISSN 1938-3207

Morgan, J. (1980) The pre-school child: diet, growth and obesity. *J Hum Nutr*, Vol. 34, No. 2, pp. 117-130, ISSN 0308-4329

Musaeva, Z.T. (1986) [Changes in the activity of lactate dehydrogenase and creatine kinase in skeletal muscle during puberty]. In: *Novye issledovaniya po vozrastnoi fiziologii [New research on the physiology of age]*, N 2, pp. 14-17, Pedagogika, Moscow, USSR (rus.)

Nagornyi, A.V., Nikitin, V.N. & Bulankin, I.N. (1963) *Problema stareniya i dolgoletiya [The problem of aging and longevity]*, Nauka, Moscow, USSR (rus.)

Nedergaard, J. & Cannon, B. (2010) The changed metabolic world with human brown adipose tissue: therapeutic visions. *Cell Metab, Vol.* 11, No. 4, pp. 268-272, ISSN 1932-7420

Nedergaard, J., Bengtsson, T. & Cannon, B. (2007) Unexpected evidence for active brown adipose tissue in adult humans. *Am J Physiol Endocrinol Metab*, Vol. 293, pp. E444–E452, ISSN 1522-1555

Petersen, S.R., Gaul, C.A., Stanton, M.M. & Hanstock, C.C. (1999) Skeletal muscle metabolism during short-term, high-intensity exercise in prepubertal and pubertal girls. *J Appl Physiol*. Vol. 87, No. 6, pp. 2151-2156, ISSN 1522-1601

Pyarnat, Ya.P. & Viru, A.A. (1975) [Age peculiarities of physical (aerobic and anaerobic) capacity]. *Fiziologiya cheloveka*, Vol. 1, No. 4, pp.692-696 (rus.) ISSN 0131-1646

Ratel, S., Bedu, M., Hennegrave, A., Dore, E. & Duche, P. (2002) Effects of age and recovery duration on peak power output during repeated cycling sprints. *Int J Sports Med*. Vol. 23, No. 6, pp. 397-402, ISSN 1439-3964

Robinson, S. (1938) Experimental studies of physical fitness in relation to age. *Arbeitsphysiol*. Vol. 10, No. 3, pp.251-323

Ross, B., Kreis, R. & Ernst, T. (1992) Clinical tools for the 90s: magnetic resonance spectroscopy and metabolite imaging. *Eur J Radiol*. Vol. 14, No. 2, pp. 128-40, ISSN 1872-7727

Round, J.M., Jones, D.A., Honour, J.W., & Nevill A.M. (1999) Hormonal factors in the development of differences in strength between boys and girls during adolescence: a longitudinal study. *Ann Hum Biol, Vol.* 26, No.1, pp. 49-62, ISSN 0301-4460

Rubner, M. (1883). Über den einfluss der körpergrösse auf stoff- und kraftwechsel. *Z. Biol.* Vol. 19, pp. 536-562.

Schmelzle, H., Schroder, C., Armbrust, S., Unverzagt, S. & Fusch, C. (2004) Resting energy expenditure in obese children aged 4 to 15 years: measured versus predicted data. *Acta Paediatr.* Vol. 93, No. 6, pp. 739-746, ISSN 1651-2227

Schmidt-Nielsen, K. (1984) *Scaling. Why is animal size so important.* Cambridge University Press, Cambridge, England

Shmal'gauzen, I.I. (1935) Rost i differencirovka. [Growth and differentiation] - In: *Rost zhivotnyh. [Growth of animals],* Ed. by Mickiewicz, M.S. pp. 74-84, Biomedgiz, Moscow, USSR (rus.)

Skinner, J.S. (Ed.) (1993) *Exercise testing and exercise prescription for special cases: theoretical basis and clinical application.* Lea & Febiger, Philadelphia, USA

Son'kin, V.D., Urakov, T.U., Pavlov, Yu.M. & Deduhova V.I. (1975) [Use of glucose load to characterize energy metabolism in children of school age]. In: *Novye issledovaniya po vozrastnoi fiziologii,* No. 2 (5), pp.58-60. Academy of Pedagogical Sciences USSR, Moscow (rus.)

Son'kin, VD. (1988) [Development of energy support for muscle activity in adolescents] *Fiziol Cheloveka.* Vol. 14, No. 2, pp. 248-255 (rus), ISSN 0131-1646

Son'kin, V.D. (2007) Physical working capacity and energy supply of muscle function during postnatal human ontogeny. Human Physiology, Vol. 33, No. 3, pp. 326-341, ISSN 0362-1197

Son'kin, V.D., Kirdin, A.A., Andreev, R.S., Akimov, E.B. (2010) Homeostatic nonshivering thermogenesis in Humans. Facts and Hypotheses. Human Physiology, Vol. 36, No. 5, pp. 599-614, ISSN 0362-1197

Son'kin, V.D. & Tambovtseva, R.V. (2011) *Razvitie myshechnoi energetiki i rabotosposobnosti v ontogeneze. [Development of muscle energetics and working capacity during ontogenesis].* Knizhnyi dom «LIBROKOM», Moscow, Russia, ISBN 978-5-397-01708-4 (rus.)

St-Onge, M.-P. & Gallagher, D. (2010) Body composition changes with aging: The cause or the result of alterations in metabolic rate and macronutrient oxidation? *Nutrition.* Vol. 26, No. 2, pp. 152–155, ISSN 1873-1244

Tambovtseva, R.V. & Kornienko I.A. (1986a) [Development of various types of muscle fibers in soleus muscle of rat postnatal ontogenesis]. *Arhiv anatomii, gistologii i embriologii. [Archive of Anatomy, Histology and Embryology],* Vol. 90, No. 1, pp. 77-81 (rus.), ISSN 0004-1947

Tambovtseva, R.V. & Kornienko, I.A. (1986b) [Development of various types of muscle fibers in the quadriceps femori and the soleus during human ontogenesis] *Arkh Anat Gistol Embriol.* Vol. 91, No. 9, pp. 96-99 (rus.), ISSN 0004-1947

Tambovtseva, R.V. & Kornienko, I.A. (1987) [Development of different types of muscle fiber in the postnatal ontogeny of guinea pig]. *Arhiv anatomii, gistologii i embriologii. [Archive of Anatomy, Histology and Embryology],* Vol. 93, No.7, pp.55-59. (rus.), ISSN 0004-1947

Tambovtseva, R.V. (2003) *Vozrastnye i tipologicheskie osobennosti energetiki myshechnoi deyatel'nosti. [Age and typological features of muscular activity energetics].* Thesis of Dissertation...Doc. Biol. Sci., Institute for Developmental Physiology Russian academy of education. Moscow, Russia (rus.)

Topol'skii, V.I. (1951) [Age peculiarities of circulation at shoulder muscles in humans]. In: *Nauch. trudy Krasnoyarskogo med. in-ta,* No.2, pp. 275-277. (rus.)

Treuth, M.S., Butte, N.F. & Herrick, R. (2001) Skeletal muscle energetics assessed by (31)P-NMR in prepubertal girls with a familial predisposition to obesity. *Int J Obes Relat Metab Disord.* Vol. 25, No. 9, pp.1300-1308, ISSN 0307-0565

Tsehmistrenko, T.A., Vasil'eva, V.A., Shumeiko, N.S. & Chernyh, N.A. (2009) [Structural changes of the cerebral cortex and cerebellum in human postnatal ontogenesis]. In: *[The development of brain and cognitive development of children].* Ed. Farber, D.A., Bezrukih, M.M., pp. 9-75, Izd. Mos. Psih.-social. Inst.; Moscow-Voronezh, Russia ISBN 978-5-9770-0361-2 (rus.)

Tverskaya, R., Rising, R., Brown, D. & Lifshitz, F. (1998) Comparison of several equations and derivation of a new equation for calculating basal metabolic rate in obese children. *J Am Coll Nutr, Vol.* 17, No. 4, pp. 333-336, ISSN 1541-1087

Ugolev, A.M., Efimova, N.V. & Skvortsova, N.B. (1976) [Functions of the intestinal hormonal (enterin) system]. *Usp Fiziol Nauk,* Vol. 7, No. 3, pp. 6-31 (rus.) ISSN 0301-1798

Van Praagh, E. & Dore, E. (2002) Short-term muscle power during growth and maturation. *Sports medicine,* Vol. 32, No. 11, pp. 701-728, ISSN 0112-1642

Van Praagh, E. (2000) Development of anaerobic function during childhood and adolescence. *Pediatr Exerc Sci,* Vol. 12, No. 2, pp. 150-173, ISSN 1543-2920

Vermorel, M., Lazzer, S., Bitar, A., Ribeyre, J., Montaurier, C., Fellmann, N., Coudert, J., Meyer, M. & Boirie, Y. (2005) Contributing factors and variability of energy expenditure in non-obese, obese, and post-obese adolescents. *Reprod Nutr Dev.* Vol. 45, No. 2, pp. 129-142, ISSN 0926-5287

Volkov, N.I. (2010) *Bioenergetics of sports activities.* Theory and Practice of Physical Culture and Sports, Moscow, Russia, ISBN 978-5-93512-054-2

Wang, Z., Heymsfield, S.B., Ying, Zh., Pierson, R.N. Jr., Gallagher, D. & Gidwani, S. (2010) A Cellular Level Approach to Predicting Resting Energy Expenditure: Evaluation of Applicability in Adolescents. *Am J Hum Biol.* Vol. 22, No. 4, pp. 476–483, ISSN 1520-6300

White, C.R. & Seymour, R.S. (2005) Allometric scaling of mammalian metabolism. *The Journal of Experimental Biology,* Vol. 208, Pt. 9, p. 1611-1619, ISSN 0022-0949

Yazvikov, V.V., Sergeev, Yu.P., Nikitina, T.V. & Bashkirov, V.F. (1978) [Histochemical characteristics of muscle fibers of different types in the performance of an untrained person intense muscular work] In: *[Proceedings of the XV All-Union scientific conference for Physiology and Biochemistry of Sport],* P.190, Moscow, USSR, November 1978 (rus.)

Young, V.R., Yu, Y.M. & Fukagawa, N.K. (1991) Protein and energy interactions throughout life. Metabolic basis and nutritional implications. *Acta Paediatr Scand Suppl.,* Vol. 373, pp. 5-24, ISSN 0300-8843

Zanconato, S., Buchthal, S., Barstow, T.J., Cooper, D.M. (1993) 31P-magnetic resonance spectroscopy of leg muscle metabolism during exercise in children and adults. *J Appl Physiol.,* Vol. 74, No. 5, pp. 2214-2218, ISSN 1522-1601

Part 2

Reviews of Bioenergetics Applied to Performance Optimization

7

Bioenergetics Applied to Swimming: An Ecological Method to Monitor and Prescribe Training

Rodrigo Zacca and Flávio Antônio de Souza Castro
Universidade Federal do Rio Grande do Sul
Brazil

1. Introduction

Systematic assessments of athletes' physiological conditions are central to monitor and prescribe swimming training according to the needs and goals. Thus, it is possible to understand the current physiological state and follow its development in order to assess the effects of training, to identify the swimmer's skills profile and to predict athletic performance (Vilas-Boas & Lamares 1997). Specifically regarding swimmers and their skills, aerobic capacity is a major determinant of these athletes performance, and it is defined as the ability to maintain a high percentage of maximal oxygen uptake (VO_{2max}) for a long period of time (DI PRAMPERO et al., 2011). Furthermore, the endurance is influenced by VO_{2max}, swimming economy (or energy cost, defined as the total energy expenditure required to move the body to a certain distance in a determined velocity) and anaerobic capacity (Dekerle & Pelayo 2011). In a group of swimmers with similar values of swimming economy and anaerobic capacity, those with greater aerobic potential (VO_{2max} and aerobic capacity) will be faster at distances of 400 m and longer. Four hundred meters, when swimming in front *crawl*, is usually suggested as a trial in which VO_{2max} is reached (Dekerle & Pelayo 2011). Thus, the longer events (800 m, 1500 m and open water marathon), which are covered primarily with energy from aerobic metabolism, are covered in a fraction of the VO_{2max}. The intensity will be lower the longer is the distance, reaching 60-65% of VO_{2max} on the 25 km open water marathon (Zamparo et al. 2005). In this sense, one of the objectives of the swimming training is to increase the aerobic capacity. Thus, a valid and reliable measure of the swimmer aerobic profile is essential to verify the benefits that the training program is or is not providing, and, also, to set training intensities according to the physiological profile of the athlete. Dekerle & Pelayo (2011) emphasize that the methodology used for this purpose cannot be considered valid unless it is reliable. Whenever possible, the degree of reliability should be assessed. The origin of the variability measurement (human error, equipment error, biological variation, or motivational factors when performing the test) needs to be taken into account. Thus, the aim of this chapter is to present a careful review of the bioenergetics contribution on the physiological assessment of the swimmer, especially related to aerobic profile.

2. Critical velocity (CV)

The performance achieved in competitions is an important setting information from training sessions in swimmers (Sweetenham & Atkinson, 2003). However, constant evaluations are necessaries during the cycles and training sessions in order to verify the effectiveness of training and ensure the best performance in the competition (Sweetenham & Atkinson, 2003). Physiological and biomechanical swimmers conditions' knowledge is crucial to implement and/or to control the training processes that surround them (Pyne et al. 2001). These assessments can be applied in the field of competitive and / or recreational swimming. Tests used to evaluate and determine swimming speeds (SS) for the development of aerobic endurance training can be divided into invasive and noninvasive (Pyne et al. 2001), based on the relationship between oxygen consumption (VO_2), blood lactate concentration ([La]), heart rate (HR) and SS (Vilas-Boas & Lamares 1997). Although the precision provided by some of these tests, which require invasive sampling, such as those using the [La], ethical conflicts may arise (Heck et al. 1985), especially when applied to children. Moreover, it is common a high number of athletes to be evaluated in a training session by only one coach, so that they may require a longer period for implementation. Another limiting factor is the high cost for each testing session (Heck et al. 1985).

Considering these difficulties, the tests that verify the SS in durations of 30 (T_{30}) and 60 (T_{60}) minutes (Olbrecht et al. 1985; Madsen 1982) or even over distances of 2000 m (T_{2000}) (Touretski 1993) and 3000 m (T_{3000}) (Madsen 1982), the perceived exertion (PE) (Lima et al. 2006), the critical velocity (CV) (Ettema 1966) and 400 m testing (T_{400}) (Wakayoshi et al. 1993a; Dekerle et al. 2006; Alberty et al. 2006; Pelayo et al. 2007) have been widely disseminated in swimming. However, T_{30}, T_{60}, T_{2000} and T_{3000} can provide very subjective information to determine training intensities in young and/or low level of experience swimmers. These protocols require the maintenance e of a given SS for a long time require psychological and physiological capacity compatible with the demands of the test (Zacca & Castro 2008, 2009). Regarding the PE, the athlete needs good training base to swim extensive sets with minimal adjustments in intensity between each repetition (Zacca & Castro 2008, 2009). In this sense, determination of SS for swimming training through the CV (Dekerle et al. 2006; Greco et al. 2008; Leclair et al. 2008; Vandewalle et al. 2008) seems to correspond to these swimmers profiles. CV's use is also justified due to the low cost and facility to apply in various populations. Another advantage is that CV is able to be gotten even during competitions (Vilas-Boas & Lamares 1997).

Since Hill (1927), it is accepted that the relationship between power output and time to exhaustion is a hyperbole. The asymptote of this relationship of power (critical power or PC) is equivalent to the slope of the regression line related to the work and time to exhaustion (time limit or *tlim*) (Monod & Scherrer 1965). Since then, CP represents, at least theoretically, the largest power that could be sustained, whose energy would be derived preferably by the aerobic metabolism without fatigue, and is suggested as a good performance index in events of long duration (Vandewalle et al. 1997).

Ettema (1966) applied the CP concept in cyclists, swimmers, speed skaters and runners. Instead of power and work, the author used speed (S) and distance limit (*dlim*), respectively. The hyperbolic relationship between S and *tlim* (Hill 1927) and the linear relationship between *dlim* and *tlim* (Equation 1), usually called critical velocity (CV), have the same physiological meaning of CP (Pepper et al. 1992; Housh et al. 2001).

In Equation 1, the slope of the regression line corresponds to CV (obtained through a two-parameter model, CV_{2par}), the y-intercept (second parameter) is mathematically defined as a finite stock of reserve power available pre-exercise (Ettema 1966; Wakayoshi et al. 1992), usually referred as "anaerobic distance capacity" (ADC_{2par}).

$$\mathrm{dlim} = CV_{2par} \cdot \mathrm{tlim} + ADC_{2par} \tag{1}$$

The non-linear SS-time limit to exhaustion ("SS-*tlim*"), the linear relationship between distance limit and time limit (*dlim-tlim*) and the linear relationship between SS and the inverse of *tlim* (Equation 2) are two-parameter models commonly used to estimate the VC (Billat et al. 1999; Housh et al. 2001; Whipp et al. 1982).

$$SS = \frac{ADC_{2par}}{\mathrm{tlim}} + CV_{2par} \tag{2}$$

Equation 2 shows that the CV can be obtained by expressing SS as a function of *tlim* (Ettema 1966). In order to revise the statement that in the hyperbolic model SS is infinite when time approaches zero, Morton (1996) proposed a mathematical model including an additional parameter representing the maximum instantaneous velocity (V_{max} obtained from a three-parameter model, $V_{max3par}$).

$V_{max3par}$ allows a time asymptote (*tlim*) which is below the x-axis where *tlim* is zero, thus providing a V_{max} in the y-intercept (Morton 1996). Equation 3 expresses SS as a function of *tlim* (Zacca et al. 2010; adapted from Morton 1996).

$$SS = \frac{ADC_{3par}}{\mathrm{tlim} + \dfrac{ADC_{3par}}{V_{max3par} - CV_{3par}}} + CV_{3par} \tag{3}$$

Where SS is the swimming speed, *tlim* is the time limit and ADC_{3par}, $V_{max3par}$ and CV_{3par} are the parameters. The fact that two-parameter model assumes that there is no upper limit for power output or SS (Morton et al. 1996; Dekerle et al. 2006) leads some authors choose three-parameter models (Gaesser et al. 1995; Bull et al. 2000; Hill et al. 2003).

However, both (two and three parameters) models have an important limitation: they do not take into account the "aerobic inertia" (τ) (Wilkie 1980; Vandewalle et al. 1989), regarding to the cardio respiratory adjustments for the VO_2 reaches the steady state or maximum value. Thus, a four-parameter model (CV_{4par}, CDA_{4par}, $V_{max4par}$ and τ) as proposed by Zacca et al. (2010) could provide more information on bioenergetics in sports (Equation 4).

$$SS = \frac{ADC_{4par}}{\mathrm{tlim} + \dfrac{ADC_{4par}}{V_{max4par} - CV_{4par}\left(1 - e^{-\frac{\mathrm{tlim}}{\tau}}\right)}} + CV_{4par}\left(1 - e^{-\frac{\mathrm{tlim}}{\tau}}\right) \tag{4}$$

Zacca et al. (2010) proposed to plot *tlim* and SS values using a four-parameter model (Equation 4). The CV was corrected on this model by an exponential factor, proposed by Wilkie (1980). This exponential factor represents the time constant of the increased aerobic involvement, called "aerobic inertia" (τ), understood as a temporary delay in the response of VO_2, caused by dissociation of O_2 absorbed in lungs and used especially by skeletal muscle. The use of CV in swimming training is suggested since 1966 (Ettema 1966). Studies by researchers about its use continue to be published (Dekerle & Pelayo 2011).

3. Intensity domains (training zones)

Some authors (Gaesser et al. 1996; Greco et al. 2008) suggest a range of intensities of three domains (sometimes referred as training zones) and others (Dekerle & Pelayo 2011) a scale of five domains and their physiological effects. According to Table 1, exercise can be conducted in three different intensity domains, resulting in very distinctive physiological effects in each of these domains (Gaesser et al. 1996; Greco et al. 2008).

Intensity domains	Effects
SEVERE	• There is no variable metabolic stabilization; • Accumulation and increase of lactate / pyruvate relationship and increase of protons concentration [H+]; • VO_2 increases toward the maximum.
HEAVY	• [La] stabilizes at high values of concentration • The efficiency appears to be lower; • High VO_2 values (development of a slow component); • It is still possible to maintain a stable physiological state and perform the exercise for a longer period.
MODERATE	• [La] stabilizes quickly; • VO_2 has a quick adjustment; • The individual can maintain this intensity for hours without exhaustion.

Table 1. Intensity domains and their physiological effects (Gaesser et al. 1996; Greco et al. 2008)

3.1 Moderate intensity domain
[La] stabilizes quickly and can be maintained almost similar to resting levels. Similarly, VO_2 shows a quick set (1-3 min) before stabilization, and the individual can maintain the intensity for hours without exhaustion. The main explanation for the "end" of the exercise refers to substrate depletion (muscle and liver glycogen), changes related to hydration and electrolytes or problems related to the process of thermoregulation (Greco et al. 2008).

3.2 Heavy intensity domain
Production and removal rates of lactate levels are high due to a high metabolic demand. Consequently, [La] tends to stabilize at higher concentrations when compared to exercise at moderate intensity. Moreover, the efficiency of the specific motor gesture seems to be smaller, generating higher VO_2 values than the linear relationship between VO_2 and exercise intensity that characterizes the Moderate intensity domain (development of a slow

component of VO$_2$). Although the metabolic stress is high, it is possible to maintain a state of physiological balance and to perform the exercise for a long period (Greco et al. 2008). However, Baron et al. (2008) found exercise performed at maximum intensity possible to maintain the stabilization of the [La], i.e., in the maximal lactate steady state (MLSS), depletion occurred while physiological reserve capacity still existed, but in association with an increase in PE assessments, as predicted by the central regulator model (Noakes & St Clair Gibson 2004; Noakes et al. 2005). The end of the exercise could then be induced by an integrative homeostatic control of peripheral physiological system to ensure specifically the maintenance of homeostasis.

3.3 Severe intensity domain
There is no stabilization in metabolic variables. Specifically, the rate of lactate production is greater than the rate of removal, with a consequent increase in the accumulation and the relationship between lactate and pyruvate and the concentration of protons ([H$^+$]) (Greco et al. 2008). At the same time, VO$_2$ increases towards to its maximum (VO$_{2max}$) and the amplitude of the slow component is much higher than those that characterize the heavy intensity exercise (Xu & Rhodes 1999). This reduces exercise tolerance, with *tlim* related to the cellular level of disturbance (metabolites production and removal rates), caused by high demand of muscle adenosine 3-phosphate (ATP) (Greco et al. 2008).

3.4 Scale of five intensity domains proposed by Dekerle & Pelayo (2011)
Dekerle & Pelayo (2011) propose a scale of five domains and their physiological effects. On this scale, lactate threshold (LT), MLSS and CV$_{2par}$ can be understood as boundaries that demarcate some intensity domains. Figure 1 shows the five intensity domains proposed by Dekerle & Pelayo (2011), in which the behavior of [La] and VO$_2$ is illustrated in each domain.

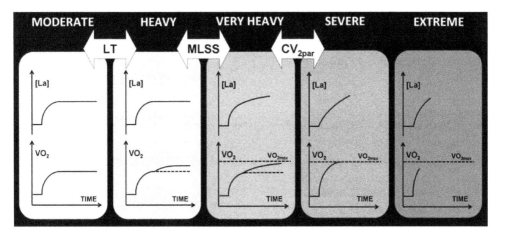

Fig. 1. Intensity domains (adapted from Dekerle & Pelayo 2011) and the response of each to [La] and VO$_2$ kinetics during exercise in different SS.

Each of the five intensity domains (Dekerle & Pelayo, 2011) is characterized by acute specific physiological responses. Dekerle & Pelayo (2011) establish the lactate threshold (LT) as the boundary between moderate and heavy domain. The LT is defined as the first increase in lactate response to an incremental test (Wasserman et al. 1990).

3.4.1 Heavy intensity domain
The exercise is performed in intensity very close to the LT, but a little higher, which causes a small increase in [La] (no more than 1 mmol·l⁻¹) in the first minutes, with subsequent stabilization close to resting levels ($\approx$ 2.1 mmol·l⁻¹). The maximum exercise intensity at which [La] stabilization occurs is defined as maximal lactate steady state (MLSS, $\approx$ 3-5 mmol·l⁻¹) (Beneke, 1995). The MLSS is the heavy intensity domain upper limit (Barstow 1994). The intensity corresponding to LT can be maintained for a very long period (e.g. aquatic marathons) and occurs at a slower speed when compared to MLSS (*tlim* $\approx$ 60 min). MLSS is located in the smaller SS than CV_{2par} (*tlim* $\approx$ 14.3 to 39.4 min). Importantly, for being difficult to detect the MLSS through the curve obtained in [La] and SS, and also to avoid any misinterpretation, the term "anaerobic threshold" should not be associated to the MLSS. Swimming in a very low SS is a difficult task (<0.4 to 0.5 m·s⁻¹ or 50-60% of V_{400} -average speed of 400 m front *crawl* in maximal effort). Thus, the lowest speed that can be adopted by swimmers using a good technique, it is almost equal to LT (Dekerle & Pelayo 2011).

3.4.2 Severe intensity domain
In SS above the MLSS (heavy intensity domain upper limit) there is an increase in [La], HR and VO_2 (occurrence of the slow component). Initially, it was suggested that the increase in VO_2 in these intensities reach the maximum (VO_{2max}) before exhaustion, which characterizes the severe intensity domain. This statement is controversial and difficult to investigate because of the low reliability of time to exhaustion obtained in constant intensity tests (variability of *tlim*) (Hinckson & Hopkins, 2005). The SS equivalent to the Severe intensity domain includes performances of approximately 2 to 60 minutes (VO_{2max} reaching the end of the exercise) with the performance of 400 m in front *crawl*, the maximum aerobic speed (MAS) and CV_{2par} lying within that domain (Lavoie & Montpetit et al 1981; Lavoie et al. 1983; Lavoie & Leone 1988; Rodrigues 2000; Pelayo et al. 2007; Billat et al. 2000; Dekerle et al. 2010).

3.4.3 Extreme intensity domain
This domain includes performances of very short duration (< 2 min). Due to the limited response of VO_2, VO_{2max} is not reached during exercise, although the task is performed to exhaustion.

3.4.4 Very heavy intensity domain
Dekerle & Pelayo (2011) suggest the subdivision of Heavy intensity domain. According to these authors, the range of effort associated to this area is wide (performances of $\approx$ 2 to 60 min) and associated with many chronic responses to training, i.e., the physiological adaptations of a training period in SS near the MLSS are different from the training adaptations induced by a training period in MAS or above.

In addition, the physiological responses to swimming at intensities equal to or above the MLSS are still unclear, since it is not certain that VO_{2max} is reached. Thus, it is justifiable to establish at least one domain between the MLSS and CV_{2par}: the "very heavy intensity domain". Thus, exercise performed in this domain (very heavy) suggests an increase in [La] and the occurrence of the VO_2 slow component, but without reaching VO_{2max} in the end of the exercise (Dekerle et al. 2010). VO_{2max} would only be achieved if the exercise was conducted in intensity above CV_{2par} and continued until exhaustion (featuring the severe domain). Thus, CV_{2par} represents the boundary between very heavy and severe intensity domain. However, Dekerle & Pelayo (2011) suggest that more experiments are needed in these models of training zones. As a result, coaches and swimmers will be able to use them with a greater degree of reliability.

Based on the information presented, it is believed that the model of five intensity domains proposed by Dekerle & Pelayo (2011) best describes the physiological responses to exercise in different intensities.

4. Physiological meaning of each parameter

4.1 Two-parameter model

4.1.1 Critical Speed (CV_{2par})

PC was used initially to determine exercise intensity that could be theoretically maintained for a long period of time without exhaustion (Monod & Scherrer 1965). CP (or CV in running or swimming) proved to be valid for aerobic capacity prediction (Dekerle et al. 2005a) and sensitive to physiological changes from aerobic training programs (Jenkins & Quigley, 1991). CP or CV determined by two-parameter model (CP_{2par} or CV_{2par}) represents the lower boundary t of the severe intensity domain (Poole et al. 1990; Hill & Ferguson 1999). Poole et al. (1990) found that when subjects performed exercise intensity on CP_{2par}, VO_2 stabilized around $75\%VO_{2max}$. In addition, studies have investigated the hyperbolic relationship between power and time to achieve VO_{2max}. The results also suggest that this relationship is the lower boundary of the severe intensity domain, or CP_{2par} (or CV_{2par}) (Hill & Smith 1999; Hill & Ferguson 1999). Thus, CV_{2par} can determine the exercise intensity equivalent to the lower boundary of the severe intensity domain.

4.1.2 Anaerobic distance capacity (ADC_{2par})

The physiological meaning of ADC_{2par} is still subject of many studies (Moritani et al. 1981; Green et al. 1994; Miura et al. 2000; Heubert et al. 2005). Evidence trying to suggest the ADC_{2par} anaerobic nature was observed in cyclists (Green et al. 1994). Also in cyclists, Heubert et al. (2005) found a decrease of 60 to 70% in ADC_{2par} values as a result of a 7 s maximal effort performed before a protocol of four exercises at constant intensity (95, 100, 110 and $115\%VO_{2max}$) and to determine the ADC_{2par} and CP_{2par}. CP_{2par} values did not change. Moritani et al. (1981) also found no differences in ADC_{2par} values in response to ischemia, hypoxia and hyperoxia. In relation to prior depletion of glycogen, Miura et al. (2000) found a decrease in ADC_{2par} values (in cycle ergometer). Jenkins & Quigley (1993) found an increase in ADC_{2par} values in response to high-intensity training in untrained individuals, but the CP_{2par} values did not change. ADC_{2par} values also showed increases in response to creatine supplementation (Miura et al. 1999) and demonstrated good correlation with predominantly anaerobic exercises (Vandewalle et al. 1989; Jenkins & Quigley 1991; Hill 1993; Dekerle et al. 2005b).

4.2 Three-parameter model
4.2.1 Critical Velocity (VC_{3par})
The oxygen supply spends a period of time to reach a steady state or maximum. This has led some researchers (Vandewalle et al. 1989; Morton 1996) questioned the "immediate" availability of CV in two-parameter models (CV_{2par}). As a result of this lapse of time, probably CV_{2par} was being overestimated. In addition, studies found that CV_{2par} could be sustained only by 14.3 to 39.4 min by swimmers (Dekerle et al. 2010). These results suggest that the concept of CV_{2par} as a speed that could be sustained infinitely would not be appropriate.
There is little information on CV and the type of mathematical model used to obtain it in sports. Morton (1996) suggests that CV_{2par} values may be overestimated. Gaesser et al. (1996) also found that three-parameter model generated CP values (CP_{3par}) significantly lower, and the subjects were able to resist in a continuous work for a long period. Thus, CV_{3par} seems not to be at the lower boundary of the severe intensity domain, requiring further investigation. Probably CV_{3par} is below the lower boundary of the severe intensity domain.

4.2.2 Anaerobic distance capacity (ADC_{3par})
Vandewalle et al. (1989) question the assumption that at exhaustion all ADC_{2par} is used, as theoretically is suggested by two-parameter models. Thus, ADC_{2par} may be underestimated (Vandewalle et al. 1989; Morton 1996).

4.2.3 Maximum instantaneous velocity ($V_{max3par}$)
As a result of the lapse of time ("immediate" availability of CV_{2par}), Morton (1996) proposed a three-parameter model (Equation 3) which the "maximum instantaneous speed" ($V_{max3par}$) was included (third parameter). With the addition of the parameter $V_{max3par}$, the three-parameter model is more accurate in estimating the CV (and therefore ADC) surpassing the initial concept of the relationship velocity-*tlim*, that when *tlim* approaches zero, velocity is infinite (Morton 1996). $V_{max3par}$ allows a time asymptote below the x-axis, where time = zero, and provides a Vmax3par value in the intercept-x (MORTON 1996).

4.3 Three-parameter model
4.3.1 Critical velocity (CV_{4par})
Both models (two and three-parameter models) have an important limitation: do not predict the "aerobic inertia" (τ) (Wilkie 1980; Vandewalle et al. 1989), related to cardio respiratory adjustments so that the VO_2 reach steady state or maximum. Thus, a four-parameter model (CV_{4par}, ADC_{4par}, $V_{max4par}$ and τ) proposed by Zacca et al. (2010) could provide more information on bioenergetics in cyclic sports. The four-parameter model proposed by Zacca et al. (2010) was based on the three-parameter model, and CV_{4par} was corrected by an exponential factor, first proposed by Wilkie (1980). This exponential factor is theoretically defined as the time constant that describes the increased aerobic involvement, the "aerobic inertia" (τ). Zacca et al. (2010) suggest that CV is sensitive to additional parameters in young swimmers (93% of the variation was explained by the mathematical model used). The effect of the models showed that CV_{2par} was higher than CV_{3par} and CV_{4par}. CV_{3par} and CV_{4par} were similar (and therefore the physiological meanings of both models are also similar). Thus, future studies are necessary to understand the physiological meaning of CV_{3par} and CV_{4par} in young swimmers and probably in other sports. Figure 2 shows the plot of the data using two, three and four-parameter models with speed and *tlim* data of 50, 100, 200, 300, 400, 800

and 1500 m from swimmers (adapted from Zacca et al. 2010). It is easy to see that the data fits more appropriately in three and four-parameter models. Thus, CV_{2par} was higher than CV_{3par} and CV_{4par}, as previously described.

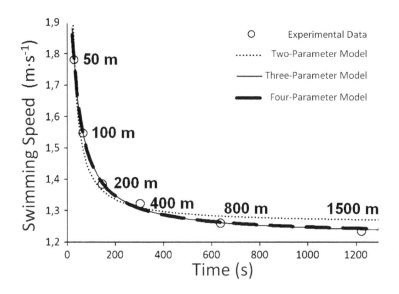

Fig. 2. Swimming speed and tlim of 50, 100, 200, 300, 400, 800 and 1500 m from sprint swimmers and fitted curves through two, three and four-parameter models (adapted from Zacca et al. 2010).

4.3.2 Anaerobic distance capacity (ADC$_{4par}$)

ADC_{2par} was originally defined as the maximum distance (m) that could be covered anaerobically (Ettema 1966). However, Costill (1994) conceptualized ADC_{2par} as the total work that can be performed by a set of limited power of the human body (phosphagen, anaerobic glycolysis and oxygen reserves) suggesting that the anaerobic energy system is predominant but not exclusive (Gastin 2001). Zacca et al. (2010) compared ADC_{2par}, ADC_{3par} and ADC_{4par} values. The results showed that ADC_{2par} (13.77 ± 2.34 m) was lower than ADC_{3par} and ADC_{4par} (30.89 ± 1.70 and 27.64 ± 0.03 m respectively). Moreover, ADC_{3par} and ADC_{4par} values were similar. These results are consistent with others that also observed an overestimation of the parameter ADC in two-parameter model (Billat et al. 2000). Dekerle et al. (2002) evaluated ten well-trained swimmers, when the objective was to verify the possibility of determining ADC_{2par}. They concluded that ADC_{2par} is not perfectly linear and is very sensitive to variations in performance. Thus, according to the authors, it is impossible to estimate the anaerobic capacity by two-parameter models. Toussaint et al. (1998) also suggest that the anaerobic capacity in swimming obtained by two-parameter model does not provide an accurate estimate of the real anaerobic capacity. It seems clear that three and four-parameter models seem more suitable to predict ADC.

4.3.3 Maximum instantaneous velocity

There are gaps in the literature regarding the prediction of V_{max} by mathematical models. Billat et al. (2000) found that $V_{max3par}$ was not different from the maximum speed obtained in 20 m at maximal effort. However, Bosquet et al. (2006) suggest that $V_{max3par}$ is smaller than the real V_{max} (obtained by the average speed of the last 10 m of a maximal 40 m effort). Zacca et al. (2010) found that V_{max} was higher in sprint than endurance swimmers (2.53 ± 0.15 m s^{-1} and 2.07 ± 0.19 m s^{-1} respectively) independent of the mathematical model used (three or four parameters). In addition, $V_{max4par}$ was greater than $V_{max3par}$ (2.42 ± 0.29 m s^{-1} and 2.18 ± 0.34m s^{-1} respectively), suggesting future studies to compare V_{max} and real V_{max}.

4.3.4 Aerobic inertia

The two-parameter model given by the relation "*SS-tlim*" (or "*Dlim-tlim*") and three-parameter model given by the relation "*SS-tlim*" have an important limitation: they do not take into account the "aerobic inertia" (τ) (Wilkie 1980; Vandewalle et al. 1989). The "τ" is a temporary delay in VO_2 response because of dissociation between O_2 absorbed in the lungs and the mainly used by skeletal muscle, lasting approximately 15 to 20 s. "τ" is associated to vasodilatation, i.e, the time it takes for the body to increase heart rate and redirect blood flow. Studies regarding oxygen kinetics during exercise with children and adolescents is limited to few articles and until recently was based on data collected with adults (FAWKNER & ARMSTRONG 2003). Invernizzi et al. (2008) suggest that the time to reach steady state in VO_2 after the beginning of the exercise depends on the characteristics of the subject: endurance swimmers reach this balance sooner than sprint swimmers, and children reach earlier than adults. Thus, "τ" could be a good tool for evaluating cardiovascular and pulmonary performance in athletes (Kilding et al. 2006; Duffield et al. 2007).

5. Swimming speeds prescription through a 400 m front *crawl* maximum effort (T_{400})

Although many distances used in swimming competition does not exceed 2 min (50, 100 and 200 m), the zone related to VO_2, commonly referred as aerobic power, is relevant in swimming (Di Prampero 2003), perhaps because T_{400} is performed in similar SS reach VO_{2max} (Rodrigues 2000). The concept of aerobic power refers to the rate of oxidative energy synthesis (i.e., the maximum power at which the oxidative system can operate, also known as maximum aerobic speed, MAS), available to the muscle work, which can be measured by VO_{2max}. Measuring VO_{2max} in swimming is always a great challenge (PELAYO et al. 2007). This is due to the fact that conventional techniques interfere in swimming biomechanics (Keskinen et al. 2003; Barbosa et al. 2010), which performs the side breathing impossible, changes can occur in hydrodynamics, and most of the times the turns are not performed (Montpetit et al. 1981). Training programs, in order to develop aerobic power in swimmers, are related to the increase in VO_{2max} and the ability to use a high percentage of VO_{2max} for a long time. Maximal aerobic power is widely used to assess aerobic fitness and training intensities prescription (Lavoie & Montpetit 1986).

In an attempt to find alternatives and make the evaluation of athletes swimming closer to reality applied in swimming pools, several studies have been conducted in order to verify the possibility to prescribe training intensities through a single test, but not so extensive such as T_{30} (Lavoie et al. 1981; Lavoie et al. 1983; Lavoie & Montpetit 1986; Rodrigues 2000; Takahashi et al. 2002; Takahashi et al. 2003, 2009;). The attainment of VO_{2max} values from the recovery curve of VO_2 (the back extrapolation method proposed by Di Prampero et al. 1976) was first tested on swimmers by Lavoie et al. back in 1983. Lavoie et al. (1983) found a high correlation between VO_{2max} and *tlim* of T_{400}. The possibility to prescribe training intensities using a single test has renewed expectations of swimming coaches and researchers. The attainment of VO_{2max} values trough the back extrapolation involves obtaining VO_2 after swimming and applying a simple regression curve between the time and the values of consumption in order to predict the value of VO_2 in time zero (Lavoie & Montpetit 1986).

It is believed that the high correlation between VO_{2max} and *tlim* T_{400} m found by Lavoie et al. (1983) is probably the first indication of the T_{400} as a non-invasive alternative. Since then, T_{400} is a reference to verify the MAS and prescribe swimming training intensities (Montpetit et al. 1981; Lavoie et al. 1983; Rodrigues 2000; Pelayo et al. 2007). However, despite many studies reporting the use of T_{400} by swimming coaches (Wakayoshi et al. 1993b; Dekerle et al. 2005a; Alberty et al. 2006; Dekerle et al. 2006; Pelayo et al. 2007), we did not find a reliable protocol for prescribe more than one swimming training zone through the T_{400}, i.e., a protocol not only able to predict aerobic power, but also another training zone.

By questioning some brazilian coaches, we find that some of them use a protocol (of unknown origin) based on the T_{400} to monitor and to prescribe three different SS for swimmers and triathletes. Table 2 presents a summary of the equations used to calculate the SS for "aerobic threshold", "anaerobic threshold" and "VO_{2max}".

OBJECTIVE	DISTANCE		EQUATION
VO_{2max} (I_{VO2})	400 m	$(t400I_{VO2})$	$= 400 / (400 / tlim\ 400\ m) \cdot k$
	800 m	$(t800I_{VO2})$	$= t400I_{VO2} \cdot 2 + 3\ s$
	200 m	$(t200I_{VO2})$	$= t400I_{VO2} / 2 - 3\ s$
	100 m	$(t100I_{VO2})$	$= t200I_{VO2} / 2 - 2\ s$
	50 m	$(t50I_{VO2})$	$= t100I_{VO2} / 2 - 1,5\ s$
ANAEROBIC THRESHOLD (I_{LA})	400 m	$(t400I_{LA})$	$= 400 / ((400 / tlim\ 400\ m) \cdot k) \cdot 0,95$
	800 m	$(t800I_{LA})$	$= t400I_{LA} \cdot 2 + 3\ s$
	200 m	$(t200I_{LA})$	$= t400I_{LA} / 2 - 3\ s$
	100 m	$(t100I_{LA})$	$= t200I_{LA} / 2 - 2\ s$
	50 m	$(t50I_{LA})$	$= t100I_{LA} / 2 - 1,5\ s$
AEROBIC THRESHOLD (I_{LAe})	400 m	$(t400I_{LAe})$	$= 400 / (((400 / tlim\ 400\ m) \cdot k) \cdot 0,95) \cdot 0,93$
	800 m	$(t800I_{LAe})$	$= t400I_{LAe} \cdot 2 + 3\ s$
	200 m	$(t200I_{LAe})$	$= t400I_{LAe} / 2 - 3\ s$
	100 m	$(t100I_{LAe})$	$= t200I_{LAe} / 2 - 2\ s$
	50 m	$(t50I_{LAe})$	$= t100I_{LAe} / 2 - 1,5\ s$

Table 2. Equations used to calculate the SS for "aerobic threshold", "anaerobic threshold" and "VO_{2max}". K is a constant: K = 0.94 if *tlim* is between 3 min 50 s to 4 min 40 s, K = 0.95 if *tlim* is between 4 min 41 s to 5 min 40 s, K = 0.96 the tlim is between 5 min 41 s to 6 min 40 s, K = 0.97 if *tlim* is above 6 min 41 s, t = time prescribed for a given distance; I_{VO2} = intensity prescribed to increase VO_{2max}, I_{LA} = intensity for anaerobic threshold and I_{LAe} = intensity prescribed for aerobic threshold.

In this protocol, the coach just needs that your athletes swim 400 m in front *crawl* under maximum intensity (in training situation, but preferably in competitive situation).

According to the protocol, the T_{400} is able to prescribe SS in three different intensities for training in swimming called (1) "aerobic threshold" (I_{LAE}) (2) "anaerobic threshold" (I_{LA}) and (3) "increased VO_{2max}" (I_{VO2}) (Olbrecht 2000; Maglischo 1999). For each intensity, the protocol suggests the time prescription for distances of 50, 100, 200, 400 and 800 m. SS prescribed by T_{400} for I_{VO2} is between 94 and 97% from the SS of 400 m (V_{400}). SS prescribed for I_{LA} is proposed as approximately 90% of the V_{400}. SS prescribed for I_{LAe} stands at approximately 84% of the V_{400}.

It can be seen throughout this review that the literature presents a wide naming to explain the [La] response to exercise. However, despite being related to the same phenomenon, the physiological responses are often different, such as LT and MLSS mentioned above, and I_{LAE} and I_{LA} used in this protocol. This means that it cannot be used interchangeably.

As Maglischo (1999) and Olbrecht (2000) suggest, sets on L_{AE} are swum in SS ranging from an intensity which is observed in the first rise in [La] above the resting level to the SS that sits comfortably below the I_{LA} of the swimmer. The total distance can vary between 2,000 and 10,000 m for adult swimmers or 20 to 120 min for young swimmers. Any distance can be used in the interval sets. Regarding the rest intervals between each repetition, it is suggested 5 to 30 s (Olbrecht 2000). Still, in I_{LA} sets, the total distance of the set can range from 2,000 to 4,000 m for adults, or approximately 30 min for younger athletes (Maglischo 1999; Olbrecht 2000). Distances between 25 and 4,000 m can be used in the interval sets (Maglischo 1999; Olbrecht 2000), with rest intervals between 10 to 30 s (Olbrecht 2000). Series aimed to increase VO_{2max}, Maglischo (1999) suggests SS slightly above the I_{LA} until 95% of best performance (Maglischo 1999) (Severe intensity domain). It is suggested distances between 25 to 2.000 m, with intervals of rest of 30 s to 120 s between each repetition (Maglischo 1999). However, the SS percentage suggested for training zones prescription have not been observed in constant speed tests until exhaustion. However, similarities were observed in *tlim* and percentage of training zones prescription between the T_{400} and the 1 mile running applied by Daniels (2005). Daniels's concepts (Daniels 2005) were based on "velocity at VO_{2max}" (vVO_{2max}).

6. Velocity at VO_{2max} (vVO_{2max})

Although VO_{2max} is accepted as the physiological variable that best describes cardiovascular and respiratory capacities (Hill & Lupton 1923; Billat & Koralsztein 1996), vVO_{2max} was measured only five decades later in order to provide a practical method to measure aerobic fitness in runners (Billat & Koralsztein 1996). In the 80's there was a growth interest in the physiological assessments in order to monitor athletic training (Billat & Koralsztein 1996). However, it is known that protocols for VO_{2max} measurement, for example, require trained professionals, special equipment and need to be conducted in a controlled environment.

The first field test used to measure vVO_{2max} was intended to replace the 12 min test Cooper (Cooper 1968) as an alternative to predict VO_{2max} in a unique effort to simplify procedures and reduce costs. Cooper (1968) reported a correlation of 0.9 between VO_{2max} and the distance covered in a 12 min test running or walking. However, the motivation and rhythm was mentioned as critical to achieve good reliability in a 12 min test (Cooper 1968). Importantly, when prescribing training intensities based on the performance test, is also considered the psychological characteristic of the race, because instead of applying laboratory tests to monitor training status of the athlete, we use the performance obtained in competitive events, which is directly affected by the willingness to deal as discomfort. Tests based on test performances reflect everything that an athlete recruited to travel any distance in a competitive situation (Daniels, 2005).

The Cooper test (1968) was based on the linear relationship between running speed and VO_2 when, while driving the subject until exhaustion, it was possible to determine VO_{2max}. Billat & Koralsztein (1996) suggest that the accuracy of prediction of VO_{2max}, or also its inaccuracy, depends on the energy cost inter-individual variation, i.e., the total energy expenditure required to move the body to a certain distance.

Daniels et al. (1984) introduced the term "velocity at VO_{2max}" (vVO_{2max}) suggesting that it is a useful variable that combines VO_{2max} and movement economy (Conley & Krähenbühl 1980) on a single factor that identifies aerobic differences among various runners or group of runners. According to Daniels (2005), vVO_{2max} explains individual differences in performance that VO_{2max} or running economy alone could not identify, i.e. individuals with the same VO_{2max} for example, may have different performance times.

Daniels et al. (1984) found in female runners who had various combinations of VO_{2max} and running economy (submaximal VO_2), that vVO_{2max} was similar to the average speed required to run 3,000 m (maintained approximately for 9 min). In a study with sub-elite distance runners, Billat et al. (1994a) measured a *dlim* at vVO_{2max} of 2,008.7 ± 496 m. However the authors suggest that there is a need to distinguish total run at vVO_{2max} and time run at VO_{2max} race only. Daniels et al. (1984) calculated vVO_{2max} extrapolating through a regression curve relating running speed and VO_2. When VO_{2max} was reached, the running speed corresponding to VO_{2max} was identified. Sub-maximal VO_2 was calculated from efforts of 6 min at speeds of 230, 248 and 268 m.min[-1] at intervals of 4 to 7 min between each effort. VO_{2max} was measured separately in a test based on the incremental pace of 5,000 m, adding 1% for the treadmill speed every minute until the test is terminated, where subjects reported that they would not be able to run more than 30 s. The highest VO_2 achieved during the maximal test was considered as VO_{2max}.

7. *tlim* that swimmers are able to keep at vVO_{2max} (tlim-vVO_{2max})

For several years, many studies have remained focusing on measuring vVO_{2max} during swimming. However, few investigations in order to determine the tlim-vVO_{2max} were carried out. This training tool which requires the swimmer to keep the exercise intensity corresponding to its vVO_{2max} has been studied mainly by the Billat et al research group. Based on the pioneering work of Hill and Lupton (1923), Billat & Koralsztein (1996) defined this parameter as the maximum time that the vVO_{2max} is maintained until exhaustion (tlim-vVO_{2max}).

The difficulties of measuring VO_2 in the aquatic environment hindered the swimming research and related modalities. The first studies were conducted in "swimming flume" (Faina et al. 1997; Demarie et al. 2001). To our knowledge, the first study in the pool, i.e., under normal swimming conditions, was performed by Renoux (2001). However, Renoux (2001) did not present results for cardio respiratory parameters such as VO_2 and ventilation. The main results obtained in studies with "swimming flume" suggested that: a) the tlim-vVO_{2max} has low inter-individual variability in swimming, unlike other sports such as running (Billat et al. 1994b), and the values are between 4 min 45 s and 6 min 15 s; b) There is an inverse relationship between tlim-vVO_{2max} and vVO_{2max}, similar to running (Billat et al. 1994c); c) There was an inverse relationship between tlim-vVO_{2max} and anaerobic threshold.

Studies carried out in swimming pool with both genders and different levels of performance shoed some results that agreed with "swimming flume" studies. Fernandes et al. (2003a, 2003b) suggest little variability in tlim-vVO$_{2max}$ between subjects at the same level of performance (Fernandes et al. 2006c), genders (Fernandes et al. 2005), or swimming techniques (Fernandes et al. 2006a). Still, there was an inverse relationship between tlim-vVO$_{2max}$ and vVO$_{2max}$ (Fernandes et al. 2003b, 2005, 2006a), and between tlim-vVO2max and anaerobic threshold corresponding blood concentrations of 3,5 mmol·l^{-1} (Fernandes et al. 2003b).

The method for obtaining vVO$_{2max}$ of swimmers in swimming pool proved to be valid by Fernandes et al. (2003a). First, each subject performed an intermittent and individualized protocol, with increments of 0.05 m.s^{-1} at each stage of 200 m and with 30 s intervals between each stage, until exhaustion. The VO$_2$ was measured directly with a ergospirometer (K4b^2, Cosmed, Rome, Italy) connected to the swimmer through a snorkel and a valve system (Keskinen et al. 2003). The concentrations of expired gases were measured *breath-by-breath*. A speed controller (*visual pacer*, TAR. 1.1, GBK-electronics, Aveiro, Portugal) with lights in the pool, was used to help the swimmers to keep their pre-determined SS. VO$_{2max}$ was considered to be reached according to primary and secondary physiological criteria: (HOWLEY et al. 1995):

a. Occurrence of a VO$_2$ plateau independent of the increase in SS;
b. [La] level (≥ 8mmol·l^{-1});
c. High respiratory exchange ratio (r ≥1,0);
d. High HR (≥90% of [220-age];
e. High value of PE (visually controlled).

Thus, vVO$_{2max}$ is equal to the SS corresponding to the first stage at which VO$_{2max}$ is reached. If a plateau lower than 2.1ml·min^{-1}·kg^{-1} could not be observed, the vVO$_{2max}$ was then calculated by the equation proposed by Kuipers et al. (1985):

$$v\dot{V}O_{2\max} = SS + \Delta S \cdot (n \cdot N^{-1})$$ (5)

where SS is the speed corresponding to the last completed stage, ΔS is the increment of speed, n indicates the number of seconds that the subjects were able to swim during the last stage, and N is the preset time (in seconds) to that stage. After determining the vVO$_{2max}$ of each swimmer, followed by an adequate recovery period, applies the test of tlim-vVO$_{2max}$ when each swimmer trying to stay in your swimming vVO$_{2max}$ (speed control) to exhaustion.

The main studies in swimming suggest that tlim-vVO$_{2max}$:

a. Correlates inversely with the energy cost, ie, it has a direct relationship with swimming economy (Fernandes et al. 2005);
b. Correlates inversely with the speed of the individual anaerobic threshold (Fernandes et al. 2006a);
c. Presents negative correlation values with the delta lactate (Δ[La]), ie, the difference found between [La] at the end and [La] at the beginning of exercise (Δ[La]) (Fernandes et al. 2008);
d. Presents negative correlation with maximum values of [La]. (Fernandes et al. 2008);
e. Shows no significant correlation with VO$_{2max}$ (Fernandes et al. 2003a; 2003b; 2005; 2006a; 2006b; 2006c);

f. Depends on the biomechanical parameters, correlating inversely with the strokes frequency and directly with the distance traveled per stroke cycle and the swimming index (product of the average SS and average distance traveled per stroke cycle) (Fernandes et al. 2006b);

g. During the protocol to obtaining tlim-vVO$_{2max}$ there is a significant increase in stroke frequency and a great decline in the distance per stroke cycle (Marinho et al. 2004, 2006).

Studies in runners and cyclists (Billat & Koralsztein 1996) found that the tlim-vVO$_{2max}$ is less than 12 minutes, and the average is about 6 minutes.

Study	n	tlim (s)	vVO$_{2max}$ (m · s^{-1})	dlim (m)
FERNANDES et al. (2003a)	10 Males (students)	325±76.5 (4min8s to 6min41s)	1.19±0.08	295.7 to 477.79
FERNANDES et al. (2003b)	15 Males (athletes)	260.2±60.73 (3min19s to 5min21s)	1.46±0.06	279.26 to 487.8
FERNANDES et al. (2006a)	8 (athletes)	243.17±30.49 (3min33s to 4min34s)	1.45±0.08	308.38 to 396.80
FERNANDES et al. (2006b)	13 Males	234.49±57.19 (4min17s to 4min51s)	1.45±0.04	257.08 to 422.93
	10 Females	231.90±52.37 (3min to 4min44s)	1.35±0.03	242.36 to 383.76
	Total = 23 (athletes)	233.37± 53.92 (3min to 4min47s)	1.40±0.06	240.44 to 419.42
FERNANDES et al. (2008)	3 Males	217.67±20.84 (3min17s to 3min58)	1.55±0.02	301.15 to 374.47
	5 Females	258.46 ± 25.10 (3min53s to 4min44s)	1.39±0.02	319.70 to 399.82
	Total = 8 (athletes)	243.20 ± 30.50 (3min32s to 4min34s)	1.45±0.08	291.54 to 418.76

Table 3. Studies that measured tlim-vVO$_{2max}$ in swimmers.

Despite these results, is not only the complexity of measuring vVO$_{2max}$ that affect the application of this concept by coaches. Because it is an abstract goal, the use of vVO$_{2max}$ and tlim-vVO$_{2max}$ in swimming training would be more attractive if a "*dlim*" was associated with tlim-vVO$_{2max}$. The studies presented in Table 3 suggest that efforts related to aerobic power (vVO$_{2max}$) have very similar *dlim* of 400 m front crawl, ranging from 4min01s (3min17s to 5min21s, elite swimmers) and 5min25s (4min8s to 6min41s, recreational swimmers). Fastest swimmers endure less time vVO$_{2max}$ likely for two reasons:

a. Higher SS imply higher energy cost (Fernandes et al. 2008);
b. Higher vVO_{2max} in best swimmers require more strenuous levels of exercise, more anaerobic system request (Fernandes et al. 2008).

8. Conclusion

Is well justified that CV_{2par} seems to be higher than CV_{3par} and CV_{4par}. CV_{3par} and CV_{4par} better represent the relationship between "SS-$tlim$" due to its better fit. CV_{3par} and CV_{4par} are similar and probably are located below the lower boundary of the severe intensity domain. However, its applicability to swimming training is questioned because of the need to conduct many maximum efforts to obtain the CV.
In this sense, obtaining vVO_{2max} through the T_{400} seems to be an interesting ecological non-invasive protocol. $tlim$-vVO_{2max} relationship should be considered during swimming training, specifically in the evaluation sessions of the training status. This parameter, together with other indicators, such as LT, MLSS, PE and general biomechanical parameters allow improving the assessment and intensity prescription of training programs. In this sense, assuming some limitations that bring non-invasive tests, vVO_{2max} can be obtained through a single effort of 400 m front *crawl* at maximum intensity (T_{400}), with the advantage of being easy to use, low cost, and have great ecological validity (i.e., reflect the real swimming condition, as it is applied in the training environment. Thus, evaluations and prescriptions for training swimmers would be more practical and accessible, not only for the shortest time spent (i.e, collected even in a competitive situation) but also because do not impact cost. The ability to prescribe more than one training zone through T_{400} still deserves further studies.

9. Acknowledgment

Thanks to Universidade Federal do Rio Grande do Sul - Programa de Pós Graduação em Ciências do Movimento Humano (http://www.esef.ufrgs.br/pos/), Grupo de Pesquisa em Esportes Aquáticos (http://www.geeaufrgs.wordpress.com), Coordenação de Aperfeiçoamento de Pessoal de Nível Superior (http://www.capes.gov.br), Grêmio Náutico União Swim Team (www.gnu.com.br), Caixeiros Viajantes Swim Team (www.caixeirosviajantes.com.br/), Grêmio Náutico Gaúcho Swim Team (www.gngaucho.com.br/gng/home.php) and Biomechanics and Kinesiology Research Group – UFRGS (www.esef.ufrgs.br/pos/gruposdepesquisa/gpbic.php). Also thanks to the Graduate Student Rodrigo Carlet. We would like to leave here a special thanks to Coach Marcelo Diniz da Costa for his great contribution to this topic.

10. References

Alberty M, Sidney M, Hout-Marchand F, Dekerle J, Gorce P, Lensel G, Pelayo P. *Effects of stroking parameters changes on time to exhaustion.* In: Vilas–Boas JP, Alves F, Marques A. (Eds). Biomechanics and medicine in swimming X: proceedings of the Xth Internacional Syposium on Biomechanics and medicine in swimming; Port J Sports Sci. 2006;6:287-289.

Barbosa T, Silva AJ, Reis AM, Costa M, Garrido N, Policarpo F, Reis VM. (2010) *Kinematical changes in swimming front Crawl and Breaststroke with the AquaTrainer® snorkel.* Eur J Appl Physiol 2010 Aug;109(6):1155-62, DOI: 10.1007/s00421-010-1459-x.

Baron B, Noakes TD, Dekerle J, et al. *Why does exercise terminate at the maximal lactate steady state intensity?* Br J Sports Med 2008;42:828–833. doi:10.1136/bjsm.2007.040444

Barstow TJ. (1994) *Characterization of VO₂ kinetics during heavy exercise.* Med Sci Sports Exerc, 26(11), 1327-34.

Beneke R. (1995). *Anaerobic threshold, individual anaerobic threshold, and maximal lactate steady state in rowing.* Med Sci Sports Exerc, 27(6), 863-7.

Billat V, Pinoteau J, Petit B, et al. (1994a) *Times to exhaustion at 90, 100 and 105% of speed at VO₂max and critical speed in elite long distance runners.* Med Sci Sports Exerc; 26 (5 Suppl.) 106S.

Billat V, Renoux JC, Pinoteau J, Petit B, Koralsztein JP. (1994b) *Reproducibility of running time to exhaustion at VO₂max in subelite runners.* Med. Sci. Sports Exerc. 2:254-257.

Billat V, Pinoteau J, Petit B, Renoux JC, Koralsztein JP. (1994c) *Time to exhaustion at 100% of velocity at VO₂max and modelling of the relation time-limit/velocity in elite long distance runners.* Eur J App Physiol. 69:271-273.

Billat V, Koralsztein JP. *Significance of the velocity at VO₂max and time to exhaustion at this velocity.* Sports Med. 1996 Aug;22(2):90-108.

Billat VL, Blondel N, Berthoin S. *Determination of the velocity associed with the longest time to exhaustion at maximal oxygen uptake.* Eur J Appl Physiol Occup Physiol. 1999; 80:159-161.

Billat VL, Morton RH, Blondel N, Berthoin S, Bocquet V, Koralsztein, Barstow TJ. (2000) *Oxygen kinetics and modeling of time to exhaustion whilst running at various velocities at maximal oxygen uptake.* Eur J Appl Physiol 82 (3) 178-187.

Bosquet L, Duchene A, Lecot F, Dupont G, Leger L. (2006) *Vmax estimate from three-parameter critical velocity models: validity and impact on 800m running performance prediction.* Eur J Appl Physiol 97: 34-42.

Bull AJ, Housh TJ, Johnson GO, Perry SR (2000) *Electromyographic and mechanomyographic responses at critical power.* Can J Appl Physiol 25(4):262–270.

Conley DL, Krahenbuhl GS. (1980) *Running economy and distance running performance of highly trained athletes.* Med Sci Sports Exerc; 12: 357-60.

Cooper KH. A *mean of assessing maximal oxygen intake.* JAMA 1968; 203: 201-4.

Costill DL, Maglischo BW, Richardson AB. (1994) *La natation.* Paris: Vigot 215.

Daniels J, Scardina N, Hayes J, et al. *Elite and subelite female middle- and long-distance runners.* In: Landers DM, editor. Sport and Elite Performers, Vol. 3. Proceedings of the 1984 Olympic Scientific Congress: 1984 Jul 19-23: Oregon. Champaign (IL): Human Kinetics, 1984: 57-72.

Daniels JT. (2005) *Daniels' Running Formula: Proven programs 800 m to the marathon.* Human Kinetic. 2nd edition ISBN 0-7360-5492-8.

Dekerle J, Sidney M, Hespel JM, Pelayo P. *Validity and Realiability of Critical Speed, Critical Stroke Rate and Anaerobic Capacity in Relation to Front Crawl Swimming Performances.* Int J Sports Med. 2002;23:93-98.

Dekerle J, Nesi X, Lefevre T, Depretz S, Sidney M, Huot-Marchand F, Pelayo P. *Stroking parameters in front crawl swimming and maximal lactate steady state speed.* Int J Sports Med. 2005a; 26: 53-58.

Dekerle J, Pelayo P, Clipet B, Depretz S, Lefevre T, Sidney M. *Critical swimming speed does not represent the speed at maximal lactate steady state.* Int J Sports Med 2005b; 26:524-530.

Dekerle J, Pelayo P, Sidney M, Brickley G. *Challenges of using critical swimming velocity. From scientists to coaches.* In: Vilas–Boas JP, Alves F, Marques A. (Eds). Biomechanics and medicine in swimming X: proceedings of the Xth Internacional Syposium on Biomechanics and medicine in swimming; Port J Sports Sci. 2006;6:296-299.

Dekerle J, Brickley G, Alberty M, Pelayo P. (2010) *Characterising the slope of the distance–time relationship in swimming.* J Sci Med Sport, May,13(3):365-70.

Dekerle J, Pelayo P. (2011) *Assessing aerobic endurance in swimming.* World book of swimming: from science to performance. Editors, Ludovic Seifert and Didier Chollet. 276-93 ISBN 978-161668-202-6.

Demarie S; Sardella F; Billat V; Magini W; Faina M. (2001) *The VO2 slow component in swimming.* Eur J Appl Physiol. 84:95-99.

DI Prampero PE, Cortili G, Magnani P, Saibene F. *Energy cost of speed skating and efficiency of work against air resistance.* Journal of Applied Physiology. 1976. 40:548-591.

Di Prampero PE. (2003) *Factors limiting maximal performance in humans.* Eur. J. Appl. Physiol. 90:420-429.

Di Prampero PE, Pendergast D, Zamparo P. (2011) *Swimming Economy (Energy Cost) And Efficiency.* World book of swimming: from science to performance / editors, Ludovic Seifert and Didier Chollet. 297-312 ISBN 978-161668-202-6.

Duffield R, Edge J, Bishop D, Goodman C. (2007) *The relationship between VO2 slow component, muscle metabolites and performance during very-heavy exhaustive exercise.* J Sci and Med Sport 10;127-134.

Ettema JH. (1966) *Limits of human performance and energy production.* Int Z Ang Physiol Einschl Arbeitphysiol. 22:45-54.

Faina M, Billat V, Squadrone R, De Angelis M, Koralsztein JP, Dal Monte A. (1997) *Anaerobic contribution to the time to exhaustion at the minimal exercise intensity at which maximal oxygen uptake occurs in elite cyclists, kayakists and swimmers.* Eur. J. Appl. Physiol. 76:13-20.

Fawkner SG, Armstrong N. (2003) *Oxygen Uptake Kinetic Response to Exercise in Children.* Sports Med 2003; 33 (9): 651-669

Fernandes RJ, Cardoso CS, Soares SM, Ascensão AA, Colaço PJ; Vilas-Boas JP (2003a) *Time limit and VO2 slow component at intensities corresponding to VO_{2max} in swimmers.* Int J. Sports Med. 24 :576-81.

Fernandes RJ, Billat V, Vilas-Boas JP. (2003b) *Time limit at vVO_{2max} and VO_{2max} slow component in swimming. A pilot study in university students.* In J.-C. Chatard (ed.), Biomechanics and Medicine in Swimming IX (pp. 331-336) Publications de l'Université de Saint-Étiene, Saint Étiene: France.

Fernandes RJ, Billat VL, Cruz AC, Colaço PJ, Cardoso CS, Vilas-Boas JP. (2005) *Has gender any effect on the relationship between time limit at VO_{2max} velocity and swimming economy?* J. Hum. Movement Stud. 49:127-148.

Fernandes R, Cardoso C, Silva JA, Vilar S, Colaço P, Barbosa T, Keskinen KL, Vilas-Boas JP. *Assessment of time limit at lowest speed corresponding to maximal oxigen consumption in the four competitive swimming strokes.* In: Vilas–Boas JP, Alves F, Marques A. (Eds). Biomechanics and medicine in swimming X: proceedings of the Xth Internacional

Syposium on Biomechanics and medicine in swimming; Port J Sports Sci. 2006a;6:128-130.

Fernandes RJ, Marinho D, Barbosa TM, Vilas-Boas JP. *Is time limit at the minimum swimming velocity of VO_{2max} influenced by stroking parameters?* Perceptual and Motor Skills, 2006b, 103, 67-75.

Fernandes RJ, Billat VL, Cruz AC, Colaço PJ, Cardoso CS, Vilas-Boas JP. (2006c) *Does net energy cost of swimming affect time to exhaustion at the individual's maximal oxygen consumption velocity?* J Sports Med Phys Fitness; 46: 373-380.

Fernandes R, Keskinen K, Colaço P, Querido A, Machado L, Morais P, Novais D, Marinho D and Vilas Boas JP. (2008) *Time limit at VO2max velocity in elite crawl swimmers.* International Journal of Sports Medicine 29, 145-150

Gaesser GA, Carnevale TJ, Garfinkel A, Walter DO, Womack CJ (1995) *Estimation of critical power with nonlinear and linear models.* Med Sci Sports Exerc 27(10):1430-1438

Gaesser GA, Poole DC. (1996) *The slow component of oxygen uptake kinetics in humans.* Exerc Sport Sci Rev 24:35-71.

Gastin PB. (2001) *Energy System Interaction and Relative Contribution During Maximal Exercise.* Sport Med 31 (10): 725 - 741.

Greco CC, Caputo F, Denadai BS. Puissance *critique et consommation maximale d'oxygène: estimation de la limite haute du domaine d'intensité severe, un nouveau challenge?* Science & Sports 23 (2008) 216-222.

Green S, Dawson BT, Goodman C, Carey MF. (1994) *Y-intercept of the maximal work-duration relationship and anaerobic capacity in cyclists.* Eur J Appl Physiol Occup Physiol. 69(6):550.

Heck H, Mader A, Hess G, Mucke S, Muller R, Hollmann W. *Justification of the 4 mmol/l lactate threshold.* Int J Sports Med. 1985;6:117-30.

Heubert RA, Billat VL, Chassaing P, Morton RH, Koralsztein JP, Di Prampero PE. (2005) *Effect of a previous sprint on the parameters of the work-time to exhaustion relationship in high intensity cycling.* Int J Sports Med. 26(7):583.

Hill AV, Lupton L. *Muscular exercise, lactic acid and the supply and utilization of oxygen.* QJ Med 1923; 16:135-71.

Hill AV (1927) *Muscular movement in man: the factors governing speed and recovery from fatigue.* McGraw-Hill, New York, pp 41-44.

Hill DW. *The critical power concept. A review.* Sports Med 1993;16:237-54.

Hill DW, Ferguson CS. *A physiological description of critical velocity.* Eur J Appl Physiol 1999; 79:290-3.

Hill DW, Smith JC. *Determination of critical power by pulmonary gas exchange.* Can J Appl Physiol 1999;24:74-86.

Hill DW, Alain C, Kennedy MD (2003) *Modeling the relationship between velocity and time to fatigue in rowing.* Med Sci Sports Exerc 35(12):2098-2105

Hincson EA, Hopkins WG. (2005) *Reliability of time to exhaustion analyzed with critical-power and log-log modeling.* Med Sci Sports Exerc, 37(4), 696-701.

Housh TJ, Cramer JT, Bull AJ, Johnson GO, Housh DJ (2001) *The effect of mathematical modeling on critical velocity.* Eur J Appl Physiol 84:469-475.

Howley ET, Basseet T, Welch HG. *Criteria for maximal oxygen uptake: review and commentary.* Med Sci Sports Exerc 1995; 27: 1292-1301

Invernizzi PL, Caporaso G, Longo S, Scurati R, Alberti G. (2008) *Correlations between upper limb oxygen kinetics and performance in elite swimmers.* Sport Sci Health 3;19-25.

Jenkins DG, Quigley BM. (1991) *The y-intercept of the critical power function as a measure of anaerobic work capacity.* Ergonomics. Jan;34(1):13-22.

Jenkins DG, Quigley BM. (1993). *The influence of high intensity exercise on the Wlim-Tlim relationship.* Medicine and Science in Sports and Exercise 25: 275 - 282.

Keskinen K, Rodriguez F, Keskinen O (2003) *Respiratory snorkel and valve system for breath-by-breath gas analysis in swimming.* Scand J Med Sci Sports 13:322–329.

Kilding AE, Winter EM, Fysh M. (2006) *A comparison of pulmonary oxygen uptake kinetics in middle and long distance runners.* Int J Sports Med 27;419-426.

Kuipers H, Verstappen FT, Keize HA, Guerten P, van Kranenburg G. *Variability of aerobic performance in the laboratory and its physiologic correlates.* Int J Sports Med 1985; 6: 197–201.

Lavoie JM, Léger LA, Montpetit RR, Chabot S. *Backward extrapolation of VO_2 from de O_2 recovery curve after a voluntary maximal 400m swim.* In:Hollander, Huijing, de Groot (Eds) Biomechanics and medicine in swimming. International series on sport science. 1983 Vol 14, pp 222-227, Human Kinetics Publishers, Champaing, Illinois, USA.

Lavoie JM, Montpetit RR. *Applied Physiology of Swimming.* Sports Med. 1986. 3:165-189.Lavoie JM, Taylor AW, Montpetit RR. *Physiological effects of training in elite swimmers as measured by a free swimming test.* J Sports Med Phys Fitness. 1981;21:38-42.

Lavoie JM, Leone M. (1988). *Functional maximal aerobic power and prediction of swimming performances.* J. Swimming Research; 4 (4), 17-19.

Leclair E, Mucci P, McGawley K, Berthoin S. *Application du concept de puissance critique à différentes populations.* Science & Sports 23 (2008) 206-215.

Lima MCS, et al. *Proposta de teste incremental baseado na percepção subjetiva de esforço para determinação de limiares metabólicos e parâmetros mecânicos do nado livre.* Revista Brasileira de Medicina do Esporte, v.12(5), p.268-274, 2006.

Madsen O. *Anaerobic training not so fast, there.* Swim Tech. 1982;19(3):13-18.

Maglisho, E. W. *Nadando ainda mais rápido.* 1ª ed. Brasileira. São Paulo. Manole, 1999.

Marinho D, Vilas-Boas J, Cardoso C, Barbosa T, Soares S, Fernandes R. (2004). *Stroke rate and stroke length in a typical swimming time limit at VO_{2max}.* In E. Van Praagh; J. Coudert; N. Fellmann; P. Duché (Eds.), Abstracts of 9th Annual Congress of the European College of Sport Science (pp. 338). Clermont-Ferrant: University of Clermont-Ferrant.

Marinho D, Ramos L, Carmo C, Vilar S, Oliveira R, Rodriguez F, Keskinen KL, Fernandes RJ, Vilas-Boas JP. (2006). *Stroke performance during front crawl swimming at the lowest speed corresponding to maximal oxygen consumption.* In: J.P. Vilas-Boas, F. Alves e A. Marques (eds), Book of Abstracts of the Xth International Symposium on Biomechanics and Medicine in Swimming. Portuguese J Sport Scienc. 6(1):45.

Miura A, Kino F, Kajitani S, Sato H, Sato H, Fukuba Y. (1999) *The effect of oral creatine supplementation on the curvature constant parameter of the power-duration curve for cycle ergometry in humans.* Jap J Physiol. 49:169-174.

Miura A, Sato H, Sato H, Whipp BJ, Fukuba Y. (2000) *The effect of glycogen depletion on the curvature constant parameter of the power-duration curve for cycle ergometry.* Ergonomics. 43:133-141.

Monod H, Scherrer J. (1965) *The work capacity of synergic muscle groups.* Ergonomics. 8:329-338.

Montpetit R, Léger L, Lavoie, JM, Cazorla G. *VO₂ Peak During Free Swimming Using the Backward Extrapolation of the O₂ Recovery Curve.* European Journal of Applied Physiology. 47, 385-391, 1981.

Moritani T, Nagata A, De Vries HA, Muro M. (1981) *Critical power as a measure of physical work capacity and anaerobic threshold.* Ergonomics. 24:339-350.

Morton RH. (1996) *A 3-parameter critical power model.* Ergonomics 39: 611-619.

Noakes TD, St Clair Gibson A. *Logical limitations to the "catastrophe" models of fatigue during exercise in humans.* Br J Sports Med 2004;38:648–9.

Noakes TD, St Clair Gibson A, Lambert EV. *From catastrophe to complexity: a novel model of integrative central neural regulation of effort and fatigue during exercise in humans: summary and conclusions.* Br J Sports Med 2005;39:120–4.

Olbrecht J, Madsen O, Mader A, Liesen H, Hollmann W. *Relationship between swimming velocity and lactic concentration during continuous and intermittent training exercises.* Int J Sports Med. 1985; 6(2):74-7.

Olbrecht, J. *The Science of Winning. Planning, periodization and optimizing swim training.* Luton: Swimshop. 2000.

Pelayo P, Alberty M, Sidney M, Potdevin F, Dekerle J. *Aerobic potencial, stroke parameters, and coordination in swimming Front-crawl performance.* International Journal of Sports Physiology and Performance. 2007 2, 347-359.

Pepper ML, Housh TJ, Johnson GO. (1992) *The accuracy of the critical velocity test for predicting time to exhaustion during treadmill running.* Int j of Sports Medicine, 13, 121-124.

Poole DC, Ward SA, Whipp BJ. *The effects of training on the metabolic and respiratory profile of high-intensity cycle ergometer exercise.* Eur J Appl Physiol 1990;59:421–9.

Pyne BD, Lee HE, Swanwick KM. *Monitoring the lactate threshold in world-ranked swimmers.* Med Sci Sports Exerc. 2001;33:291-297.

Renoux J-C. (2001). *Evaluating the time limit at maximum aerobic speed in elite swimmers. Training implications.* Arch. Physiol. Biochem. 109(5):424-9.

Rodrigues FA. *Maximal oxygen uptake and cardiorespiratory response to maximal 400-m free swimming, running tests in competitive swimmers. J.* Sports Med. Phys. 2000. 40: 87-95.

Sweetenham B, Atkinson J. *Championship swim training.* Champaign: Human Kinetics, 2003.

Takahashi S, Wakayoshi K, Hayashi A, Sakaguchi Y, Kitagawa K. *A Method for Determining Critical Swimming Velocity.* Int J Sports Med 2009; 30: 119–123.

Takahashi S, Wakayoshi K, Nagasawa S, Kitagawa K. *A simplified method for determination of critical swimming velocity as a swimming fatigue threshold for sprinters and distance swimmers.* (in Japanese with English abstract). Jap J Biom Sports Exer 2002; 6: 110–115.

Takahashi S, Wakayoshi K, Nagasawa S, Sakaguchi Y, Kitagawa K. *A simplified method for determination of critical swimming velocity as a swimming fatigue index for freestyle sprinters and distance swimmers.* In: Chatard JC. (Ed) Biomechanics and medicine in swimming IX: proceedings of the IX Internacional Syposium on Biomechanics and medicine in swimming, Université de Saint-Etiene 2003 421-427.

Touretski G. *1993 Japan official swimming coach clinic. Jap Amateur swim federation.* 1994:93-139.

Toussaint HM, Wakayoshi K, Hollander AP, Ogita F. (1998) *Simulated front crawl swimming performance related to critical speed and critical power.* Med Sci Sports Exerc. Jan;30(1):144-51.

Vandewalle H, Kapitaniak B, Grun S, Raveneau S, Monod H. (1989) *Comparison between a 30-s all-out test and a time-work test on a cycle ergometer.* Eur J Appl Physiol Occup Physiol. 58(4):375.

Vandewalle H, Vautier JF, Kachouri M, Lechevalier JM, Monod H. (1997) *Work-exhaustion time relationships and the critical power concept.* J Sports Med Phys Fitness 37: 89–102.

Vandewalle H. *Puissance critique: passé, présent et futur d'um concept.* Science & Sports 23 (2008) 223-230.

Vilas-Boas J, Lamares JP. *Velocidade Crítica: Critério para a Avaliação do Nadador e para a Definição de Objetivos.* In: XX Congresso Técnico Científico da Associação Portuguesa dos Técnicos de Natação, 1997.

Wakayoshi K, Ilkuta K, Yoshida T, Udo M, Moritani T, Mutoh Y. *Determination and validity of critical velocity as an index of swimming performance in the competitive swimmer.* Eur J Appl Physiol. 1992;64:153-7.

Wakayoshi K, Yoshida T, Ikuta Y, Mutoh Y, Miyashita M. *Adaptations to six months of aerobic swim training: changes in velocity, stroke rate, stroke length and blood lactate.* Int J Sports Med. 1993a; 14:368-72.

Wakayoshi K, Yoshida T, Udo M, Harada T, Moritani T, Mutoh Y, Miyashita M. *Does critical swimming velocity represent exercise intensity at maximal lactate steady state?* Eur J Appl Physiol. 1993b;66:90-95.

Wasserman K, Beaver WL, Whipp BJ. (1990) *Gas exchange theory and the lactic acidosis (anaerobic) threshold.* Circulation, 81 (1 Suppl), II14-30.

Whipp BJ, Huntsman DJ, Stoner N, Lamarra N, and Wasserman K. (1982) *A constant which determines the duration of tolerance to high-intensity work.* Federation Proceedings 41:1591

Wilkie DR (1980) *Equations describing power input by humans as a function of duration of exercise.* In: Cerretelli P, Whipp B (eds) *Exercise bioenergetics and gas exchange.* Elsevier North Hollands, Amsterdam, pp 75-80.

Xu F, Rhodes EC. *Oxygen uptake kinetics during exercise.* Sports Med 1999; 27:313-27.

Zacca R, Castro FAS (2009) *Comparison between different models to determine the critical speed in young swimmers.* Brazilian J Exerc Physiol 8(2):52–60.

Zacca R, Castro FAS. *Velocidade Crítica em nadadores juvenis: determinação por meio de diferentes combinações de distância de prova.* In: XII Congresso de Ciências do Esporte e Educação Física dos Países de Língua Portuguesa, Porto Alegre, 2008.

Zacca R, Wenzel BM, Piccin JS, Marcilio NR, Lopes AL, Castro FAS. (2010) *Critical velocity, anaerobic distance capacity, maximal instantaneous velocity and aerobic inertia in sprint and endurance young swimmers.* Eur J Appl Physiol (2010) 110:121–131 Doi 10.1007/s00421-010-1479-6.

Zamparo P, Bonifazi M, Faina M, Milan A, Sardella F, Shena F, Capelli C. (2005) *Energy cost of swimming of elite distance swimmers.* Eur J Appl Physiol, 94, 697-704.

8

Optimisation of Cell Bioenergetics in Food-Associated Microorganisms

Diego Mora and Stefania Arioli

University of Milan, Dipartimento di Scienze e Tecnologie Alimentari e Microbiologiche,
Milano,
Italy

1. Introduction

Microorganisms display a considerable versatility, with mechanisms that govern cell bioenergetics and a large number of redox active molecules being used as electron donors or acceptors. We will not review the basis of microbial bioenergetics here, but instead focus attention on the metabolic systems that microorganisms have evolved to optimise the efficiency of cell catabolism and cell energy homeostasis. The mechanisms that act in the regulation of cell bioenergetics belong to the complexity of biological systems in which large networks of metabolic pathways interact to govern the life and responsiveness of cells towards environmental fluctuations. During growth, all microorganisms determine considerable changes in the environmental concentration of nutrients, organic acids and other molecules generated by cell catabolism. As a consequence, microorganisms are constantly faced with different environmental stimuli and stresses. The natural habitats of some microorganisms may fluctuate erratically, whereas others which are more predictable offer the opportunity to prepare in advance for the next environmental change. In this context, microorganisms may have evolved the bioenergetic machinery to anticipate environmental fluctuations by adapting to their temporal order of appearance. Food matrixes represent an example of 'predictable' fluctuating environments, generated by anthropic activities and able to drive the speciation of several microorganisms. The nutrient's richness, and specifically the abundance of mono- and disaccharides that characterise several food matrixes (such as milk and grape juice), have allowed the speciation of lactic acid bacteria (LAB) and yeasts with a high fermentation capacity. The bakers' yeast *Saccharomyces cerevisiae* degrades sugars to two-carbon components – in particular, ethanol – even in the presence of excess oxygen, thus using a fermentation metabolism instead of the energetically favourable respiration metabolism (2 mol versus about 32 mol of ATP per mol of glucose respectively). *S. cerevisiae* alcoholic fermentation has been exploited for several millennia throughout the world in a variety of food processes of crucial importance for humans, such as the making of beer, wine and bread. Moreover, LAB species have partially lost the genetic information need in order to carry out a respiratory metabolism on behalf of a homofermentative pathway in which lactic acid is the primary product, or a heterofermentative pathway in which lactic acid, CO_2, acetic acid and/or ethanol are produced (Kandler, 1983). The seemingly simplistic metabolism of LAB has been exploited throughout history for the preservation of foods and beverages in nearly all societies, and dates back to the origins of agriculture. The domestication of LAB strains passed down through various culinary traditions and the continuous passage of food stuffs has resulted in modern-day cultures that are able to carry out these fermentations (Makarova *et al.*, 2006).

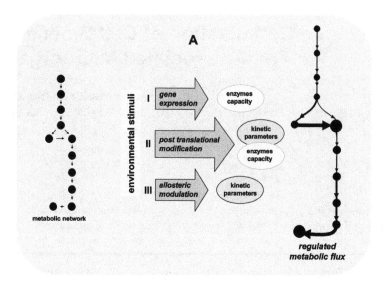

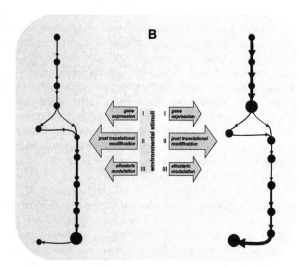

Fig. 1. View of the overlapping regulatory mechanisms modulating metabolic fluxes. A) Example of a metabolic network and schematic representation of the three layers of cellular regulatory mechanisms. The metabolic flux and metabolite pools' concentrations are subjected to the three layers of regulation. The regulation mechanisms act as a response to environmental stimuli. B) Different environmental stimuli (blue and green areas) affect the metabolic fluxes thereby determining the accumulation or depletion of intermediate metabolites.

Both LAB and *S. cerevisiae* have definitely evolved their energetic metabolism to reach maximum fitness in a defined environmental niche characterised by a high carbohydrates concentration. Milk, the proposed evolutionary environmental niche for the LAB *Streptococcus thermophilus, Lactobacillus delbrueckii* subsp. *bulgaricus,* and the "domesticated" strains of *Lactococcus lactis* (Bolotin *et al.,* 2004; van de Guchte, 2006; Passerini *et al.,* 2010), and the man-made niches, grape juice, has driven the evolution of the domesticated strains of *S. cerevisiae* (Martini, 1993, Fay & Benavides, 2005). These two are environments in which mono- and disaccharides resources are both large and dense. In these environmental contexts, fast sugar consumption, lactic acid or ethanol production, accumulation and tolerance, and the ability to propagate without oxygen are some of the 'winning' traits, and they have apparently evolved and become specialised to perfection in these fermenting microorganisms. In other words, energetic limitation is an important factor for organisms in their natural environment and therefore the ATP-production pathways have been under strong selection pressure during evolution (Pfeiffer *et al.,* 2001). Similarly, we can hypothesise that all mechanisms acting in the regulation and optimisation of the ATP-production pathways are subjected to the same selection pressures.

The complexity of the understanding of how metabolic fluxes are modulated arises from multiple overlapping regulatory mechanisms and metabolic feedback into regulatory networks (Figure 1). The *in vivo* capacity of an enzyme to govern and modulate a metabolic flux is a function of its abundance and kinetic properties. Both abundance and the kinetic properties of enzymes are governed by three layers of cellular regulatory mechanisms: i) *gene expression*, acting on enzyme abundance, ii) *post-translational modification*, modulating enzyme abundance and kinetic parameters, and iii) *allosteric modulation*, exclusively affecting the kinetic parameters. Moreover, the *in vivo* metabolic flux depends also on the *in vivo reactant concentrations* (Gerosa & Sauer, 2011) which are function of thermodynamics and reaction kinetics, i.e. parameters that a cell may modulate only indirectly.

This chapter examines the mechanisms regulating the primary metabolism by using as model organisms the dairy species *L. lactis* among prokaryotes, and the bakers' yeast *S. cerevisiae* among eukaryotes. Moreover, some enzymatic activities and metabolic pathways are described and their physiological role is revisited, taking into consideration the optimisation of the cellular bioenergetics as a result of an environment-dependent selection pressure.

2. The regulation of the energetic metabolism in lactic acid bacteria

Despite the wide use of LAB in food production and the role of some species for their health benefits for the human gastro intestinal tract, the regulatory mechanisms that govern the main energetic metabolism of these bacteria have still not been completely disclosed. Most of the studies have been carried out on the 'domesticated' *L. lactis* species, a member of the LAB widely used in the industrial manufacture of milk-fermented products. The most important industrial application of *L. lactis* is based on its energetic metabolism, which leads mainly to the production of high amounts of lactic acid. Anaerobic glycolysis is the principal energy-generating process of *L. lactis*, it is thus considered exclusively as a fermenting microorganism. Nevertheless, in aerobic conditions and in presence of an exogenous source of heme, *L. lactis* may be able to carry out oxidative phosphorylation (Duwat *et al.,* 2001). This cofactor-dependent respiration capacity has also been discovered in other LAB species (Lechardeur *et al.,* 2011). Although named and used for their capacity to produced lactic acid, numerous LAB can be induced through a respiratory metabolism, thereby improving the population size and its survival. It follows that *L. lactis* is currently industrially produced as biomass using a heme-dependent respiration, while in the manufacture of fermented milk and cheeses the homolactic fermentation has a key role in the food matrix's transformation or preservation as a consequence of the sizable production lactic acid. Due to the relatively recent timing of studies on the respiratory behaviour of *L. Lactis*, most of the work carried out in order to elucidate the intricate regulation of the energetic metabolism in this species has focused on lactic fermentation. A detailed description of the dynamics of metabolic pools were obtained through *in vivo* measurements, and kinetic analysis by using cell extracts and the techniques of nonlinear systems modelling (Voit *et al.*, 2006). The monitoring of the glycolytic intermediates made, at first glance, intuitive sense. During homolactic fermentation, glucose was taken-up by lactococcal cells and converted into glucose 6-phosphate and then fructose 1,6-biphosphate. The latter is converted into trioses, which ultimately form lactate (Figure 2).

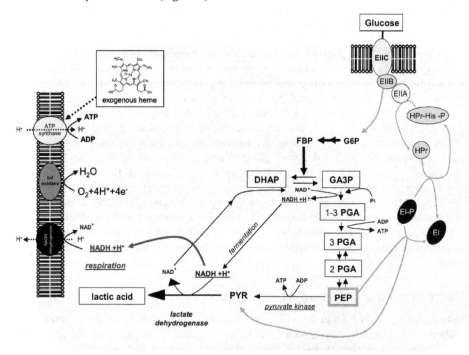

Fig. 2. Simplified representation of glycolysis, homolactic fermentation and heme-dependent respiration in *L. lactis*. The Black arrows show the metabolic fluxes. The red arrow shows the regeneration of NAD+ occurring during the heme-dependent respiration. Glucose-6-phosphate (G6P), dihydroxyacetone phosphate (DHA-P), glyceraldehyde-3-phosphate (GA3P), 1,3-biphosphoglycerate (1-3PGA), 3-phosphoglycerate (3PGA), 2-phosphoglycerate (2PGA) phosphoenolpyruvate (PEP), pyruvate (PYR). The green arrows show the role of PEP as a phosphate donor in the glucose uptake, PTS-dependent phosphoenolpyruvate phosphotransferase system (PTS). EIIC, EIIB, EIIA, HPr, HPr-His-P, EI, and EI-P are components of the PTS system.

While it is usually assumed that the accumulation of intermediates in a linear pathway is disadvantageous because their storage is chemically costly, in L. lactis it a strongly persistent accumulation of trioses (3-phosphoglycerate and phosphoenolpyruvate) at relatively high concentrations (6-20 mM) after glucose consumption was observed (Voit et al., 2006). The reason for the accumulation of trioses in the glycolytic pathway was identified when the overall primary metabolism was considered (Figure 2) together with the nutritional characteristics of the environments where many homofermentative bacteria, including L. lactis, live. These environments are characterised by the availability of glucose, which fluctuates widely between high concentrations and extended periods of starvation. As long as glucose is available, the glycolytic pathway is efficiently fed so as to obtain energy production and population growth. During glucose starvation, it becomes crucial to be well-prepared for future carbohydrate availability, when the cell must use them as fast as possible in order to restart the flux of the glycolytic pathway and grow. The maintenance of the high concentration of trioses is, therefore, necessary because glucose transport across the membrane depends upon phosphoenolpyruvate (PEP) as phosphate donor through a PTS system (Figure 2) (Voit et al., 2006). As such, it can be speculated that L. lactis and other homofermentative LAB have evolved regulatory mechanisms to be able to control the level of PEP in order to bridge normal periods of starvation.

The maintenance of the 'necessary' concentration of PEP (PEP holding pattern) during starvation periods requires a fine tuning of downstream reactions in the pathway. If pyruvate kinase is closed too rapidly, unnecessary amounts of materials are stored in the form of trioses. Otherwise, if pyruvate kinase is deactivated too slowly, the glycolytic flux is accelerated towards the production of lactate. In other words, this regulatory mechanism has evolved to use the phosphotransferase system rather than ATP for glucose phosphorylation, thereby having most of the glycolytic process short-circuited through the PTS system. The main ecological advantage of such metabolic control is that cells use the first available glucose directly in order to produce pyruvate and than lactate, thereby acidifying the local environment when potential competitors attempt to take up glucose (Voit et al., 2006).

Beside the PEP holding pattern, a further interesting metabolic control mechanism developed by L. lactis is represented by the 'feed-forward activation', which is quite rare in metabolic systems. The observation of a transient high concentration of fructose 1,6-biphosphate (FBP) during glucose consumption led to the hypothesis of a regulatory role for this glycolytic intermediate. It was suggested that FBP represents a strong activator of the pyruvate kinase (PK), thereby facilitating the very quick conversion of PEP into pyruvate and lactate while glucose is available. On the other hand, the reduction of glucose availability and, therefore a drop in FBP concentration, allows the decrease of PK activity until an effective stop when glucose is no longer available. The specific activation of PK by FBP has also affected the tuning of PEP concentration. This complex regulation of the energetic metabolism was strictly driven by environmental and ecological constraints. In a more general view, the 'PEP holding' strategy and the FBP 'feed-forward activation' represent an adaptive prediction of environmental changes (in this case related to the availability of carbohydrates). The increasing concentration of PEP during glucose starvation represents a metabolic anticipation of the next environmental stimulus (i.e. new glucose availability). The anticipation of environmental change is considered an adaptive trait because pre-exposure to the stimulus – which typically appears early in the ecology – improves the organism's fitness when it encounters a second stimulus (Mitchell et al., 2009). In the regulatory mechanisms described above, carbohydrates-availability and carbohydrates-starvation represent two consecutive and predictable environmental stimuli for fermentative domesticated LAB. In L. lactis the FBP 'feed-forward activation' represents a clear example of the relevance of allosteric regulatory mechanisms (Figure 1) on the modulation of the energetic metabolism.

A study done to explain the ability of L. lactis to grow, retain an active metabolism and survive at low pH highlights the complexity and the interplay of the overlapping regulatory mechanisms that operate in the regulation of the energetic metabolism. Culturing the microorganisms at low environmental pH sees the biomass yield diminished and the energy dedicated to maintenance increased as a response to the organic acid inhibition and cytoplasmic acidification (Even at al., 2003). The request for energy for maintenance in acid conditions resulted in an increase in glucose consumption and the glycolytic rate with a significant reduction of biomass yield relative to ATP production. The adjustment of the metabolic flux in response to a low environmental pH was determined by an increase in the enzymes' capacity and by a specific modulation of the enzyme activities of the glycolytic pathway. A transcription profile and regulation analysis were effective in evaluating the contribution of each layer of regulatory mechanisms in the observed phenomena, highlighting the primary contribution of translational regulation to the increased concentration of glycolytic enzymes in acidic conditions, and confirming that the translation apparatus of L. lactis was optimised under acid stress conditions (Even et al., 2003). In this case, the decrease of intracellular pH due to the acidity of the extracellular environment determines an important decrease in enzyme activity that was compensated for by an increase in the enzyme capacity through the efficiency of the translation machinery. In this context, it should be underlined that the enzyme concentration results from the rate of protein synthesis, corrected by dilution coefficient, which is affected by protein turnover (normally negligible except under conditions of stress) and the rate of cell division (at each cell division, the enzymes cellular content will be halved). It follows that cells growing at different rates will have substantially different rates of protein synthesis, even though the specific activities may remain similar (Even et al., 2001). More recently, the primary role of allosteric regulatory mechanisms in controlling the glycolytic flux of L. lactis has been questioned, underlining the predominant regulatory role of the enzymes' concentration. This statement was supported by a new methodology whereby experimental measurements of fluxes and enzyme concentrations can be integrated into flux functions capable of predicting the 'fulsome' from the proteome (Rossel et al., 2011). Nevertheless, by such an approach the understanding of the role of each layer of regulation can only be partially addressed. In this case, the approach of regulation analysis is more informative in delineating which regulatory layer is responsible for establishing fluxes through a given enzyme (Gerosa & Sauer, 2011).

Concerning the hierarchical (i.e. expression and post-translational modification) regulation of the energetic metabolism of LAB, the little information available are related to the catabolite control protein CcpA, the major regulator of the carbon metabolism in L. lactis and other Gram-positive bacteria. CcpA belongs to the LacI/GalR family of bacterial regulator proteins, and the disruption of the *ccpA* gene reduces the carbon catabolite repression (CR) of several genes involved in carbohydrate metabolisms. CcpA-mediated regulation depends basically on three elements: i) a specific *cis*-acting DNA sequence, termed catabolite-responsive element (*cre*) which is present near the promoter region of genes affected by CR, ii) the HPr protein, a phosphotransferase protein of the PTS system, and iii) the concentration of glycolytic intermediates (such as FBP). A metabolite-activated kinase has been shown to phosphorylate HPr on residue serine 46. This phosphorylated form of HPr [HPr(Ser-P)] interacts with CcpA, and this interaction enhances the binding of CcpA to *cre* on the promoter region of genes, so affecting the level of their expression. Indirectly, the phosphorylation of HPr on serine residue, enhanced by high level of glycolytic intermediates (*e.g.* FBP), reduces the number of HPr molecules that can be phosphorylated on histidine residue so as to ensure the functionality of the PTS system in sugar uptake across the membrane (Figure 1). Besides the role of CcpA in the control of sugar metabolism (mainly the sugar uptake), it was demonstrated that the role of this protein in the transcriptional activation of the glycolytic *las* operon, encoding the enzymes phosphofructokinase, pyruvate kinase and L-lactate dehydrogenase (Luesink *et al.*, 1998). Specifically, the disruption of the *ccpA* gene lowered the activity of pyruvate kinase and L-lactate dehydrogenase, resulting in the production of metabolites that are characteristic of a mixed-acid fermentation. It was, therefore, speculated that homolactic fermentation in *L. lactis* is maintained by CcpA-mediated repression of mixed-acid fermentation (Luesink *et al.*, 1998).

The regulatory function of CcpA on the energetic metabolism was further confirmed when its primary role in the regulation of aerobic and respirational growth of *L. lactis* was described (Gaudu *et al.*, 2003). CcpA was found to repress NADH oxidase activity, thus maintaining a correct $NADH/NAD^+$ ratio that directed the metabolism in favour of respiration. Moreover, it was proposed that a CcpA-mediated repression of the heme transportation system thereby prevented the oxidative damage provoked by precocious heme uptake at the start exponential growth. CcpA thus appears to govern a regulatory network that coordinates oxygen, iron and the energetic metabolism.

3. The regulation of the glycolytic pathway in *Saccharomyces cerevisiae*

The *S. cerevisiae* metabolism has been exploited by humans for several millennia through a variety of food processes in order to produce alcoholic beverages and leavened bread. Alcoholic fermentation began due to the presence of indigenous yeast in grapes, must, wort and dough, and with total ignorance regarding the existence of microorganisms and their fermentative role. In practice, humans started to apply microbiology before the role of yeast in beer, wine and bread production was formally proven by Pasteur in 1860 (Pasteur, 1860). Starting with the work of Emil Christian Hansen at the Carlsberg Laboratory in Copenhagen, in the early 1880s, the control of the *S. cerevisiae* metabolism became of crucial importance to enhance the efficiency of fermentation processes as well as the quality of the various products. Alcoholic fermentation is not the unique energetic metabolism in *S. cerevisiae* since it can use the more energetically favourable respiration, which sees a significant increase of ATP being produced per mole of glucose (Figure 3).

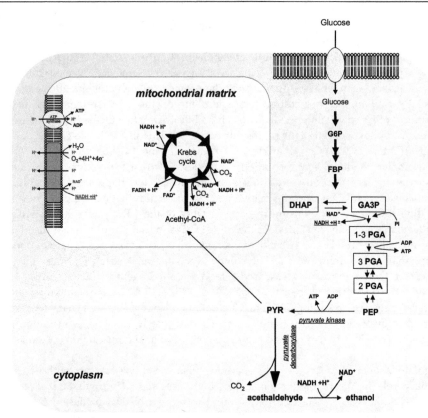

Fig. 3. Simplified representation of the glycolysis, alcoholic fermentation and respiration metabolism in *S. cerevisiae*. Black arrows show the metabolic fluxes. Glucose-6-phosphate (G6P), dihydroxyacetone phosphate (DHA-P), glyceraldehyde-3-phosphate (GA3P), 1,3-biphosphoglycerate (1-3PGA), 3-phosphoglycerate (3PGA), 2-phosphoglycerate (2PGA) phosphoenolpyruvate (PEP), pyruvate (PYR).

A fundamental characteristic of *S. cerevisiae* is the ability to ferment sugars, even in the presence of oxygen in aerobic conditions. This phenomenon is called the Crabtree effect, in honour of Herbert Grace Crabtree who first described the reversible switch between the glycolytic and oxidative metabolism in some cancer cells (Crabtree 1929). In more general terms, the duality of the *S. cerevisiae* metabolism allows this microorganism to use two different strategies for exploiting resources: the 'selfish' strategy and the 'cooperative' strategy. According to the 'selfish' strategy the individuals quickly consume resources and increase their own reproduction rate, whilst according to the 'cooperative' strategy the individuals exploit resources slowly but efficiently. A high rate of ATP production per unit of time is associated with a high reproduction rate and is considered to be a 'selfish' strategy (e.g., fermentation), whereas a high yield of ATP production (the number of units of ATP per unit of resource consumed) is associated with a low reproduction rate but with high biomass production, and is therefore considered to be a 'cooperative' strategy (e.g., respiration) (Pfeiffer, 2001). Given that, resource supply is one of the most important ecological factors that drive the evolution of organisms, the presence in *S. cerevisiae* of two different metabolic strategies for exploiting resources (fermentation and respiration) represents an ecological advantage that has allowed this species to survive under different environmental conditions. The duality of the *S. cerevisiae* metabolism increases the complexity of the regulatory mechanisms interacting with each other to control the energetic metabolism under different environmental stimuli.

Despite *S. cerevisiae* has been extensively studied with regard to several of its characteristics, little information is available concerning the complexity of the regulatory mechanisms acting on the glycolytic pathway, i.e. the common pathway for fermentative and respiratory metabolism (Figure 3). Glycolysis is a highly conserved pathway from bacteria to yeast and humans, and presumably it has been under intense evolutionary pressure for its robust efficiency. It therefore represents an interesting model for investigating the correlation between the different levels of gene expression. As stated by the central dogma of molecular biology (DNA encodes mRNA and mRNA encodes proteins), a strong correlation was anticipated amongst mRNA concentrations, protein concentrations and metabolic fluxes. However, all attempts to verify these correlations – starting from the data on mRNA and protein levels, enzyme activities and *in vivo* fluxes – were far from perfect. A recent study developed a method to dissect the hierarchical regulation of *S. cerevisiae* glycolysis into contributions by transcription, translation, protein degradation and post-translational modification (Daran-Lapujade *et al.*, 2007). The authors propose the calculation of two coefficients, the hierarchical regulation coefficient ρ_h and the metabolic regulation coefficient ρ_m. ρ_h quantifies to what extent the local flux through the enzyme is regulated by a change in enzyme capacity which is affected by a cascade of gene expression, from transcription to post-translational modification. ρ_m quantifies the relative contribution of changes in the interaction of the enzyme with the rest of the metabolism to the regulation of the enzyme's local flux. While ρ_h can be measurable, ρ_m is calculated assuming that $\rho_h + \rho_m = 1$. It follows that a reaction that is purely regulated by a cascade of gene expression would have a ρ_h of 1, whereas a reaction that is solely metabolically regulated would have ρ_h of 0 and ρ_m of 1. A study by Daran-Lapujade compared different cultivation conditions in order to compare a fully respiratory metabolism with a fully anaerobic fermentative metabolism. Moreover, the anaerobic fermentative metabolism was studied by increasing the carbon fluxes in glycolysis by adding to the culture the non-metabolisable weak acid benzoic acid. The comparison of the three different cultivation conditions, carried out using a glucose-limited chemostat at the same dilution rate, highlights an increase of carbon fluxes (5- to 11-fold) in anaerobic rather than in aerobic cultures, with a further increase in the presence of benzoic acid. The dissection analysis revealed that in most cases the fluxes resulted from both hierarchical and metabolic regulatory mechanisms (ρ_h between 0.2 and 0.5). Surprisingly, the increase of glycolytic fluxes stimulated by benzoic acid revealed a dominant contribution of metabolic regulation because most of the reactions showed small ρ_h values and ρ_m values which were close to 1 (with the exception of the reactions governed by phosphofructokinase, fructose-bisphosphate aldolase, triose-phosphate isomerase and pyruvate kinase) (Daran-Lapujade, 2007). A further dissection approach was useful for analysing the contribution of transcription, mRNA degradation, translation, protein degradation or post-translational modification, to the hierarchical regulation of enzymes' capacities. The main conclusion was that fluxes through glycolytic enzymes were only marginally regulated by mRNA levels, whereas most of the observed gene-expression regulation was exerted at the level of protein synthesis and/or degradation and the post-translational level. It was, therefore, speculated that in *S. cerevisiae*, the whole glycolytic regulation is an interplay of purely hierarchical regulation (ρ_h close to 1), purely metabolic regulation (ρ_m close to 1), cooperative regulation (ρ_m and ρ_m between 0 and 1) and antagonistic regulation (both ρ_h and ρ_m negative). The nature and the role of post-translational modification, which appeared to be relevant in the control of glycolytic fluxes, has not yet been investigated systematically for all glycolytic enzymes, even though phosphorylation seems to be the predominant mechanism of protein modification.

The ability of *S. cerevisiae* to switch from respiratory to fermentative metabolism is an important characteristic in the evolutionary and ecological context and for many of its industrial applications. In the natural – evolutionary – context, this ability may have helped this organism to quickly recover sugars and create a hostile environment for competing microorganisms. Concerning the industrial application of *S. cerevisiae*, yeast biomass starved of glucose during storage must rapidly adapt to a high sugar concentration when it is added to bread dough or wort. As has been reported, the shift from respiratory to fermentative metabolism resulted in a rapidly increase of the yeast glycolytic flux in order to compensate the differences in the ATP yield of the two metabolisms. The dynamics of glycolytic regulation during the adaptation of *S. cerevisiae* to fermentative metabolism have been investigated with the aim of understanding the time-dependent, multilevel regulation of glycolytic enzymes during the metabolic switch just described (van de Brink *et al.*, 2008). It was reported that within 45 min of the switch from respiratory to fermentative metabolism, the glycolytic flux increases eightfold without any changes in the glycolytic enzymes' capacities, thereby highlighting an increase of the enzymes activities via metabolic regulation (i.e. the regulation of activities by interaction with low-molecular-weight substrates, products and effectors). By prolonging the incubation during the fermentative metabolism under anaerobic, glucose-excessive conditions, a hierarchical regulation of enzymes was also observed. Specifically, the capacity of the kinases of the upper part of the glycolysis remained unaffected, whereas the enzymes' capacities of the lower part of the glycolysis increased, establishing a new homeostasis of glycolytic metabolites. The delay of the transcriptional regulation compared to the metabolic regulation of glycolytic enzymes observed after the metabolic switch was ascribed to the dramatic change in the rate of ATP production. While the glucose consumption rate increased more than 12-fold during the 2 hours after the switch, the rate of ATP decreased during the first 15 minutes as a result of the reduced ATP yield under fermentative conditions. It was, therefore, speculated that cells energy levels influence the induction of the enzymatic capacity in glycolysis. Due to the fact that an increased level of glycolytic enzymes was only observed 45 minutes after the metabolic switch, and given that the majority of the relevant transcripts were induced after 10 minutes, the step was severely affected by the cellular energetic status which was identified in the translation machinery.

4. Alkalising reactions and cell bioenergetics

Food associated bacteria, and in particular LAB, have been selected and used by humans in several food processes because of their ability to acidify milk or vegetables in order to obtain a more stable and safer food products. Acidification occurs in homofermentative LAB through lactic acid production as the final product of their energetic metabolism. It follows that the growth of LAB determines a significant change in the environmental chemical composition, with a progressive decrease in sugar concentration and a simultaneous increase in lactic acid concentration thereby determining a decrease in environmental pH. Consequently, LAB during each fermentation process are faced with 'predictable' environmental changes, ending with the cessation of growth due to carbon source starvation and, mostly, with an environmental pH which is incompatible with the metabolic processes of the microorganism. The exposure to low pH for long period times determines the arresting of growth and a dramatic decrease of glycolytic fluxes, structural damage to the cell membrane and macromolecules (such as DNA and proteins), and a progressive loss in viability. Weak acids, such as lactic acid, have potent antimicrobial activity because the undissociated forms of weak acids pass freely through the cell membrane. Since the cytoplasmic pH is generally higher than that of the growth medium, inside the cell the weak acid dissociates by releasing a proton and leading to the acidification of the cytoplasm. Due to environmental constraints, LAB have developed through evolution a 'make-accumulate-consume' metabolic strategy (Pfeiffer *et al.*, 2001; Rozpedowska *et al.*, 2011) in order to have a faster sugar consumption, lactic acid production and accumulation. This strategy is aimed at rapidly monopolising sugars and creating an unfavourable environment so as to out-compete other microorganisms by the rapid secretion of fermentation products. In order to survive themselves, LAB have therefore developed a series of mechanisms to counteract low environmental pH and the negative effects of weak organic acids produced by their own metabolism. Several of these mechanisms have been extensively studied (Cotter & Hill, 2003) to understand how LAB protect themselves from the challenge posed by low-pH environments, such as food and gastric juice, and how they develop the strategies by which they can be aided or impeded. Nevertheless, the role of these mechanisms in the regulation of the energetic metabolism was barely investigated even though the loss of the activity of the relatively acid-sensitive glycolytic enzymes (which severely affects the ability to produce ATP) was well known. Indeed, even if LAB species are acid-tolerant bacteria, they cannot be considered to be acidotrophic, and the optimum pH of the highly conserved glycolytic enzymes is close to neutral-alkaline values (Hutkins & Nannen, 1993).

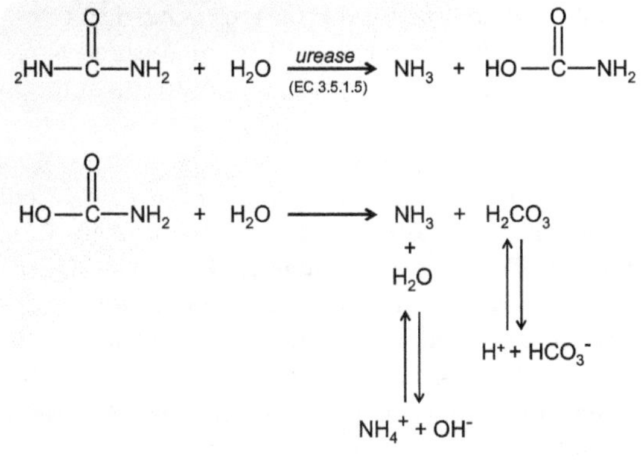

Fig. 4. Reaction catalysed by urease (EC 3.5.1.5) and the spontaneity of carbamate in ammonia and carbonic acid.

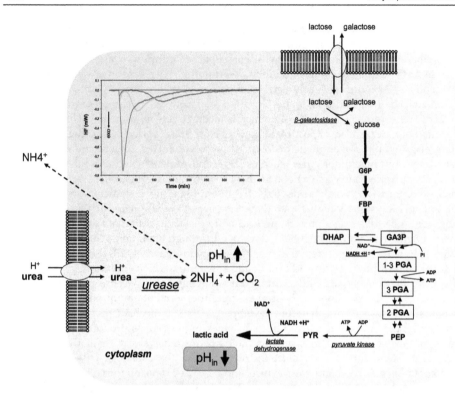

Fig. 5. Simplified representation of glycolysis, homolactic fermentation and urease activity in *S. thermophilus*. The inset represents the raw isothermal titration calorimetry data (heat flux versus time) of *S. thermophilus* lactose metabolism either alone (blue line) or in the presence of ammonia (green line) or urea (red line) (for the detailed experimental procedure see Arioli *et al.*, 2010).

Quite recently, the urease activity – an enzymatic reaction known as a stress response to counteract environmental acidic pH in several bacteria – has been described as a metabolic regulatory mechanism of the energetic metabolism in the dairy bacterium *Streptococcus thermophilus* (Arioli *et al.*, 2010). Urease is a multi-subunit urea amidohydrolase (EC 3.5.1.5) that catalyses the hydrolysis of urea to yield ammonia and carbamate, which spontaneously decomposes to yield a second molecule of ammonia and carbonic acid (Figure 4). The released carbonic acid and the two molecules of ammonia are in equilibrium with their deprotonated and protonated forms respectively, and the net effect of these reactions is an increase in intracellular (pH_{in}) and extracellular (pH_{out}) pH (Figure 5). Urea hydrolysis increases the catabolic efficiency of *S. thermophilus* by modulating the intracellular pH and thereby increasing the activity of β-galactosidase, glycolytic enzymes, and lactate dehydrogenase. Moreover, urease increases the overall change in enthalpy generated by the microbial metabolism as a consequence of an increased glycolytic flux (Figure 5).

In light of these considerations, urease activity – which is stimulated when environmental pH is weakly acidic (pH 5.8-6) (Mora *et al.*, 2005) – should be considered as a regulatory system that has evolved to optimise the activity of the glycolytic enzymes. These enzymes are exposed to an increasingly acidic intracellular environment and must maintain cell energy homeostasis when the pH_{out} and pH_{in} decrease as a result of lactic acid production. Urease biogenesis is only important when the cells are actively growing, since it increases the fermentative capacity of *S. thermophilus* and leads to rapid growth and an increased acidification rate in milk (i.e. urease favour a cytoplasmic background suitable for a 'make-accumulate-consume' strategy). If we consider that energetic limitation is an important factor for organisms in their natural environment, we then expect that the properties of ATP-production pathways have been under strong selection during evolution (Pfeiffer *et al.*, 2001). Similarly, the regulatory mechanisms which act in optimising the efficiency of the ATP-production pathway should be under the same evolutionary selection. In this context, it is notable that eleven genes are necessary in order for the maintenance of an active urease, which accounts for 0.9% of the estimated core genome of *S. thermophilus*. This enzyme has been found in all the previously characterised *S. thermophilus* strains, and urease-negative mutants are not common in nature. The *S. thermophilus* genome has mainly evolved following divergent evolution from the phylogenetically related pathogenic streptococci bacteria. Loss-of-function mutations, counterbalanced by the acquisition of relevant traits (e.g. lactose utilisation) have resulted in a *S. thermophilus* genome that is well-adapted for dairy colonisation (Bolotin *et al.*, 2004). Because urease is not common in pathogenic streptococci (Mora *et al.*, 2005), its acquisition and maintenance within the *S. thermophilus* genome is likely to be dependent upon its contribution to the environmental fitness of this microorganism when linked to the environmental availability of urea. Urea is the major nitrogenous waste product of most terrestrial animals. Urea is produced in the liver, carried in the bloodstream to the kidneys and excreted in urine. Urea is also present in milk and in the secretions of the major and minor exocrine glands at concentrations approximately equivalent to serum, so a large proportion of circulating urea is translocated onto epithelial surfaces by secretory systems or else in tissue exudates. In this context, it is not surprisingly that urease is present in a high number of human pathogenic bacteria and represents an important factor in infection and disease (Burne and Chen, 2000; Mora *et al.*, 2005).

Since the activity of the bioenergetic machinery is modulated by the intracellular pH, the mechanism of metabolism regulation in other urease-positive bacteria, including human pathogens, should be further analysed. All of the metabolic reactions that result in the alkalisation of the cytosol of acidogenic organisms (such as those involved in the arginine deiminase (ADI) pathway, the citrate metabolism or else those involved in malolactic conversion) should be analysed in light of these novel findings. Indeed, and not surprisingly, all previous pathways act by subtracting protons from the cytoplasm and are strongly induced by an acidic environmental pH (Magni *et al.*, 1999; Cotter & Hill, 2003; Broadbent *et al.*, 2010). The conserved role of alkalising reactions across acidogenic bacteria is also supported by the data obtained for *L. lactis* IL1403-945 and *S. pneumoniae* SP292-945 in the presence of glucose and cellobiose as a carbon source (Arioli *et al.*, 2010). In both cases, the rate of ATP produced during the sugar catabolism was increased, alkalising with the ammonia the cytoplasm. Besides the selfish utility of urease for cells harbouring this enzymatic activity, the cooperative behaviour of urease in an ecological context in which different microbial species share the same environment was also underlined. Urea hydrolysis results in a rise of both pH_{in} and pH_{out} due to the rapid diffusion of ammonia outside the cell. It follows that in the presence of urea and a urease-positive microorganism, (or a urease-negative microorganism) sharing the same micro-environment, there will be benefits from the local transient increase of pH (Arioli *et al.*, 2010).

5. Conclusions and perspectives

The regulation and control of metabolic fluxes in microbes is based on our knowledge of regulatory networks topology, on input-output regulatory logics and metabolic feedback, and on the quantitative effect of control exerted by regulation events. No less important is our understanding of how metabolic regulatory circuits have evolved and what the significance of the impact of environmental constraints on the regulatory configuration will be. It has recently been described that microbes can 'learn' form exposure to a series of new environmental changes and rearrange some regulatory networks so as to predict the new environmental stimuli (Mitchell *et al.*, 2009). The ecological forces and the molecular mechanisms that govern this ability are not clear but it is evident that the regulatory networks that link environmental stimuli to microbial responses are complex and can evolve rapidly (Cooper, 2009). The origin of the adaptability of regulatory networks could be ascribed to microbial cell individuality and the underlying sources of heterogeneity. This heterogeneity is related to stochastic fluctuations in transcription or translation, despite a genetically homogeneous background and constant environmental conditions. Heterogeneity at single-cell level is typically masked in conventional studies of microbial populations, which are based on the average behaviour of thousands or millions of cells, but it has the potential to create variant subpopulations better equipped to persist during environmental perturbation (Avery, 2006). In other words, a population might enhance its fitness by allowing individual cells to make a stochastically transition amongst multiple phenotypes, thus ensuring that some cells are always prepared for erratic, unpredictable environmental fluctuations. It can be therefore be concluded that the regulatory mechanisms that act in the optimisation of the bioenergetics of food-associated bacteria should be analyzed by always taking into consideration the 'predictable' succession of environmental stimuli that have driven their domesticated speciation and evolution.

6. References

Arioli, S., Ragg, E., Scaglioni, L., Fessas, D., Signorelli M., Karp, M., Daffonchio, D., De Noni, I., Mulas, L., Oggioni, M., Guglielmetti, S., Mora, D. (2010). Alkalizing reactions streamline cellular metabolism in acidogenic microorganisms. *Plos One*, Vol.5, No.11 (November 2010) e1520. doi:10.1371/journal.pone.0015520, ISSN 1932-6203

Avery, S.V. (2006). Microbial cell individuality and the underlying sources of heterogeneity. *Nature Review Microbiology*, Vol.4, No.8 (August 2006) pp.577-587, ISSN 1740-1526

Bolotin, A., Quinquis, B., Renault, P., Sorokin, A., Ehrlich, S.D., Kulakauskas, S., Lapidus, A., Goltsman, E., Mazur, M., Pusch, G.D., Fonstein, M., Overbeek, R., Kyprides, N., Purnelle, B., Prozzi, D., Ngui, K., Masuy, D., Hancy, F., Burteau, S., Boutry, M., Delcour, J., Goffeau, A., & Hols, P. (2004). Complete sequence and comparative genome analysis of the dairy bacterium *Streptococcus thermophilus*. *Nature*, Vol.22, No.12 (December 2004) pp. 1554-1558, ISSN 1087-0156

Broadbent, J.R., Larsen, R.I., Deiel, V. & Steele, J.L. (2010). Physiological and transcriptional response of *Lactobacillus casei* ATCC 334 to acid stress. *Journal of Bacteriology*, Vol.192, No.9 (May 2010) pp. 2445-2458, ISSN 0021-9193

van de Brink, J., Canelas, A.B., van Gulik, W.M., Pronk, J.T., Heijnen, J.J., de Winde, J. & Daran-Lapujade, P. Dynamics of glycolytic regulation during adaptation of *Saccharomyces cerevisiae* to fermentative metabolism. *Applied and Environmental Microbiology*, Vol.74, No.18 (September 2008) pp. 5710-5723, ISSN 0099-2240

Burne, R.A. & Chen, Y.Y. (2000). Bacterial ureases in infection diseases. *Microbes and Infection*, Vol.2, No.5 (April 2000) pp. 533-542, ISSN 1286-4579

Crabtree, H.G. (1929). Observations on the carbohydrate metabolism of tumors, *Biochemical Journal*, Vol.23, No.3 (1929) pp. 536–545, ISSN 0264-6021

Cooper, T.F. (2009). Microbes exploit groundhog day. *Nature*, Vol. 460, No.7252 (July 2009) pp. 181, ISSN 0028-0836

Cotter, P.D. & Hill, C. (2003). Surviving the acid test: responses of Gram-positive bacteria to low pH. *Microbiology and Molecular Biology Reviews*, Vol.67, No.3 (September 2003) pp. 429-453, ISSN 1092-2172

Daran-Lapujade, P., Rossel, S., van Gulik, W.M., Luttik, M.A.H., de Groot, M.J.L., Slijper, M., Heck, A.J.R., Daran, J.M., de Winde, J.H., Westerhoff, H.V., Pronk, J.T. & Bakker, B.M. (2007). The fluxes through glycolytic enzymes in *Saccharomyces cerevisiae* are predominantly regulated at posttranscriptional levels. *Proceedings of the National Academy of Science of the USA*, Vol.104, No.40, (October 2007) pp. 15753-15758, ISSN 0027-8424

Duwat, P., Sourice, S., Cesselin, B., Lamberet, G., Vido, K., Gaudu, P., Le Loir, Y., Violet, F., Loubiere, P. & Gruss, A. (2001). Respiration capacity of the fermenting bacterium *Lactococcus lactis* and its positive effect on growth and survival. *Journal of Bacteriology*, Vol.183 No.15 (August 2001) pp. 4509-4516, ISSN 0021-9193

Even, S., Lindley, N.D. & Cocaign-Busquet, M. (2001). Molecular physiology of sugar catabolism in *Lactococcus lactis* IL1403. *Journal of Bacteriology*, Vol.183, No.13 (July 2001) pp. 3817-3824, ISSN 0021-9193

Even, S., Lindley, N.D. & Cocaign-Busquet, M. (2003). Transcriptional, translational and metabolic regulation of glycolysis in *Lactococcus lactis* subsp. *cremoris* MG 1363 grown in continuous acidic cultures. *Microbiology*, Vol.149, No.7 (July 2003) pp. 1935-1944, ISSN 1350-0872

Kandler, O. (1983). Carbohydrate metabolism in lactic acid bacteria. *Antonie Van Leeuwenhoek*, Vol.49, No.3 (May 1983), pp. 209-224

Fay, J.C. & Benavides, J.A. (2005). Evidence for domesticated and wild populations of *Saccharomyces cerevisiae*. *Plos Genetic*, DOI: 10.1371/journal.pgen.0010005

van de Guchte, M. (2006). The complete genome sequence of *Lactobacillus bulgaricus* reveals extensive and ongoing reductive evolution. *Proceedings of the National Academy of Science of the USA*, Vol.103, No.24, (April 2011) pp. 9274-9279, ISSN 0027-8424

Gerosa, L. & Sauer, U. (2011). Regulation and control of metabolic fluxes in microbes. *Current Opinion in Biotechnology*, Vol.22, No.4 (August 2011) pp. 566-575, ISSN 0958-1669

Gaudu, P., Lamberet, G., Poncet, S. & Gruss, A. (2003). CcpA regulation of aerobic and respiration growth in *Lactococcus lactis*. *Molecular Microbiology*, Vol.50, No.1 (October 2003) pp. 183-192, ISSN 0950-382X

Hutkins, R.W. & Nannen, N.L. (1993). pH homeostasis in lactic acid bacteria. *Journal of Dairy Science*, Vol.76, No.8 (August 1993) pp. 2354-2365, ISSN 0022-0302

Lechardeur, D. *et al.* (2011). Using heme as an energy boost for lactic acid bacteria. *Current Opinion in Biotechnology*, Vol.22, No.2 (April 2011) pp. 143-149, ISSN 0958-1669

Luesink, E.J., van Herpen, R.E.M., Grossiord, B.P., Kuipers, O.P. & de Vos, W.M. (1998). Transcriptional activation of the glycolytic *las* operon and catabolite repression of the *gal* operon in *Lactococcus lactis* are mediated by the catabolite control protein

CcpA. *Molecular Microbiology*, Vol.30, No.4 (November 1998) pp.789-798, ISSN 0950-382X

Makarova, K., Slesarev, A., Wolf, Y., Sorokin, A., Mirkin, B., Koonin, E., Pavlov, A., Pavlova, N., Karamychev, V., Polouchine, N., Shakhova, V., Grigoriev, I., Lou, Y., Rohksar, D., Lucas, S., Huang, K., Goodstein, D.M., Hawkins, T., Plengvidhya, V., Welker, D., Hughes, J., Goh, Y., Benson, A., Baldwin, K., Lee, J.-H., Diaz-Muniz, I., Dosti, B., Smeianov, V., Wechter, W., Barabote, R., Lorca, G., Altermann, E., Barrangou, R., Ganesan, B., Xie, Y., Rawsthorne, H., Tamir, D., Parker, C., Breidt, F., Broadbent, J., Hutkins, R., O'Sullivan, D., Steele, J., Unlu, G., Saier, M., Klaenhammer, T., Richardson, P., Kozyavkin, S., Weimer, B. & Mills, D. (2006). Comparative genomics of the lactic acid bacteria. *Proceedings of the National Academy of Science of the USA*, Vol.103, No.42 (October 2006), pp. 15611-15616

Magni, C., Mendoza, D., Konings, W.N. & Lolkema, J.S. (1999). Mechanism of citrate metabolism in *Lactococcus lactis*: resistance against lactate toxicity at low pH. *Journal of Bacteriology*, Vol.181, No.5 (March 1999) pp. 1451-1457, ISSN 0021-9193

Martini, A. (1993). Origin and domestication of the wine yeast *Saccharomyces cerevisiae*. *Journal of Wine Research*, Vol.4, No.3 (September 1993) pp. 165-176, ISSN 0957-1264

Mitchell, A., Romano, G.H., Groisman, B., Yona, A., Dekel, E., Kupiec, M., Dahan, O. & Pilpel, Y. (2009). Adaptive prediction of environmental changes by microorganisms. *Nature*, Vol. 460, No.7252 (July 2009) pp. 220-224, ISSN 0028-0836

Mora, D., Monnet, C., Parini, C., Guglielmetti, S., Mariani, A., Pintus, P., Molinari, F., Daffonchio, D. & Manachini, P.L. (2005). Urease biogenesis in *Streptococcus thermophilus*. *Research in Microbiology*, Vol.156, No.9 (November 2005) pp.897-903, ISSN 0923-2508

Mora, D., Monnet, C., Daffonchio, D. (2005). Balancing the loss and acquisition of pathogenic traits in food-associated bacteria. *Microbiology*, Vol.151, No.12 (December 2005) pp. 3814-3816, ISSN 1350-0872

Pasteur, L. (1860). Mémoire sur la fermentation alcoolique. *Annales de Chimie et the Physique*, Vol.58 pp. 323-426

Pfeiffer, T., Schuster, S. & Bonhoeffer, S. (2001). Cooperation and competition in the evolution of ATP-producing pathways. *Science*, Vol.292, No.5516, (April 2001), pp. 504-507, ISSN 0036-8075

Passerini, D. Beltramo, C., coddeville, M., Quentin, Y., Ritzenthaler, P., Daveran-Mingot, M.-L. & Le Bourgeois, P. (2010). Genes but not genomes reveal bacterial domestication of *Lactococcus lactis*. *Plos One*, Vol.5, No.12 (December 2010), e15306. doi:10.1371/journal.pone.0015306, ISSN 1932-6203

Rossel, S., Solem, C., Jensen, P.R. & Heijnen, J.J. (2011). Towards a quantitative prediction of the fluxome from the proteome. *Metabolic Engineering*, Vol.13, No.3 (May 2011), pp. 253-262, ISSN 1096-7176

Rozpędowska, E., Hellborg, L., Ishchuk, O.P., Orhan, F., Galafassi, S., Merico, A., Woolfit, M., Compagno, C. & Piškur, J. (2011). Parallel evolution of the make-accumulate-consume strategy in *Saccharomyces* and *Dekkera* yeast. *Nature Communications*, Vol.2, No.302 (May 2011) DOI: 10.1038/ncomms1305, ISSN 2041-1723

Voit, E., Neves, A.R. & Santos, H. (2006). The intricate side of systems biology. *Proceedings of the National Academy of Sciences of the USA*, Vol.103, No.25, (June 2006), pp. 9452-9457, ISSN 0027-8424

9

Invertebrates Mitochondrial Function and Energetic Challenges

Oliviert Martinez-Cruz[1], Arturo Sanchez-Paz[2],
Fernando Garcia-Carreño[3], Laura Jimenez-Gutierrez[1],
Ma. de los Angeles Navarrete del Toro[3] and Adriana Muhlia-Almazan[1]

[1]*Molecular Biology Laboratory, Centro de Investigacion en Alimentacion
y Desarrollo (CIAD), Hermosillo, Sonora,*
[2]*Laboratorio de Sanidad Acuicola, Centro de Investigaciones Biologicas
del Noroeste (CIBNOR), Hermosillo, Sonora,*
[3]*Biochemistry Laboratory, Centro de Investigaciones Biologicas
del Noroeste (CIBNOR), La Paz,*
Mexico

1. Introduction

The term "invertebrate" recalls all animal species lacking a backbone or a bony skeleton. Although "invertebrate" is not a scientific term that encloses a taxonomic rank, this group includes animal species represented by over 30 phyla and it includes the first animals that successfully inhabited the earth, an event that – according to the fossil evidence – dates back to around 600 million years ago. This group is composed of several different phyla, such as annelids, molluscs, sponges, cnidarians, echinoderms, and all species from the phylum Arthropoda – which is the largest among invertebrates and is comprised by insects, arachnids and crustaceans (nearly reaching 1,033,160 species).

Since they appeared for the first time during the Cambrian period, invertebrates have played an important ecological role since they are frequently the key constituents of many trophic chains and they occupy virtually every available ecosystem on Earth, being characterised by notable variations in temperature, oxygen concentrations, food availability and food quality. Also, many species occupy highly specific and important roles in nature as pollinators, parasites or vectors for parasitic diseases affecting human and animal health.

It is clear that the ability of invertebrates to inhabit almost every ecosystem – as well as the diverse array of morphological and behavioural strategies used to obtain nutrients from the environment – is an accurate reflection of the enormous ability of these organisms to solve their most basic energetic requirements. From blood-suckers such as mosquitoes, intestinal nematodes and leeches (hirudin), to small plankton marine feeders such as cnidarians and marine benthic bivalves, all species face changes in food availability throughout their life cycle which affect their energy stores and growth rates (Peck, 2002; Popova-Butler & Dean, 2009). A beautiful example of highly specific energy stores – crucial during invertebrates' life cycle and important to human health – is that of the female mosquito (*Anopheles gambiae*), which usually feeds on sugar to gain energy to fly and to cope with metabolic requirements; however, anautogenous mosquitoes require the energy resulting from blood digestion in order to produce eggs, and it is during blood sucking that *Plasmodium vivax* (the parasite from infected females) enters into the vertebrate host to produce Malaria, a major health problem around the world (Das et al., 2010).

Large energetic demands during external work are observed throughout the life of several invertebrate species, and a clear example may be found in insect flight, which is considered to be one of the most energetically demanding processes of animal locomotion (Harrison & Roberts, 2000). Besides this, being an aerobic process that requires a permanent oxygen supply and depends upon ATP cellular production, the high energetic cost of flying is related to the frequency of the flight muscles' contraction (Vishnudas & Vigoreaux, 2006). In vertebrate species, the existence of high-energetic molecules in the muscle (phosphocreatine) during its exercise has been well documented (Jubrias et al., 2001); however, in invertebrate species, the presence of phosphagen-kinases that catalyse the synthesis of these high-energetic phosphorylated molecules has not been widely distributed (Ellington & Hines, 1991). The insect flight muscle seems to lack such molecules, but some flying species are able to surpass such energy needs by the proximity of mitochondria to muscle myofibrils, thus facilitating the export of energy rich nucleotides – such as ATP – to myofibrils (Vishnudas & Vigoreaux, 2006).

Some other invertebrate phyla – such as crustaceans – are able to synthesise phosphagens differently from that of vertebrates, like phosphocreatine. Phosphoarginine – a phosphagen of L-arginine found in the tail muscle of shrimp and crabs as well as in the flight muscle of flying insects – is the chemical energy storage system of these tissues, and thus animals are able to rapidly produce ATP when it is required (Wegener, 1996; Kotlyar et al., 2000). The enzyme responsible for the synthesis of phosphoarginine from ATP and L-arginine in invertebrates is named 'arginine kinase' and it is also considered to be a major allergen protein for shrimp-allergic individuals (Garcia-Orozco et al., 2007).

Since energetics are considered to be a key factor in limiting organisms' adaptation to extreme temperatures, several invertebrate species inhabiting marine polar environments are known to show a remarkable plasticity as regards their cellular system. Such adaptations may include an increasing number of mitochondria per cell as the temperature decreases as well as differences in the mitochondrial characteristics relating to the species' lifestyle, from motile species to sedentary ones (Peck, 2002). Studies in the mitochondrial function of the eurythermal polychaete *Arenicola marina* have concluded that invertebrates inhabiting higher latitudes – and consequently exposed to cold temperatures – showed higher oxygen consumption, mitochondrial densities and mitochondrial capacities when compared with those organisms living at lower latitudes with higher temperatures (Sommer & Portner, 1999; Peck, 2002). This adaptation of cold-acclimatised organisms is thought to occur in order to equate the level of metabolic activity present at warmer temperatures.

Among other important environmental factors affecting the bioenergetic state of organisms, marine invertebrates face large daily fluctuations in the dissolved oxygen concentrations of water, as well as wide salinity changes between open ocean and coastal waters - where many species live at least during one specific stage of their life cycle - (Dall et al., 1990). Such variations can adversely affect some species whose physiological mechanisms usually do not allow them to cope with low oxygen levels (as oxyregulators) or to handle salinity changes (as osmoregulators). However, several species are able to swim or move from one place to other, searching for a suitable site to grow, reproduce and survive (Hochachka &

Somero, 2002; Abele et al., 2007). Nevertheless, other invertebrate species are highly adapted to live in extreme conditions such as those living in hypoxic or even anoxic environments, like the brine shrimp *Artemia franciscana* (Eads & Hand, 1999; 2003).

As has previously been stated, this chapter reviews the current state of knowledge of the mitochondrial function of invertebrate species. It asks two central questions: 1) How are invertebrates able to adapt to such diverse environmental conditions by using a common set of structures and mechanisms – their mitochondrial machinery – to fulfil their energy requirements along their entire life cycle? 2) Is it really important to understand the role of mitochondria in the life history of invertebrates? This chapter also includes original data on crustacean responses to the external factors affecting such mitochondrial functions as hypoxia, starvation and the energetically expensive molt cycle.

2. The highly conserved mitochondrial machinery of invertebrates: Same functions, different challenges

Following the endosymbiotic origin from primitive bacteria – at least 2 billion years ago – when atmospheric oxygen levels rose and subsequently remained relatively steady, mitochondria have experienced large changes among species, from α-proteobacteria to mammals. During the adaptation process of organisms to their new dynamic environment, some mitochondrial characteristics have remained highly conserved even among distantly related species, such as their rod shape - the overall structure including two phospholipid membranes – and, with some exceptions, their conserved characteristic genome content of 22 tRNAs, 2 rRNAs, and 13 genes encoding protein subunits of the enzymes from the oxidative phosphorylation system (OXPHOS) (Boore, 1999; Gray et al., 1999).

Besides mitochondrial encoded proteins, a significant fraction of the original mitochondrial genes have moved to the nucleus. Thus, in the mammalian mitochondria, approximately 76 subunits – which are part of the respiratory chain – are encoded by nuclear genes, and all of them must be imported into the mitochondria. The complete protein machinery involved in mtDNA replication, transcription and translation (including all of the ribosomal protein subunits) is encoded by nuclear genes (St. John et al., 2005; Falkenberg et al., 2005). Furthermore, several of these imported proteins are highly conserved among species, some of them accomplishing key roles as subunits alpha and beta of the ATP-synthase, which are part of the catalytic sites of the enzyme (Martinez-Cruz et al., 2011).

In addition to those key proteins that maintain a conserved function, hundreds of new proteins have been described among invertebrate species as being imported to mitochondria, each presumed to participate in at least one of the large number of metabolic pathways occurring in this organelle. However, its major conserved function allows mitochondrion to produce – from food assimilated compounds via oxidation – the proton motive force that drives ATP synthesis (Rich & Marechal, 2010). This complex process produces 95% of the cellular ATP that cells need for biosynthesis, transport and motility (Wilson et al., 1988; Dudkina et al., 2008; Diaz, 2010), and any significant change in the system could result in deleterious consequences for the whole cell metabolism and – consequently – reduce its efficiency or provoke its death (Mayevsky & Rogatsky, 2007).

Throughout the years (and mostly based in the study of human pathologies) researchers have found that mitochondria are involved in various critical functions – such as thermoregulation – in the synthesis of essential molecules – such as phospholipids and heme – in the programmed cell death or apoptosis of mediating multiple cellular signalling pathways (Ryan & Hoogenraad, 2007). Mitochondria are also essential in the cholesterol metabolism and the detoxification of ammonia in the urea cycle. In addition, there is a close relationship between mitochondria and different cell types. It is well known that the number of mitochondria in individual cell types varies according to their function and energy requirements (St. John et al., 2005; Chen & Chan, 2009). Thus, highly energetic tissues as the flight muscle of flying insects and the midgut gland of crustaceans are known to contain a large number of mitochondria, just as occurs in the skeletal muscles of vertebrates during endurance training (Harrison & Roberts, 2000).

Mitochondria are known as dynamic organelles that cannot be made *de novo*, and instead they divide through a highly regulated process called mitochondrial fission, mediated by a defined set of new proteins recruited from the cytoplasm, which are added to pre-existing sub-compartments and protein complexes to a point whereby the organelle grows and divides (Ryan & Hoogenraad, 2007). Furthermore, mitochondria are now seen as a set of organelles that are able to migrate throughout the cell, to fuse and divide regulating mitochondrial function (Chen & Chan, 2009).

Recent findings have also confirmed the existence of dynamic mitochondrial supercomplexes – defined as the association of protein complexes distributed along the inner mitochondrial membrane – on mammals, plants, yeasts (*Yarrowia lipolytica*), and bacteria (Nübel et al., 2009; Wittig & Schägger, 2009; Dudkina et al., 2010). Complexes I, III and IV are able to associate in order to promote electron transport as single OXPHOS complexes or else as a supercomplex called respirasome (I + III$_2$ + IV$_{1-2}$) both of which can autonomously carry out respiration (Wittig et al., 2006). Furthermore, complex V – the mitochondrial F_1F_OATP-synthase – is associated to form dimeric, trimeric and tetrameric organisations (Dudkina et al., 2008). Unfortunately, to our knowledge, there are no reports confirming the existence of these mitochondrial protein associations from invertebrate species.

A general description of the most recent advances covering mitochondrial enzymes participating in the electron transport chain and the OXPHOS, including some particular findings on the enzymes of some invertebrate species, is presented below:

2.1 Complex I, NADH: Ubiquinone-oxidoreductase (EC. 1.6.5.3)

Is an enzyme which provides the input to the respiratory chain by catalysing the transfer of two electrons from NADH from - glycolysis - to ubiquinone, and which utilises the free energy released in this redox reaction for the translocation of four protons across the membrane, from the matrix to the intermembrane space. The proton translocation from the mitochondrial matrix generates the proton-motive force required for ATP synthesis at the end of the respiratory chain during oxidative phosphorylation (Friedrich & Weiss, 1997; Dudkina et al., 2008). However, this proton-pumping enzyme is the largest, most complicated and least-well understood of the respiratory chain (Zickermann et al., 2008). Another unconventional function of complex I is the generation of reactive oxygen species (ROS) – such as the superoxide ion (O_2^-) – and, even if it is not a strong oxidant, it is a precursor of most other ROS and, consequently, contributes significantly to cellular oxidative stress. In mammalian mitochondria, the superoxide production is predominantly produced by complex I (Turrens, 2003).

The scarce information available concerning mitochondrial complex I from invertebrates includes basic descriptive reports of the nucleotide sequences of the NADH subunits - most of them from the mitochondrial genome-, their proteins, and an interesting study of site-directed mutagenesis aiming to understand the subunits' function in model insect species such as *Drosophila spp.* (Tovoinen et al., 2001; Sanz et al., 2010).

In addition, the existence of an alternative oxidase (AOX) in the animal mitochondria has been confirmed. Previously, this enzyme – which catalyses the O_2-dependent oxidation of ubiquinol, producing ubiquinone and H_2O – was thought to be limited to plants, some fungi and protists. The major difference between complex I and AOX is that the electron flow from ubiquinol to AOX is not coupled to the generation of a proton motive force, decreasing energy conservation in oxidative phosphorylation. The complementary DNA sequence that encodes AOX in invertebrate species from the phyla Porifera, Cnidaria, Nematoda, Anellida, Mollusca, and Echinodermata, has been characterised and it has been suggested that it may contribute on the acclimation of animals to stress conditions, mainly when the cytochrome pathway is inhibited (McDonald et al., 2009).

2.2 Complex II, Succinate: Ubiquinone- Oxidoreductase (EC 1.3.99.1)

Also called Succinate Dehydrogenase (SDH), is a functional member of the Krebs cycle and the aerobic respiratory chain, and it couples the oxidation of succinate to fumarate with the reduction of quinone to quinol (QH_2). Most probably, this enzyme presents the most striking differences among the mitochondrial complexes in the electron transport chain and OXPHOS (Rich & Marechal, 2010). It must be noticed that the oxidation of succinate to fumarate is the only Krebs reaction that takes place in the mitochondrial inner membrane itself; this reaction does not participate in proton translocation from one side to the other of the inner mitochondrial membrane. The energy carrier flavin adenine dinucleotide (FAD) forms a part of complex II, and succinate oxidation begins after the binding of succinate to the enzyme. This covalent binding of FAD to the enzyme increases the redox potential to a level that allows succinate oxidation (Rich & Marechal, 2010).

Contrary to the four human and yeast mitochondrial complexes, which include subunits that are encoded by the mitochondrial genome, the four subunits of SDH are encoded in the nuclear genome (SDH1 to SDH4; Figueroa et al., 2002).

Early studies of complex II (SDH) from invertebrates reported the isolation of mitochondrial fractions from the body muscles of the worm *Nereis virens* and from the tail muscle of the lobster *Homarus gammarus*, and reported high activity in both enzymes (Mattisson, 1965). Unfortunately, there is scarce new information available concerning complex II in invertebrates. However, the study of mitochondria from parasite species – used as animal models – can be considered a framework that has guided our knowledge in the understanding of such critical endogenous processes as aging, mitochondrial dysfunction and the role of the organelle in apoptosis (Grad et al., 2008; Wang & Youle, 2009). Thus, it has been suggested that mitochondria may influence the longevity of the nematode *Caenorhabditis elegans* through the rate of ROS production and by the stress-evoked signals that are known to act in a cell-non-autonomous manner during mitochondrial protein regulation (Durieux et al., 2011). Furthermore, *C. elegans* has been used as a model to investigate the mitochondrial mechanisms of human aging and tumourigenesis by studying the catalytic effects of mutation in the genes encoding the SDH iron-sulphur subunit. Promising results suggest that the SDH ubiquinone-binding site can become a source of superoxide and that the pathological consequences of SDH mutations can be diminished with antioxidants, such as ascorbate and N-acetyl-l-cysteine (Huang & Lemire, 2009).

2.3 Complex III, Ubiquinol: Cytochrome C Oxidoreductase or Cytochrome BC₁ (EC 1.10.2.2)

Is a multimeric enzyme complex involved in the transfer of electrons from ubiquinol to cytochrome C, and it is also coupled to electrons' transfer across the inner mitochondrial membrane. This bovine enzyme is formed by 10 nuclear encoded subunits, with only one encoded in the mtDNA (Xia et al., 1997). The catalytic mechanism of the enzyme includes the complex mechanism of the protonmotive Q-cycle that provides the additional efficiency of the energy conservation of the electrons transferred (Mitchell, 1976; Rich & Marechal, 2010). In such species as mammals and yeasts it has been observed that as the rate of electron transfer is reduced, the enzyme may leak electrons to molecular oxygen, promoting the formation of the superoxide ion. This mitochondrial dysfunction has been widely studied, and its role in the O_2 sensing pathway has been investigated because the increasing production of reactive oxygen species (ROS) is the result of organisms in hypoxic/anoxic conditions (Guzy et al., 2007). New evidence suggests that ROS generated by the mitochondrial complex III are required for the hypoxic activation of transcription factors such as HIF (Hypoxia Inducible Factor); however, this topic will be more extensively discussed below.

The mitochondrial complex III from invertebrates has been poorly studied, but recent reports about these species confirm the importance of studying its basis and applications. An interesting example is the study about the control of Chagas disease, which severely affects the health of the human population in Latin America and which is caused by the protozoan parasite *Trypanosoma cruzi*. Genes et al. (2011) reported such bacteria species as *Serratia marcescens* biotype A1a, which is regularly found in the gut of the vector insect *Rhodnius prolixus*, and which demonstrates the trypanolytic activity conferred by prodigiosin. Prodigiosin is a potent bacterial tripyrrolic compound with various biological activities. This study suggests the abnormal mitochondrial function of *T. cruzi* since prodigiosin inhibits the mitochondrial complex III, affecting subsequent oxidative phosphorylation.

2.4 Complex IV, Cytochrome C oxidase (EC 1.9.3.1)

Is the terminal enzyme of the electron transport chain and it catalyses the reduction of molecular oxygen to water. The reduction of oxygen by this enzyme – which is responsible for biological energy conversion in mitochondria (Belevich et al., 2010) – is also linked to the translocation (pumping) of four protons across the membrane. This movement of electrons is subsequently coupled to ATP synthesis by the ATP-synthase (Khalimonchuk & Rödel, 2005). The cytochrome C oxidase (CO) has been described as one of the electron transport chain elements which is highly affected by changes in oxygen levels – since cytochrome C reduction is oxygen-dependent – and becomes more reduced when oxygen levels increase (Wilson et al., 1988).

The CO from eukaryotes consists of 11-13 subunits, depending on the species. It belongs to the family of heme-cooper enzymes, some of them suggested as hypoxia sensors. The enzyme is highly regulated by transcription factors, hormones, lipid membranes and the second messengers that control its activity (Ripamonti et al., 2006; Semenza, 2007; Fontanesi et al., 2008). As observed in other mitochondrial complexes, CO also includes mitochondrial encoded genes as subunits CO1, CO2, and CO3 which form the functional core of the enzyme; the rest are nuclear-encoded subunits and their functions – even in the most studied animal models – remain unclear, although they are assumed to participate in the assembly, stability and regulation of the enzyme (Rich & Marechal, 2010). Moreover, CO is also regulated by the existence of various isoforms from each nuclear-encoded subunit which is known to be tissue- and specie-specific (e. g. CO5a and CO5b, CO6a, CO6b and CO6c, and CO7a, CO7b, CO7c, etc.; Diaz, 2010).

The CO genes' expression and the activity of the enzyme are known to be affected by external factors. In crustacean species, such as the grass shrimp *Palaemonetes pugio*, the gene expression of subunits CO1 and CO2 is positively or negatively regulated by low dissolved oxygen concentrations in water (Brouwer et al., 2008). References and further reading may be available for this article.

In insects, as with the sweet potato hornworm *Agrius convolvuli*, diapause – the delay in development in response to regularly and recurring periods of adverse environmental conditions – is induced by low temperatures. During this physiological state, the neurological activity, oxygen consumption rate and metabolic levels are low compared to undiapause animals; and it has been found that the genetic expression of the CO1 subunit is down-regulated. When the organism terminates diapause, CO1 is up-regulated and the enzyme activity also increases (Uno et al., 2004). Other insect species, such as the cotton boll worm *Helicoverpa armigera*, show diverse responses during diapause: the levels of CO1 mRNA and enzyme activity are low, suggesting that the diapause state is different in each species (Yang et al., 2010).

In some species, CO participates in organism detoxification, as observed in the polychaetes *Hediste diversicolor* and *Marenzelleria viridis* which inhabit eutrophicated regions with low oxygen levels and high sulphide concentrations - where CO functions as an alternative pathway of oxidation - (Hahlbeck et al., 2000). In addition, when sulphide becomes hydrogen sulphide (HS) – a weak acid that occurs in marine and aquatic environments such as hydrothermal vents, mudflats and marshes – HS is known to reversibly inhibit CO activity, affecting the aerobic metabolism of certain species, such as the worm *Urechis caupo* (Julian et al., 1998).

2.5 Complex V, ATP synthase (EC 3.6.3.14)

Is a multimeric enzyme that transforms the kinetic energy of the protons' electrochemical gradient to synthesise the high energy phosphate molecule ATP. Nowadays, it is well-known that the enzyme can also hydrolyse ATP, functioning as an ATPase (Boyer, 1997; Tuena de Gomez-Poyou et al., 1999). This mitochondrial enzyme comprises a catalytic sector F_1 (composed by $\alpha_3\beta_3\gamma\delta\varepsilon$ subunits), and a transmembrane hydrophobic sector F_O (composed of at least three subunits: a, b_2 and c_{10-12}), both linked by a central and a peripheral stalk (Mueller et al., 2004). As in other mitochondrial complexes, this enzyme includes subunits encoded in both the nuclear and mitochondrial genomes, in a tightly coordinated process to assemble this multimeric complex (Itoi et al., 2003; Muhlia-Almazan et al., 2008).

During the oxidative phosphorylation process in mitochondria, the electron transport chain generates a proton gradient that is proposed to drive the rotation of Fo, a central rotor located in the inner mitochondrial membrane. This rotation movement is believed to reverse the rotation of the F_1 nanomotor, inducing – via a conformational change – the sequential release of ATP from three identical catalytic sites followed by the sequential synthesis of newly formed ATP from Pi +ADP at these sites (Cardol et al., 2005). Biochemical and structural studies of the F_1 sector from bovine enzymes have demonstrated that catalytic sites are integrated mainly by three β subunits that alternate with three α subunits. The three catalytic sites formed by these three pairs of α/β subunits are grouped in segments forming a sphere, which is connected to the γ subunit which connects F_1 to Fo (Lai-Zhang & Mueller, 2000).

Due to its complex structure and the dual role that the ATP synthase plays in cells, the current state of research concerning this mitochondrial enzyme is both abundant and relevant; however, for the majority of invertebrate taxa, the information regarding this enzyme appears to be almost non-existent, restricted to some insect species for the more studied models. Analyses of the mitochondrial transcriptome and proteome from these species – which have been exposed to different environmental conditions – have shown that the ATP-synthase subunits can be affected in their expression, and that specific subunits of this multimeric complex can also play additional roles in the mitochondrial function. These findings suggest that invertebrates are able to respond by changing their metabolism to maintain cell homeostasis.

In the fruit fly *Drosophila melanogaster* and the California purple sea urchin *Strongylocentrotus purpuratus* the gene expression of the ATP-synthase subunit alpha (*atpα*) was measured at early developmental stages, and it was found that the amount of mRNA varies throughout development in both species. Contrary results showed that during the larval stage the nuclear and mitochondrially encoded ATP synthase genes appear to be temporally co-regulated in *Drosophila*, although in the sea urchin this development pattern was not observed (Talamillo et al., 1998). In 2005, Kidd et al. analysed null mutants of the ATP-synthase subunit ε in *Drosophila* spp., and a dramatic delay in the growth rate of the first instar larvae that finally died was reported. In addition, in fly embryos the ATP-synthase activity had a six-fold reduction.

Most likely, the first two studies concerning the ATP synthase of crustacean species were published in 2001. The authors characterised the enzymatic properties of F_1 and evaluated its sensitivity to specific inhibitors and modulators in the gills of the freshwater crayfish *Orconectes virilis*; they included, as an important contribution, the standardised methods for isolating mitochondria from crustacean tissues and some results about their enzyme stability at different temperatures and pH conditions (Li & Neufeld, 2001a, 2001b).

Recent reports on the most-studied shrimp species – *Litopenaeus vannamei* – have characterised and studied several mitochondrial and nuclear encoded subunits from tissues such as muscles, gills, pleopods and the midgut gland (Muhlia-Almazan et al., 2008; Martinez-Cruz et al., 2011). The complementary DNA sequences of the *atp6* subunit encoded in the mtDNA and the *atp9* (a nuclear encoded subunit) were characterised and their deduced proteins, as major components of the F_O sector, were included in a molecular model which predicted that in the shrimp F_OF_1 ATP synthase the *atp9* oligomeric ring may contain 9-10 proteins (Figure 1; Muhlia-Almazan et al., 2008).

Over the last decade, the effects of a viral agent which provokes shrimp death have been deeply studied. The white spot syndrome virus (WSSV) is perhaps the most devastating shrimp disease, causing massive mortalities in global aquaculture systems (Sanchez-Paz, 2010). In 2006, Wang et al. analysed the gene expression profile of the fleshy prawn *Fenneropenaeus chinensis* in response to WSSV infection through cDNA microarrays. Genes including the ATP-synthase A chain and arginine kinase were found to be down-regulated during WSSV infection. Additional studies in other shrimp species, reported thirty additional genes which are involved in the antiviral process as part of the shrimp's defence system. One of the most interesting findings of these studies was that the interferon-like protein (IntlP) – known as an antiviral factor – showed increased expression in virus-resistant shrimp (He et al., 2005). Later, Rosa & Barraco (2008) suggested that the shrimp interferon-like protein (IntlP) is rather a region of the insect mitochondrial b subunit of the ATP-synthase, due to the high identity between both proteins (60–73%). Recently, Liang et al. (2010) have suggested the ATP-synthase subunit β (*atp* β) - earlier called BP53 – as a protein involved in the WSSV binding to shrimp cells that may play an important role in the antiviral defence system of shrimp against WSSV.

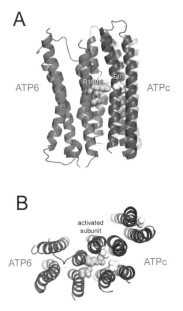

A) Ribbon lateral view, and B) Ribbon front view of the subunit ATP6 complex with three ATP9 subunits. The predicted functional residues are marked in both subunits, R160 from ATP6, and E99 from ATP9. (Taken from Muhlia-Almazan et al., 2008).

Fig. 1. Molecular Model of the ATP9- ATP6 Subcomplex from the Shrimp *L. vannamei*.

Transcriptomes and proteomes have provided a lot of information, not only about the characteristics of specific sequences of nucleotides or amino acids, but also about the proteins' structure and function in invertebrate organisms under diverse environmental conditions (Clavero-Salas et al., 2007). Moreover, novel proteins have been reported as accessories to the mitochondrial protein complexes in invertebrates species, such as the ticks *Ornithodoros moubata* and *O. erraticus*, where six novel proteins similar to the ATP synthase subunit 6 (*atp6*) were identified in the salivary glands. These proteins are attractive targets for controlling ticks and tick-borne pathogens (Oleaga et al., 2007).

Actually, and based in the mitochondrial highly conserved function, generic models of the electron transport chain in mitochondria have been constructed using bioinformatic tools to predict how the rate of oxygen consumption through the system – and the redox states of some intermediates such as NAD/NADH, ubiquinone, and cytochromes – respond to physiological stimuli such varying oxygen levels and other rapid energy demands (Banaji, 2006).

Ultimately, it is remarkable that the mitochondrial function has remained in all animal species through its long and peculiar evolutionary history and under the influence of variable selective pressures. Moreover, structural and biochemical adaptations promoting highly effective mitochondrial functions have allowed organisms to inhabit unusual environments.

3. The Invertebrates mitochondrial genome

The study of the mitochondrial genome has provided enormous amounts of information from which it has become feasible to infer the origin of species by using comparative and evolutionary genomics (Jiang et al., 2009) in order to understand the ancient phylogenetic relationships among species, to comprehend population genetics (Boore et al., 1995; Boore, 1999), and to recognise the mechanisms coordinating the nuclear and mitochondrial genomes so as to synthesise a large number of functional proteins located in this organelle.

To date, the mtDNA of several invertebrates has been sequenced and characterised, including ascidians (Yokobori et al., 1999), echinoderms (Jacobs et al., 1988; Asakawa et al., 1995), insects (Clary & Wolstenholme, 1985), nematodes (Okimoto et al., 1992), molluscs (Yu & Li, 2011; Cheng et al., 2011), and various crustacean species such as shrimp and crabs (Staton et al., 1997; Shen et al., 2007; Peregrino-Uriarte et al., 2009). Several reports have shown that the mitochondrial genome of invertebrate species varies, and ranges between 12 and 20 kbp. This may be due to contrasting ecological habitats or it may be a response to different selective pressures (Table 1).

Phylum	Species	mtDNA size (bp)	GenBank Acc. No.	References
Porifera	*Plakinastrella sp.*	19,790	NC_010217	Lavrov et al., 2008
	Negombata magnifica	20,088	NC_010171	Belinky et al., 2008
	Aphrocallistes vastus	17,427	NC_010769	Rosengarten et al., 2008
Cnidaria	*Hydra oligactis*	16,314	NC_010214	Kayal & Lavrov, 2008
	Aurelia aurita	16,937	NC_008446	Shao et al., 2006
	Fungiacyathus stephanus	19,381	NC_015640	---
Platyhelminthes	*Symsagittifera roscoffensis*	14,803	NC_014578	Mwinyi et al., 2010
	Clonorchis sinensis	13,877	JF729304	Cai et al., 2011
	Taenia taeniaeformis	13,647	NC_014768	Liu et al., 2011
Rotifera	*Brachionus plicatilis*	12,672	NC_010484	Suga et al., 2008
Acanthocephala	*Leptorhynchoides thecatus*	13,888	NC_006892	Steinauer et al., 2005

Phylum	Species	mtDNA size (bp)	GenBank Acc. No.	References
Nematoda	*Caenorhabditis elegans*	13,794	NC_001328	Wolstenholme et al., 1994
	Necator americanus	13,605	AJ417719	Hu et al., 2002
Onychophora	*Oroperipatus sp.*	14,493	NC_015890	Segovia et al., 2011
Brachiopoda	*Laqueus rubellus*	14,017	AB035869	Noguchi et al., 2000
Echinodermata	*Acanthaster planci*	16,234	NC_007788	Yasuda et al., 2006
	Strongylocentrotus purpuratus	15,650	NC_001453	Qureshi & Jacobs, 1993
	Cucumaria miniata	17,538	NC_005929	Arndt & Smith, 1998
Mollusca	*Crassostrea gigas*	18,225	EU672831	Ren et al., 2010
	Cepaea nemoralis	14,100	NC_001816	Terrett et al., 1996
	Octopus minor	15,974	HQ638215	Cheng et al., 2011
Annelida	*Platynereis dumerilii*	15,619	AF178678	Boore & Brown, 2000
	Lumbricus terrestris	14,998	NC_001673	Boore & Brown, 1995
Arthropoda				
Subphylum Chelicerata	*Centruroides limpidus*	14,519	NC_006896	Davila et al., 2005
Subphylum Crustacea	*Litopenaeus vannamei*	15,989	DQ534543	Shen et al., 2007
Subphylum Myriapoda	*Scutigera coleoptrata*	14,922	NC_005870	Negrisolo et al., 2004
Subphylum Hexapoda	*Apis mellifera*	16,343	NC_001566	Crozier & Crozier, 1993

Table 1. Invertebrates' mitochondrial genome size of the species of different phyla.

Because of the wide variability of environmental conditions in which a large number of invertebrate species are distributed, several specific mtDNA-rearrangements have been found when compared with those observed in the mtDNA of mammals. Such novel arrangements include the mitogenome from the blue mussel *Mytilus edulis* (Hoffmann et al., 1992), and that of the fruit fly *Drosophila melanogaster* (Clary & Wolstenholme, 1985; Garesse, 1988) and the horseshoe crab *Limulus polyphemus* (Staton et al., 1997).

Also, some species – or groups of species – may lack some genes, such as nematodes whose mtDNA lacks a gene for ATP8 (Keddie et al., 1998), or cnidarians like the coral *Sarcophyton glaucum* which includes an unusual gene encoding an extra tRNA (Beaton et al., 1998). Moreover, major changes have been found in invertebrates' mtDNA, such as the mitochondrial genes of *Lumbricus terrestris*, which are all known to be encoded in the same strand and, unlike others, the genes coding A8 and A6 are separated by a long 2700 nucleotides fragment (Boore & Brown, 1995).

In 2006, the description of the mtDNA of the moon jellyfish (*Aurelia aurita*) was reported. It was surprising to find that mitochondria of this organism contain a linear genome, which became the first non-circular genome described in a Metazoan. Besides its linearity, its organisation involves two additional sequences of 324 and 969 nucleotides, the last (ORF969) encodes a putative family B-DNA polymerase, tentatively identified as *dnab*, which was previously only reported in algae mtDNAs (Shao et al., 2006). Subsequently, the linear mitogenome of Cnidarians of the genus *Hydra* was also reported, although it was found that it is fragmented as two linear mitochondrial "chromosomes" (mt1 and mt2) where all genes are unidirectionally-oriented (Voigt et al., 2008).

In addition, the invertebrate's mitochondrial genetic code differs from the universal/standard genetic code, and it is suggested that this is species-specific since several studies have identified some changes in animal mitochondrial code, as shown by Table 2 (taken from Watanabe, 2010). As observed in this table, invertebrate mtDNAs are largely represented by different changeable codons – depending upon the species. This is the case for the AUA codon which usually codes Ile in the standard genetic code but in the mitochondria of some species of Nematoda, Mollusca, Platyhelminthes and Vertebrata it encodes a Met (Himeno et al., 1987; Bessho et al., 1992). Also, in several species, the start codon differs from the AUG but still codifies a methionine, and in most of the species the stop codon is an incomplete codon, such as UA or U (Watanabe, 2010).

Codon (Universal code)	AUA (Ile)	AAA (Lys)	AGA (Arg)	AGG (Arg)
Vertebrates (human, bovine, rat, mouse, chicken, frog)	Met	Lys	Term	Term
Prochordates (ascidian, asymmetron)	Met	Lys	Gly	Gly
Echinoderms (sea urchin, starfish)	Ile	**Asn**	Ser	Ser
Arthropods	Met	Lys	Ser	Ser
Most (shrimp, daphnia)	Met	Lys	Ser	Ser
Insect (Drosophila)	Met	Lys	Ser	-
Molluscs (squid, octopus, Liolophura, Mesogastropoda)	Met	Lys	Ser	Ser
Nematodes (nematodes, ascaris)	Met	Lys	Ser	Ser
Platyhelminthes	Met	**Asn**	Ser	Ser
Most (Echinostomida, Trematoda)	Ile	**Asn**	Ser	Ser
Rhabditophora (Planaria)	Ile	Lys	Arg	Arg
Coelenterates (jellyfish, coral, sea anemone, hydrozoa)	Ile	Lys	Arg	Arg

Table 2. The relationships between the genetic codes of animal mitochondria. Modified from: Watanabe, 2010. Bold letter: non-universal codon; Term: termination codon.

Although, to date, the mitochondrial genes expression mechanisms are not fully understood, and the evolutionary processes by which the mitogenome suffers a rearrangement are not clear. It is proposed that a new order in genes' arrangements must preserve or facilitate those signals or mechanisms required for the transcription and processing of RNAs to accomplish the mitochondrial function in animal species (Boore, 1999).

The mitochondrial DNA from animal cells is known to be easily affected, since it is not protected by DNA-binding proteins or histones such as nuclear DNA. Several studies have found that mtDNA can be affected by aging, hypoxia and random events of mutation or insertion/deletion (rates of mutation for mitochondrial genomes are known to be much higher than those in the nuclear DNA) that can produce increased oxidative stress and high levels of ROS in this organelle. Defective proteins which result from altered mtDNA molecules cause defective mitochondrial function, as an impaired respiratory chain and increased electron leaks so as to finally generate larger amounts of ROS (Wei et al., 1998).

Insects' mitogenomes are known to be affected at the transcriptional level by chemicals, since the mtDNA copy number has been shown to increase to meet the bioenergetic demands of the organism, as observed in the fly *D. melanogaster* when exposed to tetracycline. Treatment with this antibiotic causes an energetic deficiency, promoting an up-regulation of the mtDNA copy number (Moraes, 2001; Ballard & Melvin, 2007).

4. Invertebrate challenges and how marine species spend energy

In most animal species, high energy levels in their bodies reveal fast growth, adequate energy storage, effective reproduction strategies and viable descendants with characteristic short life spans; however, reduced energy levels in a biological system results in affected gene expression, low survival rates and reduced metabolic rates and, therefore, a need on the part of physiological mechanisms to slow the ageing rate until environmental conditions are enhanced and higher energy levels are again reached (Stuart & Brown, 2006). In their natural habitat, many invertebrate species must undergo endogenous physiological processes during their life cycle, such as molting, starvation, quiescence and metamorphosis, among others. Many of these processes imply high energetic expense, causing a low energy status that reduces their ability to reach the adult stage (Hochachka & Somero, 2002).

The role of metamorphosis – one of the most amazing physiological endogenous processes in nature – becomes strikingly important when considering the large number of animal species that undergo metamorphic changes. Frequently, the energetic balance of holometabolous insects during metamorphosis is negative, because there is no energy gain and species must face all these changes by using any energetic reserves previously stored (Nestel et al., 2003).

During their larval stages, insects – such as Lepidopterans – show fast growth rates, as observed in the tobacco worm larvae of *Manduca sexta* which increases its mass 10,000-fold in just 16 days at the final larval instar (Goodman et al., 1985). The midgut epithelium of this species is a highly aerobic tissue that digests and absorbs nutrients, and transports ions at high rates. During metamorphic changes, the midgut epithelium is programmed to die and the larval midgut should maintain structural and functional integrity until the pupal epithelium is formed. During this process, ATP synthesis and mitochondrial function must be obligatorily maintained. Thus, organisms resolve this by reducing mitochondrial substrate oxidation, a clear indication that the electron transport chain may be a site of modulation during metamorphosis (Chamberlin, 2004).

Quiescence and estivation are also two responses that some species may display during unfavourable environmental conditions in which insufficient energy is available to grow and breed. These dormant states allow species to survive by reversibly down-regulating their metabolism to low levels for up to several years. Among invertebrates, many species show quiescent states at stress conditions, including nematodes, crustaceans such as the brine shrimp *Artemia franciscana* (Hand, 1998), the estivating pulmonate snail *Helix aspersa* (Pedler et al., 1996), and various insect species entering in diapause, such as *Helicoverpa armigera*. Studies have proposed that a coordination mechanism is required when animals enter into the dormant state so as to maintain cellular homeostasis by both energy-consuming and energy-producing pathways. During quiescence, *A. franciscana* can reduce its metabolism essentially to zero, this metabolic-rate suppression affects the mitochondrial respiratory capacity and the rates of ATP-consuming processes (Barger et al., 2003). In the embryos of *Artemia franciscana*, anoxia provokes the organism to enter into a quiescent state. During experimental gradual oxygen removal, various biochemical responses are observed, such as a pH decrease, the reduction of heat production and the depression of ATP levels. Also, genetic responses, such as the down-regulation of RNA transcription, are observed during quiescence (Hand, 1998).

Often, metabolic rates have been inversely related to the life span of mammals. Moreover, when mitochondrial respiration has been inhibited by RNAi techniques, the life span extends in *C. elegans* (Lee et al., 2003), and long-lived mutants of this nematode concomitantly show decreased metabolic rates (Stuart & Brown, 2006).

The process by which mitochondrial respiration affects or extends life span has been studied in several organisms, including yeasts, worms, flies and mice (Lee et al., 2010). Electron transport in mitochondria is the main producer of superoxide anion (O^-), which in turn generates several types of reactive oxygen species (ROS), as has been mentioned (mitochondrial Complex III). In fact, according to various studies, ROS are not only undesirable toxic metabolites promoting organism oxidative stress, but they are also molecules that participate in the mitochondria-nucleus's signalling pathways (Storz, 2006). Emerging data on *C. elegans* suggests a new described pathway where superoxide serves as an intracellular messenger, whereby with increasing superoxide concentration a signal transduction pathway is triggered, resulting in changes in the pattern of gene expression of nuclear proteins and which finally results in an increased life span (Yang & Hekimi, 2010). However, different mechanisms have also been proposed as being implicated in the aging process, such as diet restriction, ubiquinone deficiency and the hypoxic response (Klimova & Chandel, 2008).

At this point, this chapter would not be complete if the energetic costs of flying for insect species were to be omitted. This activity is probably the most expensive process recorded in nature. It is by now a well-known and remarked-upon fact that the metabolic rate during insect flight increases over 50-100 fold above the resting rate (Ellington, 1985). Thus, it is clear that the flight muscle of insects is the model tissue that many researchers have adopted in order to understand mitochondrial function since it is capable of effectively producing and hydrolysing large amounts of ATP (Sherwood et al., 2005). Insect flight is a highly oxygen-dependent process, and the flight muscle metabolism is fully aerobic; thus, it has been suggested that the amazing aerobic capabilities of insects are based on a highly efficient mode of oxygen delivery that includes their oxygen transport system in a well distributed system of tracheae and tracheoles (Wegener, 1996).

In addition, several studies have demonstrated that the function and energy needs of certain tissues are highly correlated with the number of mitochondria per cell (Robin & Wong, 1988). This agrees with the large quantities of mitochondria with pronounced cristae and large surface areas that are found in the flight muscle cells of the honey bee *Apis mellifera* (Suarez et al., 2000). To date, it is well-known that oxygen uptake rates in mitochondria cristae are much higher in the flying muscle of *A. mellifera* than that observed in mammals' mitochondria – this can explain the higher electron transport rates observed in such enzymes as cytochrome c oxidase, whose maximum catalytic capacity was recorded in this species during flight - (Suarez et al., 2000).

Besides the increase on the ATP hydrolysis rate during flight, other mitochondrial adaptations to the highly and continuous energy requirements of flying species have been reported, such as the remarkable dependence on the synthesis of energy-rich phosphate compounds like phosphoarginine. Phosphoarginine, as mentioned above, constitutes a usable pool of high energy phosphate (Hird, 1986) so as to maintain the high rate of ATP turnover in flying insects (Wegener, 1996).

In addition to the various metamorphic changes in their life, crustaceans undergo a frequent and cyclic process: molting. During the molt cycle, crustaceans are exposed to a temporary scarcity of food since they lack the ability to handle food until their new exoskeleton is synthesised. Several adaptive strategies have been recognised as being employed by these organisms so as to avoid the adverse effects of starvation, such as the storage of fuel compounds in their midgut gland (Sanchez-Paz et al., 2007), changes in locomotor activity (Hervant & Renault 2002), and a decrease in oxygen consumption (Morris et al., 2005). However, little attention has been paid to the bioenergetic consequences of starvation in shrimp; since the composition of food plays an important role in oxidative phosphorylation, the nutritional status of shrimp species, such as *Litopenaeus vannamei*, may affect its major bioenergetic functions.

In our lab, we have hypothesised that, due of its central role in the cell energy metabolism, the expression of genes encoding the different polypeptide subunits that compose ATP synthase during unpredictable episodes of food shortage may ultimately be modulated. Thus, we experimentally evaluate the effect of starvation in the gene expression of subunits *atpα*, *atpβ* and *atp*in the shrimp midgut gland, during a period of short-term food deprivation (5 days). Our results (Figure 2) show that the mRNA amounts from subunits *atpα* and *atpβ* which directly participate during ATP synthesis decreased as starvation time increased; however, no significant changes were observed in the mRNA amounts of *atp9*, which forms the oligomeric ring from Fo in the shrimp ATP-synthase.

Sanchez-Paz et al., (2007) reported a gradual decrease of glycogen in the midgut gland of the white shrimp as starvation progressed. After a 24 h starvation period, the glycogen content dropped by about 50%, which correlates with an increase of the *atp9* subunit after 24 h of starvation, suggesting that glycogen may be used as fuel to generate ATP and pyruvic acid. As glycogen stores become depleted, the organism must increasingly rely on fatty acid catabolism as a source for ATP synthesis. In general, starved shrimp showed a sharp decrease in their midgut gland lipidic constituents for up to 120 h (more noticeable in acylglycerides).

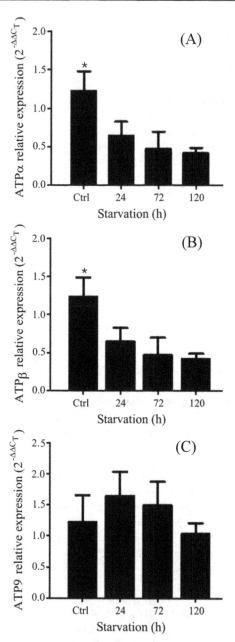

Fig. 2. The relative expression of A) ATPα, B) ATPβ and C) ATP9 mRNA in the midgut gland of the white shrimp *Litopenaeus vannamei* in response to a short-term starvation period. Expression values are given based on normalisation to L8. The data is represented as the mean and standard deviation of triplicate determinations. (*) Statistical significance was considered at $P < 0.05$.

Various studies have shown that during starvation-induced lipolysis there is a decrease in the amount of ATP, which was accompanied by a fall in some subunits of the FoF_1-ATP synthase (Vendemiale et al., 2001). It is well-known that starvation tampers with cellular detoxification systems and may expose cells to oxidative injury (Di Simplicio et al., 1997; Vendemiale et al., 2001), leading to an impaired production of ATP and a reduced uptake of substrates for mitochondrial metabolism. The results from our study, together with results from previous studies, prompt us to suggest that shrimp are capable to satisfy their energy demands through a complex combination of mechanisms that enables them to survive the adverse effects of food scarcity.

Due to its density, viscosity (800 times more dense and 50 times more viscous than air) and low oxygen solubility, water – as a respiratory medium – imposes difficulties for aquatic breathers in obtaining the necessary supply of oxygen from their surrounding environment so as to keep breathing and bringing oxygen into their systems. This process becomes more complicated when considering additional parameters (such as temperature, salinity and depth) affecting the dissolved oxygen concentration of seawater, causing additional constraints on marine species' development (Sherwood et al., 2005). All the species inhabiting marine environments should face these dynamic environmental conditions, which in over the last few decades have been seriously affected by a wide variety of anthropogenic activities, such as industrial and agricultural runoffs (Wu, 2002).

Several studies have found that marine invertebrates may respond to stress conditions by changes at the transcriptional level. In crustacean species such as the crab *Eriocheir sinensis*, different gene expression profiles from gills were characterised during acclimation to high cadmium concentrations in water. Analyses have revealed over-expressed genes, such as disulphide isomerase, thioredoxin peroxidase and glutathione S-transferase. Under the same conditions, ATP synthase beta, alpha tubulin, arginine kinase, glyceraldehyde-3-phosphate dehydrogenase and malate dehydrogenase were down-regulated. The results demonstrated that acute and chronic exposure to waterborne cadmium induced a decreased abundance of the transcript-encoding enzymes involved in energy transfer; this suggests that chronic metal exposure induced an important metabolic reorganisation (Silvestre et al., 2006).

Some other species which face high cadmium concentrations are marine intertidal molluscs, such as oysters, which live in estuaries were fluctuating temperatures and levels of trace metals are known to directly affect mitochondrial function. Isolated mitochondria from the oyster *Crassostrea virginica* which were exposed to low cadmium concentrations (1 $\mu mol \cdot L^{-1}$) resulted in a progressive uncoupling that increased with the increasing dose of cadmium; this response agrees with that observed in mammals. However, unlike mammals, molluscs are ectotherms and the exposure to the combined effects of high temperatures and cadmium concentrations severely affected mitochondrial function since elevated temperatures increased the sensitivity of this organelle to cadmium and promoted an increase in the rate of ROS production (Sokolova, 2004). These results highlight the key role of temperature in the mitochondrial system of ectotherm species.

Most invertebrates are described as ectotherm species because their body temperatures vary with the environment. At very low temperatures, polar marine invertebrates were expected to show low metabolic rates, as previously observed in Antarctic fish; however, in 1999 Sommer & Portner found important intraespecific differences in the mitochondrial function of the polychaete *Arenicola marina* from the North Sea and the colder White Sea. Their results concluded that invertebrate life is more costly at higher latitudes, where oxygen uptake, tissues mitochondrial densities and mitochondrial capacities were higher.

Remarkable abilities have been recorded in invertebrate species inhabiting extreme environments. The term "metabolic plasticity" perfectly describes such organisms as the intertidal periwinkle snail *Littorina littorea*, which has the ability to deal with very low temperatures and also to tolerate the changing environmental conditions imposed by the tidal cycle, implying continuous oxygen deprivation (Storey, 1993). Besides the biochemical and physiological mechanisms previously identified in this species, the over-expressed gene encoding a metallothionein (MT) was recently found during the exposure to low temperature and anoxic conditions of the tissues of *L. littorea*. Since thermogenesis is a process that requires high oxygen consumption and since it is also accompanied by a sharp rise in reactive oxygen species (ROS) generation, the authors describe the ability of MT to function as an antioxidant and as a reservoir of essential metals that contributes to survival under these conditions (English & Storey, 2003).

The deep sea hydrothermal vents are a different type of extreme environment where thermophilic species such as the Pompeii worm *Alvinella pompejana* inhabit. Shin et al. (2009) studied the structure and biochemical characteristics of the Cu,Zn-superoxide dismutase (SOD) of this species and found striking similarities between this enzyme and that of humans, but with an enhanced stability and catalysis – characteristics that may mean that this enzyme is potentially suitable for scientific and medical application. Other mitochondrial proteins have been proposed as a part of gene therapy for devastating human diseases by preventing the cell damage caused by oxidative stress. AOX – the mitochondrial alternative oxidase previously mentioned – is suggested to work in any cell, becoming chemically active only when it is required. AOX is provided to the cell by engineering a gene from a marine invertebrate snail *Ciona intestinalis*; this protein is under analysis as a therapeutic tool tested in mammalian disease models (Hakkaart et al., 2006).

5. How do invertebrates face hypoxia?

Hypoxia is probably one of the most studied factors affecting the central metabolic pathways of living organisms, including invertebrates. Aquatic species usually face hypoxic events in freshwater or marine environments as a daily cyclic routine in the shallow waters of lagoons, estuaries and mangroves during the dark hours, when plants and algae do not produce oxygen and organic matter is continuously oxidised (Dall et al., 1990). However, nowadays the frequency, abundance and severity of hypoxic events in coastal waters have increased due to anthropogenic activities resulting in deteriorating environments affecting marine organisms (Diaz, 2001). It is well known that hypoxia depresses the growth rate of marine animals, as it disturbs metabolic pathways and promotes the reallocation of energy resources (Wei et al., 2008; Wang et al., 2009).

Several studies have examined the physiological responses of invertebrate species to hypoxia, such as growth, stress resistance and even behaviour patterns in aquatic species able to vertically and horizontally migrate through the water column to reach more oxygenated zones (Eads & Hand, 2003; Burgents et al., 2005; Abe et al., 2007; Seibel, 2011). In fact, among invertebrates there are hypoxia-tolerant species, such as bivalve molluscs and annelids, with highly adapted structures and mechanisms to deal with hypoxia, and some others, such as crustaceans, whose tolerance to hypoxia depends on their habitat, food, and energy needs. Unfortunately, the responses to hypoxic conditions – at the molecular and biochemical levels – of the mitochondrial proteins and enzymes that participate in the respiration process are still poorly studied for most invertebrate species.

The main physiological responses from invertebrates to hypoxia are somewhat similar to those from vertebrates since in the reduction or absence of oxygen, animal cells are not able to produce enough energy to survive. Such general responses are clearly a legacy of the evolutionary past from ancestral forms and they serve adaptive ends. In marine species, such as crustaceans and molluscs, reduced oxygen consumption and metabolic rates have been confirmed during hypoxia; in addition, glucose utilisation and lactate accumulation as indicators of a switch to anaerobic metabolism have been detected at low oxygen concentrations in water (Racotta et al., 2002; Martinez-Cruz, 2007; Soldatov et al., 2010). In the brine shrimp *A. Franciscana*, the intracellular pH falls at anoxia, heat production is reduced and ATP concentrations are also depressed to low levels (Hand, 1998; Eads & Hand, 2003).

A large amount of information is now available about the changes at the transcriptional level promoted by hypoxia in invertebrates, most of it concerning aquatic species. In our lab, we have evaluated the effects of hypoxia in the gene expression of F_OF_1 ATP synthase subunits, such as *atp9, atp6, atpα, atpβ, atpγ, atpδ, and atpε*, in different tissues of the white shrimp *L. vannamei*. Results show a general trend towards increase the amount of mRNA as oxygen concentrations decrease (Martinez-Cruz, 2007; Martinez-Cruz et al., 2011; Martinez-Cruz et al. in preparation). Also, significant changes in the amount of mRNA from the mitochondrial- and nuclear- encoded subunits of the ATP synthase were detected at different molt stages and tissues, according to the energy requirements of each stage and the specific requirements of the function of each tissue (Muhlia-Almazan et al., 2008). Chronic exposure to severe hypoxia (1.5 mg/mL during 7 days) also causes the increased transcription of mitochondrial-encoded genes, such as the 16S, CO1, and CO2 subunits from the cytochrome C oxidase in the grass shrimp *Palaemonetes pugio* (Brouwer et al., 2008). To date, microarray technologies have revealed a set of genes that are up- and down-regulated in *P. pugio* during chronic, acute and moderate hypoxia; the results revealed that various genes encoding mitochondrial proteins were affected (Li & Brouwer, 2009).

In the absence of oxygen, animal cells activate transcription factors – such as the well-studied vertebrates hypoxia-inducible factor (HIF) – which has been reported in invertebrates from worms to flies (Semenza, 2007). When activated, HIF leads the organism to exhibit metabolic adaptation to hypoxia by regulating the genetic expression of some proteins and enzymes involved in central biological processes such as glycolysis, erythropoiesis, breathing and angiogenesis so as to maintain cell homeostasis (Klimova & Chandel, 2008). In the shrimp *P. pugio*, a homolog protein to HIF-α called gsHIF was found in this hypoxia-tolerant species. It includes all the conserved domains of vertebrates' HIF proteins, and an additional polypeptide sequence of 130 residues that has not been found in databases, and its participation in the functional properties of the protein has not yet been determined (Li & Brouwer, 2009). In the white shrimp *L. vannamei*, HIF-1 is a heterodimer formed by two subunits: HIF-1β, which is constitutively expressed in shrimp cells and HIF-1α, which is differentially expressed in hypoxic conditions. HIF-1 is suggested in crustaceans to be the master regulator that senses decreased oxygen availability and transmits signals promoting the physiological responses mentioned above (Soñanez-Organis et al., 2009). Additional functions have been attributed to HIF in coral species, such as *Acropora millepora*, where the diel cycle in the central metabolism appear to be governed by the circadian clock and regulated by the HIF system operating in parallel (Levy et al., 2011).

As a part of the HIF-regulated metabolic responses to hypoxia in invertebrates, the activities of specific enzymes – most of them part of the central metabolism – are known to increase. In bivalves such as *Anadara inaequivalvis*, the increased activities of enzymes – such as malate and lactate dehydrogenases – were detected at hypoxia (Soldatov et al., 2010). Also, increases in the catalase and GST activities during anoxia in the estuarine crab *Chasmagnathus granulate* have been observed. It has been suggested that such responses may be a strategy to prepare the organisms for oxidative stress in an effort to protect tissues against oxidative damage during re-oxygenation. An important decrease in SOD activity (which occurred after aerobic recuperation) was also detected; and it could have been caused by the accumulation of hydrogen peroxide production during re-oxygenation (de Oliveira et al., 2005).

At normoxia, the small levels of ROS produced by the metabolism in normal animal mitochondria come from carrying electrons along the mitochondrial complexes I, II, and III (Turrens, 2003). However, when oxygen levels are reduced, the presence of the final electron acceptor in the mitochondrial respiratory chain fails, producing a reduction in the rate of electron transport and a decrease in oxygen consumption. Under these conditions, the membrane potential increases as does ROS production (Guerrero-Castillo et al., 2011).

It has been reported that in invertebrate species considered to be hypoxia-tolerant, the absolute rate of H_2O_2 production is at least an order of magnitude less per mg of mitochondrial protein than that measured on mammalian species (Abele & Puntarulo, 2004). However, some other species which are not tolerant to hypoxia tend to produce higher levels of ROS at low oxygen levels; thus, it is suggested that they display alternate pathways in order to maintain the mitochondrial respiratory rate and avoid an over-production of ROS (Guerrero-Castillo et al., 2011).

Nowadays, the alternative mechanism of proton sinks has been evidenced in invertebrates since uncoupling proteins (UCPs) have been identified in these species (Abele et al., 2007). Such proteins have been involved in various functions, including thermoregulation, body composition, antioxidant defence and apoptosis. UCPs are thought to dissipate the proton gradient across the inner mitochondrial membrane and may help in controlling ROS production (Yu et al., 2000).

In *Drosophila*, an UCP5 protein over-expressed in a heterologous system has shown to have similar functional abilities to an uncoupling protein (Fridell et al., 2004), while in the marine eastern oyster, *Crassostrea virginica*, UCP5 is represented by two transcript forms: UCP5S (small) and UCP5L (large). However, their function has not been determined since its gene expression is not affected by hypoxia, cadmium exposure or different temperatures (Kern et al., 2009). In addition, a novel protein (UCP6) in invertebrates is considered to be an ancestral form of the vertebrates UCP1, UCP2, and UCP3 (Sokolova & Sokolov, 2005).

In mammals, it is known that less-severe hypoxia induces protective mechanisms. This phenomenon – called hypoxic preconditioning (HP) – appears in two forms: immediate preconditioning (which occurs only a few minutes after a sub-lethal hypoxic episode and declines after 4 h) and delayed preconditioning (which requires gene expression changes and takes place 12 to 24 h later and can last for days) (Dirnagl et al., 2009). In the nematode *C. elegans*, the delayed form of HP has been found to induce unfolded protein response pathways – at this point, misfolded proteins serve as early hypoxic sensors that trigger signalling pathways to induce a hypoxia protective response (Mao & Crowder, 2010).

6. The role of mitochondria in invertebrate programmed cell death (Apoptosis)

Besides the various functions just described, mitochondria also acts as the arsenal of the cell. Numerous and complex processes, still poorly understood, can trigger the release of mitochondrial components into the cytoplasm and subsequently induce cellular apoptosis of the organelle (Hengarter, 2000). It is not our intent here to provide exhaustive coverage of all the issues relating to apoptosis in great detail, but rather to give the reader a basic description of the process – to highlight its importance and to show the challenges that those interested in this topic will face.

As has been mentioned, studies in invertebrate biology are paramount to an understanding of biodiversity and to the search for potential uses for their metabolic capabilities and products for biotechnologies. Besides, comparative sciences may facilitate the use of invertebrate models in understanding the biology and pathology of farmed animals and humans. This is due – in spite of differences in the biochemical, physiological, and cellular characteristics that make invertebrates and vertebrates so obviously different – to the fact that most parts of such grades of their biology have remained similar in both groups through their evolution. For example, invertebrate cells – whether wounded by harsh environments or by the expression of abnormal proteins – die as do vertebrate cells, indicating that the powerful advantages of invertebrate molecular genetics might be successfully used for testing specific hypotheses about human diseases, for the discovery of drugs and for non-biased screens for suppressors and enhancers of maladies (Driscoll & Gerstbrein, 2003). The same criteria apply for all cellular functioning, as for apoptosis.

Apoptosis (from the Greek: "falling off") – or programmed and regulated cell death and elimination – is a pivotal process in embryogenesis, the orderly elimination of wounded or infected cells, and the maintenance of tissue homeostasis. The process is so important that it is estimated that on a daily basis the human body must get rid of approximately 10^{10} cells. Through apoptosis, cells die quietly in a controlled, regulated fashion; while in another forms of cell death – such as in necrosis – a series of uncontrolled events occur leading to serious and irreversible damage. Given the proper conditions, apoptosis destroys the cell swiftly and neatly. In contrast, necrosis causes the rupture of the cell, releasing its content into the surrounding tissue. Tampering with apoptosis may result in devastating health problems, such as cancers, immune diseases, neurodegenerative disorders and the proliferation of viruses. Apoptosis is executed by a variety of membrane, organelle, cytoplasmic and nucleus signalling, and initiator and effector molecules, including a subfamily of cysteine proteases known as caspases (Jiang & Wang, 2004).

In mammals, the active role of mitochondria in apoptosis induction has been well-established. In invertebrate models of apoptosis, such as the fly *Drosophila melanogaster* and the worm *C. elegans*, the role that mitochondria play during apoptosis and, in particular, during apoptosis initiation is less clear (Rolland & Conradt, 2006). While key regulators of apoptosis in *Drosophila* and *C. elegans* have been found in association with mitochondria, the significance of these associations has not been rigorously tested.

The regulated destruction of a cell is a basic process in Metazoa, as multicellular animals are obligated to remove damaged or harmful cells. During apoptosis, cells die in an orderly, regulated sequence of molecular, biochemical, and cellular processes. According to the endosymbiotic theory, the origin of apoptosis is currently regarded as the result of molecular interactions in which some components of a signal transduction pathway affects other pathways through interaction of some initiator and effector proteins. Accordingly, apoptosis could have arisen simultaneously with – and as a by-product of – endosymbiosis (Kroemer, 1997). However, it has also been proposed that apoptosis may be the result of the acquisition of the aerobic metabolism by early eukaryotes (Frade & Michaelidis, 1997).

Apoptosis is a unique phenomenon of tissue kinetics as it can be said that life is critically controlled by the operational centre of cell, the nucleus. Instead, death is a process controlled by the power house of the cell, the mitochondria. Thus, even cells lacking nucleus commit apoptosis. In general, the two-step membrane depolarisation and free radical release taking place in the mitochondria may trigger apoptosis. This in fact is not so peculiar if we understand that mitochondria were once free-living bacteria which did not need an external gene control for achieving their functions. Once each came into symbiosis forming a eukaryotic cell, it retained some capacity to operate partially independently.

There are several major apoptotic pathways, but the most well-known and studied are the extrinsic and the intrinsic pathways, which respond to different environmental and cellular challenges in vertebrates. The intrinsic pathway is also called the mitochondrial pathway because of the involvement of mitochondria. There are mitochondrial proteins that induce this process (proapoptotic) and others that limit cell death (antiapoptotic). Both proteins interact so as to cooperate and govern the cell's fate. Also, the origin of the activation signals of apoptosis taking place on the mitochondria is a clue molecule, cytochrome C (Cyt C), which is released from the mitochondria to form the apoptosome complex. The intrinsic pathway – with some differences – is a mostly conserved pathway among metazoans (for a comprehensive review look at Wang & Youle, 2009). Cyt C is a key component of the apoptosome complex for activating the initiator caspase-9 after its release from mitochondria. Under non-apoptotic conditions, Cyt C is kept inside the respiratory chain. Against some cellular challenges, like the alteration of the DNA in the mitochondria or the nucleus, Cyt C is released from its membrane, crossing the external membrane and initiating the formation of the apoptosome complex. In essence, mitochondrial proteins – like Cyt C and caspases – are not hired guns and during non-apoptotic conditions they are responsible for various basic mitochondrial roles for normal cell functioning. The compartmentalisation of such mitochondrial proteins isolates them from interacting with partners or targets, a mechanism to prevent the unwanted activation of apoptosis in normal cells. Only after their appropriate release into the cytoplasm do such proteins play the role of triggers to initiate the cell's suicide.

The classical invertebrate model organisms for the study of apoptosis are *C. elegans* and *Drosophila*. In spite of the fact that the regulators of apoptosis have been found in such model organisms, the involvement of mitochondria in apoptosis is not conclusive. So far, no irrefutable evidence of the release of Cyt C from the intermembrane space has been found. Also, the involvement of Cyt C in the apoptosome formation in *Drosophila* is controversial, and some evidence suggests that Cyt C is not necessary (Rolland & Conradt, 2006).

The current evidence indicates that the whole process of apoptosis -including the involved proteins and the regulation mechanisms- in crustaceans is far more diverse than has been assumed from the studies with model organisms. Recent studies have shown that several proteins in the apoptotic network are quite conserved between mammals and arthropods; however, it is clear that the integration of such homologous proteins in the physiology and pathophysiology of crustaceans needs further experimental assessment. Some unresolved questions regarding this topic are: how does the regulation of the process occur? Is crustacean apoptosis transcriptionally regulated, as in *Drosophila* (RHG ´killer´ proteins)? Or is it controlled by pro- and anti-apoptotic Bcl-2 family proteins, as in vertebrates? The issues that should be investigated in the short-term are whether the calcium-induced opening of the mitochondrion permeability transition pore (MPTP), commonly found on vertebrate species, also occurs in crustaceans. Furthermore, the study of the differences in the regulation of the intrinsic pathway of crustacean apoptosis will lead to an understanding of their adaptation to challenging environments; this is because marine organisms have to deal with seasonal as well as circadian changes in environmental variables. Some examples are UV radiation, temperature and dissolved oxygen, and even some biological stresses such as toxins that may vary over time. But this is not all: other variables that may inhibit apoptosis must be considered. "Characterisation of the players, pathways, and their significance in the core machinery of crustacean apoptosis is revealing new insights for the field of cell death"(Menze et al., 2010).

Apoptosis is a key host response to viral infection. Viruses that can modulate a host's apoptotic responses are likely to gain important opportunities for transmission. Here, we review recent studies that demonstrate that the particles of Invertebrate Iridescent Virus6 (IIV-6) (Iridoviridae, genus Iridovirus), or an IIV-6 virion protein extract, are capable of inducing apoptosis in lepidopteran and coleopteran cells, at concentrations 1000-fold lower than that required to shut-off the host's macromolecular synthesis (Williams et al., 2009). Throughout the process of pathogen–host coevolution, viruses have developed a battery of distinct strategies to overcome the biochemical and immunological defences of the host. Thus, viruses have acquired the capacity to subvert host cell apoptosis, control inflammatory responses, and evade immune reactions. Since the elimination of infected cells via programmed cell death is one of the oldest defence mechanisms against infection, disabling host cell apoptosis might represent an almost obligatory step in the viral life cycle. Conversely, viruses may take advantage of stimulating apoptosis, either to kill uninfected cells from the immune system or else to induce the breakdown of infected cells, thereby favouring viral dissemination (Galluzzi et al., 2008).

7. Conclusion and future perspectives

As stated by Van der Giezen in 2009 "over the last 5–10 years, it has become apparent that the organelle known as the mitochondrion is a much more fluid entity than generally believed," so "why should mitochondrion be the same in all eukaryotes while other cellular structures show such great evolutionary malleability?"

It is our belief that since natural selection has given invertebrates the opportunity to evolve in quick steps, a large window is opening in the field of mitochondrial research among these species, giving an outstanding opportunity to researchers to contribute to an increase in knowledge, not only because there is scarce information, but also because many species have shown special and unique characteristics that need to be explained.

At this point, the information reviewed clearly shows that invertebrates display remarkable physiological capabilities, including highly specialised mechanisms for adjusting mitochondrial functions to solve their energetic demands under the stressful conditions they usually face. These species also include within their systems ancient and novel molecules and structures acting to reach an adaptive state, from the increasing number of mitochondria per cell to the highly complex function of the HIF system.

It is also remarkable that the number of invertebrate species considered as potential models in the study of mitochondrial function has increased. New data on marine invertebrates, such as molluscs and crustaceans and non-*Drosophila* species, are emerging. Since there is still an immense lack of knowledge about invertebrates, important efforts in new animal models should focus on i) the description of mitochondrial systems in species inhabiting extreme environments, ii) the recognition and understanding of the causes and effects of mitochondrial disorders, and iii) the development of unsolved phylogenetic relationships among species and phyla. This may also open important opportunities for new biotechnological applications to better face the effects of global changes such as warming, hypoxic conditions and chronic stressors that specifically affect the central metabolic pathways in such species.

If the regulation of apoptosis in crustaceans is as varied as their diversity as a species, or at least their Families, then the potential for discovering novel biomolecules is immense. Such molecules may find uses in biotechnologies across diverse industries, including pharmacology. We endorse the hypothesis that an advanced knowledge in apoptosis will provide some clues about how crustaceans deal with viral infections and enable the proposal of feasible strategies to protect farmed crustaceans.

8. References

Abe, H.; Hirai, S. & Okada, S. (2007). Metabolic Responses and Arginine kinase Expression under Hypoxic stress of the Kuruma prawn *Marsupenaeus japonicus*. *Comparative Biochemistry and Physiology. Part A*, Vol.146, No.1, (January 2007), pp. 40-46, ISSN 0300-9629.

Abele, D. & Puntarulo, S. (2004). Formation of Reactive Species and Induction of Antioxidant Defence Systems in Polar and Temperate Marine Invertebrates and Fish. *Comparative Biochemistry and Physiology Part A*, Vol. 138, No. 4, pp. 405-415, ISSN 0300-9629.

Abele, D.; Phillip, E.; Gonzalez, P.M. & Puntarulo, S. (2007). Marine Invertebrate Mitochondria and Oxidative Stress. *Frontiers in Bioscience*, Vol.12, (January 2007), pp. 933-946, ISSN 1093-9946.

Arndt, A & Smith, M.J. (1998). Mitochondrial Gene Rearrangement in the Sea Cucumber Genus Cucumaria. *Molecular Phylogenetics and Evolution*, Vol.15, No.8, (August 1998), pp. 1009-1016, ISSN 1055-7903.

Asakawa, S.; Himeno, H.; Miura, K. & Watanabe, K. (1995). Nucleotide Sequence and Gene Organization of the Starfish *Asterina pectinifera* Mitochondrial Genome. *Genetics*, Vol.140, (July 1995), pp.1047-1060, ISSN 0016-6731.

Ballard, J.W.O. & Melvin, R.G. (2007). Tetracycline Treatment Influences Mitochondrial Metabolism and mtDNA Density two Generations after Treatment in Drosophila. *Insect Molecular Biology*, Vol.16, No.6, (December 2007), pp. 799–802, ISSN 0962-1075.

Banaji, M. (2006). A Generic Model of Electron Transport in Mitochondria. *The Journal of Theoretical Biology*, Vol.243, No.4, (December 2009), pp. 501-516, ISSN 0022-5193.

Barger, J.L.; Brand, M.D.; Barnes, B.M., & Boyer, B.B. (2003). Tissue-Specific Depression of Mitochondrial Proton Leak and Substrate Oxidation in Hibernating Arctic ground Squirrels. *American Journal of Physiology Regulatory, Integrative and Comparative Physiology*, Vol.284, No.5, (May 2003), pp. R1306–R1313, ISSN 0363-6119.

Beaton, M.J.; Roger, A.J. & Cavalier-Smith, T. (1998). Sequence Analysis of the Mitochondrial Genome of *Sarcophyton glaucum*: Conserved Gene Order Among Octocorals. *Journal of Molecular Evolution*, Vol.47, No.6, (December 1998), pp. 697-708, ISSN 0022-2844.

Belevich, I.; Gorbikova, E.; Belevich, N. P.; Rauhamäki, V.; Wikström, M. & Verkhovsky, M. I. (2010). Initiation of the Proton Pump of Cytochrome c Oxidase. *Proceedings of the National Academy of Sciences U.S.A.*, Vol.107, No.43, (October 2010), pp. 18469-18474, ISSN 1091-6490.

Belinky, F.; Rot, C.; Ilan, M. & Huchon, D. (2008). The Complete Mitochondrial Genome of the Demosponge *Negombata magnifica* (Poecilosclerida). *Molecular Phylogenetics and Evolution*, Vol.47, No.3, (January 2008), pp. 1238-43, ISSN 1055-7903.

Bessho, Y.; Ohama, T. & Osawa, S. (1992). Planarian mitochondria II. The Unique Genetic Code as Deduced from Cytochrome c Oxidase Subunit I Gene Sequences. *Journal of Molecular Evolution*, Vol.34, No.4, (April 1992), pp. 331-335, ISSN 0022-2844.

Boore, J.L. & Brown, W.M. (1995). Complete DNA Sequence of the Mitochondrial Genome of the Annelid Worm, *Lumbricus terrestris*. *Genetics*, Vol.141, No.1, (September 1995), pp. 305-319, ISSN 0016-6731.

Boore, J.L; Collins, T.M.; Stanton, D.; Daehler, L.L.; Brown, W.M. (1995). Deducing the Pattern of Arthropod Phylogeny from Mitochondrial DNA Rearrangements. *Nature*, Vol.376, No.6536, (July 1995), pp.163-165, ISSN 0028-0836.

Boore, J.L. (1999). Animal Mitochondrial Genomes. *Nucleic Acids Research*, Vol.27, No.8, (April 1999), pp. 1767-1780, ISSN 0305-1048.

Boore, J.L. & Brown, W.M. (2000). Mitochondrial Genomes of *Galathealinum*, *Helobdella*, and *Platynereis*: Sequence and Gene Arrangement Comparisons Indicate that Pogonophora is not a Phylum and Annelida and Arthropoda are not Sister Taxa. *Molecular Phylogenetics and Evolution*, Vol.17, No.1, (January 2000), pp. 87-106, ISSN 1055-7903.

Boyer, P.D. (1997). The ATP synthase- A Splendid Molecular Machine. *Annual Review of Biochemistry*, Vol.66, (May 1995), pp. 717–749. ISSN 0066-4154.

Brouwer, M.; Brown-Peterson, N.J.; Hoexum-Brouwer, T.; Manning, S. & Denslow, N. (2008). Changes in Mitochondrial Gene and Protein Expression in Grass shrimp, *Palaemonetes pugio*, Exposed to Chronic Hypoxia. *Marine Environmental Research*, Vol.66, No.1, (July 2008), pp. 143-145, ISSN 0141-1136.

Burgents, J.E.; Brunett, K.G. & Burnett, L.E. (2005). Effects of Hypoxia and Hypercapnic Hypoxia on the Localization and the Elimination of *Vibrio campbellii* in *Litopenaeus vannamei*, the Pacific White Shrimp. *Biological Bulletin*, Vol.208, No.3, (June 2005), pp.159-168, ISSN 0006-3185.

Cai, X.Q.; Liu, G.H.; Song, H.Q.; Wu, C.Y.; Zou, F.C.; Yan, H.K.; Yuan, Z.G.; Lin, R.Q. & Zhu, X.Q. (2011). Sequences and Gene Organization of the Mitochondrial Genomes of the Liver Flukes *Opisthorchis viverrini* and *Clonorchis sinensis* (Trematoda). *Parasitology Research*, (May 2011), ISSN (electronic) 1432-1955.

Cardol, P.; Gonzalez-Halphen, D.; Reyes-Prieto, A.; Baurain, D.; Matagne, R. F. & Remacle, C. (2005). The Mitochondrial Oxidative Phosphorylation Proteome of *Chlamydomonas reinhardtii* Deduced from the Genome Sequencing Project. *Plant Physiology*, Vol.137, No.2, (February 2005), pp. 447–459, ISSN 0032-0889.

Chamberlin, M.E. (2004). Control of Oxidative Phosphorylation during Insect Metamorphosis. *American Journal of Physiology- Regulatory, Integrative and Comparative Physiology*, Vol.287, No.2, (April 2004), pp. R314-R321, ISSN 0363-6119.

Chen, H. & Chan, D.C. (2009). Mitochondrial Dynamics–Fusion, Fission, Movement, and Mitophagy–in Neurodegenerative Diseases. *Human Molecular Genetics*, Vol.18, No.2, (October 2009), pp. R169–R176, ISSN 0964-6906.

Cheng, R.; Zheng, X.; Lin, Z.; Yang, J. & Li, Q. (2011). Determination of the Complete Mitochondrial DNA Sequence of *Octopus minor*. *Molecular Biology Reports*, (June 2011) On line version. ISSN (electronic) 1573-4978.

Clary, D.O. & Wolstenholme, D.R. (1985). The Mitochondrial DNA Molecule of *Drosophila yakuba*: Nucleotide Sequence, Gene Organization and the Genetic Code, *Journal of Molecular Evolution*, Vol.22, No.3, (February 1986), pp. 252-271, ISSN 0022-2844.

Clavero-Salas,A.; Sotelo-Mundo, R.; Gollas-Galván, T.; Hernandez-Lopez, J.; Peregrino-Uriarte, A.; Muhlia-Almazán, A.; Yepiz-Plascencia, G. (2007). Transcriptome Analysis of Gills from the White Shrimp *Litopenaeus vannamei* Infected with White Spot Syndrome Virus. *Fish & Shellfish Immunology*, Vol.23, No.2, (August 2007), pp. 459-472, ISSN 1050-4648.

Crozier, R.H. & Crozier, Y.C. (1993). The Mitochondrial Genome of the Honeybee *Apis mellifera*: Complete Sequence and Genome Organization. *Genetics*, Vol.133, No.1, (January 1993), pp. 97-117, ISSN 0016-6731.

Dall, W.; Hill, B.J.; Rothlisberg, P.C. & Staples, D.J. (1990). The Biology of Penaeidae, In: *Advances in Marine Biology*, J.H.S. Blaxter & A.J. Southward, (Ed.), 1-488, Academic Press, ISBN 0-12-026127-8, San Diego, CA.

Das, S.; Radtke, A.; Choi, Y.J.; Mendes, A.M.; Valenzuela, J.G. & Dimopoulos, G. (2010). Transcriptomic and Functional Analysis of the *Anopheles gambiae* Salivary Gland in Relation to Blood Feeding. *BMC Genomics*, Vol.566, (October 2010), pp. 566, ISSN 1471-2164.

Davila, S.; Piñero, D.; Bustos, P.; Cevallos, M.A. & Davila, G. (2005). The Mitochondrial Genome Sequence of the Scorpion *Centruroides limpidus* (Karsch 1879) (Chelicerata; Arachnida). *Gene*, Vol.360, No.2, (November 2005), pp. 92-102, ISSN 0378-1119.

de Oliveira, U.O.; da Rosa-Araujo, A.S.; Bello-Klein, A.; da Silva, R.S.M & Kucharski, L.C. (2005). Effects of Environmental Anoxia and Different Periods of Reoxygenation on Oxidative Balance in Gills of the Estuarine Crab *Chasmagnathus granulata*. *Comparative Biochemistry and Physiology. Part B*, Vol.140, No.1, (January 2005), pp. 51-57, ISSN 1096-4959.

Diaz, R.J. (2001). Overview of Hypoxia around the World. *Journal of Environmental Quality*, Vol.30, No.2, (March-April 2001), pp. 275-281, ISSN 0047-2425.

Diaz, F. (2010). Cytochrome c Oxidase Deficiency: Patients and Animal Models. *Biochimica et Biophysica Acta*, Vol.1802, No.10, (January 2010), pp. 100-110, ISSN 0925-4439.

Dirnagl, U.; Becker, K. & Meisel, A. (2009). Preconditioning and Tolerance Against Cerebral Ischaemia: from Experimental Strategies to Clinical Use. *Lancet Neurology*, Vol.8, No.4, (April 2009), pp. 398–412, ISSN 1474-4422.

Di Simplicio, P.; Rossi, R.; Falcinelli, S.; Ceserani, R. & Formento, M.L. (1997). Antioxidant Status in Various Tissue of the Mouse after Fasting and Swimming Stress. *European Journal of Applied Physiology*, Vol.76, No.4, pp.302-307, ISSN 0301-5548.

Driscoll, M. & Gerstbrein, B. (2003). Dying for a Cause: Invertebrate Genetics Takes on Human Neurodegeneration. *Nature Reviews*, Vol.4, No.3, (March 2003), pp. 181-194, ISSN 471-0056.

Dudkina N.; Sunderhaus, S.; Boekema, E. & Braun, H. (2008). The Higher Level of Organization of the Oxidative Phosphorylation System: Mitochondrial Supercomplexes. *Journal of Bioenergetics and Biomembranes*. Vol.40, No.5, (October 2008), pp. 419-424. ISSN 1573-6881.

Dudkina, N. V.; Kouril, R.; Peters, K.; Braun, H.P. & Boekema, E.J. (2010). Structure and Function of Mitochondrial Supercomplexes. *Biochimica et Biophysica Acta*, Vol.1797, No.6-7, (June-July 2010), pp. 664-70, ISSN 0006-3002.

Durieux, J.; Wolff, S. & Dillin, A. (2011). The Cell-Non-Autonomous Nature of Electron Transport Chain-Mediated Longevity. *Cell*, Vol.144, No.1, (January 2011), pp. 79-91, ISSN 0092-8674.

Eads, B.D. & Hand, S.C. (1999). Regulatory Features of Transcription in Isolated Mitochondria from *Artemia franciscana* Embryos. *American Journal of Physiology*, Vol.277, No.6, (December 1999), pp. R1588-R1597, ISSN 0363-6119.

Eads, B.D. & Hand, S.C. (2003). Mitochondrial mRNA Stability and Polyadenylation during Anoxia-Induced Quiescence in the Brine Shrimp *Artemia franciscana*. *The Journal of Experimental Biology*, Vol.206, No.20, (October 2003), pp. 3681-3692, ISSN 0022-0949.

Ellington, C.P. (1985). Power and Efficiency of Insect Flight Muscle. *The Journal of Experimental Biology*, Vol.115, (March 1985), pp. 293-304, ISSN 0022-0949.

Ellington, W.R. & Hines, A.C. (1991). Mitochondrial Activities of Phosphagen Kinases are not Widely Distributed in the Invertebrates. *The Biological Bulletin*, Vol.180, No.3, (June 1991), pp. 505-507, ISSN 1062-3590.

English, T.E. & Storey, K.B. (2003). Freezing and Anoxia Stresses Induce Expression of Metallothionein in the Foot Muscle and Hepatopancreas of the Marine Gastropod *Littorina littorea*. *The Journal of Experimental Biology*, Vol.206, No.14, (July 2003), pp. 2517-24, ISSN 0022-0949.

Falkenberg, M.; Larsson, N.G. & Gustafsson, C.M. (2005). DNA Replication and Transcription in Mammalian Mitochondria. *Annual Review of Biochemistry*, Vol.76, (July 2007), pp. 679-699, ISSN 0066-4154.

Figueroa, P.; Leon, G.; Elorza, A.; Holuigue, L.; Araya, A. and Jordana, X. (2002). The Four Subunits of Mitochondrial Respiratory Complex II are encoded by Multiple Nuclear Genes and Targeted to Mitochondria in *Arabidopsis thaliana*. *Plant Molecular Biology*, Vol. 50, No. 4-5, (Noviember 2002), pp. 725-734, ISSN 0735-9640.

Fontanesi, F.; Soto, I. & Barrientos, A. (2008). Cytochrome c Oxidase Biogenesis: New Levels of Regulation. *The International Union of Biochemistry and Molecular Biology*, Vol.60, No.9, (September 2008), pp. 557-568, ISSN 1521-6543.

Frade, J. M. & Michaelidis, T. M. (1997). Origin of Eukaryotic Programmed Cell Death: A Consequence of the Aerobic Metabolism. *BioEssays*, Vol.19, No. 9, (September 1997), pp. 827-832, ISSN 0265-9247.

Fridell, Y.-W.; Sanchez-Blanco, A.; Silvia, B.A. & Helfand, S.L. (2004). Functional Characterization of a Drosophila Mitochondrial Uncoupling Protein. *Journal of Bioenergetics and Biomembranes*. Vol. 36, No. 3, (June, 2004), pp. 219-228, ISSN 1573-6881.

Friedrich, T. & Weiss, H. (1997). Modular Evolution of the Respiratory NADH: Ubiquinone Oxidoreductase and the Origin of its Molecules. *The Journal of Theoretical Biology.* Vol.187. (September 1997). pp 529-540, ISSN 0022-5193.

Galluzzi L.; Brenner, C.; Morselli, E.; Touat, Z. & Kroemer, G. (2008). Viral Control of Mitochondrial Apoptosis. *PLoS Pathogens*, Vol.4, No.5, (May 2008), ISSN 1553-7366.

Garcia-Orozco, K.D.; Aispuro-Hernandez, E.; Yepiz-Plascencia, G.; Calderon-de-la Barca, A.M. & Sotelo-Mundo, R. (2007). Molecular Characterization of Arginine Kinase, an Allergen from the Shrimp *Litopenaeus vannamei*. *International Archives of Allergy and Immunology*, Vol.144, No.1, (May 2007), pp 23-28, ISSN 1018-2438.

Garesse, R. (1988). *Drosophila melanogaster* Mitochondrial DNA: Gene Organization and Evolutionary Considerations. *Genetics*, Vol.118, No.4, (April 1988), pp. 649-663, ISSN 0016-6731.

Genes, C.; Baquero, E.; Echeverri, F.; Maya, J.D. & Triana, O. (2011). Mitochondrial Dysfunction in *Trypanosoma cruzi*: the Role of *Serratia marcescens* Prodigiosin in the alternative Treatment of Chagas Disease. *Parasites & Vectors*, Vol.4, (May 2011), pp. 66, ISSN 1756-3305.

Goodman, W.G.; Carlson, R.O. & Nelson, K.L. (1985). Analysis of Larval and Pupal Development in the Tobacco Hornworm (Lepidoptera: Sphingidae), *Manduca sexta*. *Annals of the Entomological Society of America*, Vol.78, No.1, (January 1985), pp. 70-80, ISSN: 0013-8746.

Grad, L.I.; Sayles, L.C. & Lemire, B.D. (2008). Isolation and Functional Analysis of Mitochondria from the Nematode *Caenorhabditis elegans*. In: *Mitochondria: Practical Protocols*, D. Leister & J.M. Herrmann, (Ed.), Methods in Molecular Biology, Vol.372, 51-66. Humana Press Inc., ISSN 1064-3745, Totowa, NJ. USA.

Gray, M. W.; Burger, G. & Lang, B.F. (1999). Mitochondrial Evolution. *Science*. Vol.283, No.5407, (March 1999), pp.1476-1481, ISSN 0036-8075.

Guerrero-Castillo, S.; Araiza-Olivera, D.; Cabrera-Orefice, A.; Espinasa-Jaramillo, J.; Gutierrez-Aguilar, M.; Luevano-Martinez, I.A.; Zepeda-Bastida, A. & Uribe-Carbajal, S. (2011). Physiological Uncoupling of Mitochondrial Oxidative Phosphorylation. Studies in Different Yeast Species. *Journal of Bioenergetics and Biomembranes*, Vol. 43, (May 2010), pp. 323-331, ISSN1573-6881.

Guzy, R.D.; Mack, M.M. & Schumacker, P.T. (2007). Mitochondrial Complex III is required for Hypoxia-induced ROS Production and Gene Transcription in Yeast. *Antioxidants and Redox Signaling*, Vol.9, No.9, (September 2007), pp. 1317-1328, ISSN 1523-0864.

Hakkaart, G. A.; Dassa, E. P.; Jacobs, H. T. & Rustin, P. (2006). Allotropic Expression of a Mitochondrial Alternative Oxidase Confers Cyanide Resistance to Human Cell Respiration. *EMBO Reports*, Vol.7, No.3, (March 2006), pp. 341-5, ISSN 1469-221X

Hand, S.C. (1998). Quiescence in *Artemia franciscana* Embryos: Reversible Arrest of Metabolism and Gene Expression at low Oxygen Levels. *The Journal of Experimental Biology*, Vol.201, No.8, (April 1988), pp. 1233–1242, ISSN 0022-0949.

Hahlbeck, E.; Arndt, C. & Schiedek, D. (2000). Sulphide Detoxification in *Hediste diversicolor* and *Marenzelleria viridis*, Two Dominant Polychaete Worms within the Shallow Coastal Waters of the Southern Baltic Sea. *Comparative Biochemistry and Physiology Part B*, Vol.125, No.4, (April 2000), pp. 457-71, ISSN 1096-4959.

Harrison, J.F. & Roberts, S.P. (2000). Flight Respiration and Energetics. *Annual Review of Physiology*, Vol.62, (March 2000), pp. 179-205, ISSN 0066-4278.

He, N.; Qin, Q. & Xu, X. (2005). Differential Profile of Genes Expressed in Haemocytes of White Spot Syndrome Virus-Resistant Shrimp (*Penaeus japonicus*) by Combining Suppression Subtractive Hybridization and Differential Hybridization, *Antiviral Research*, Vol.66, No.1, (April 2005), pp. 39-45, ISSN 0166-3542.

Hengarter, M.O. (2000). The Biochemistry of Apoptosis. *Nature*, Vol.40, No.6805, (October 2000), pp. 770-776, ISSN 0028-0836.

Hervant, F. & Renault, D. (2002). Long-term Fasting and Realimentation in Hypogean and Epigean Isopods: A Proposed Adaptive Strategy for Groundwater Organisms. *The Journal of Experimental Biology*, Vol.205, No.14, (July 2002), pp. 2079-2087, ISSN 0022-0949.

Himeno, H.; Masaki, H.; Ohta, T.; Kumagai, I.; Miura, K.I. & Watanabe, K. (1987). Unusual Genetic Codes and a Novel Genome Structure for tRNA SerAGY in Starfish Mitochondrial DNA. *Gene*, Vol.56, No.2-3, pp. 219-230, ISSN 0378-1119.

Hird, F.J. (1986). The Importance of Arginine in Evolution. *Comparative Biochemistry and Physiology. Part B*, Vol.8, No.2, pp. 285-8, ISSN 0305-0491.

Hochachka, P.W. & Somero, G.N. (2002). Influence of Oxygen Availability, In: *Biochemical Adaptation*, P.W. Hochachka &, G.N. Somero, (Ed.), 107-157, Oxford University Press, ISBN 0-19-511702-6, NY, USA.

Hoffmann, R.J.; Boore J, L. & Brown W.M. (1992). A Novel Mitochondrial Genome Organization for the Blue Mussel *Mytilus edulis*. *Genetics*, Vol.131, No.2, (June 1992), pp. 397-412, ISSN 0016-6731.

Hu, M.; Chilton, N.B. & Gasser, R.B. (2002). The Mitochondrial Genomes of the Human Hookworms, *Ancylostoma duodenale* and *Necator americanus* (Nematoda: Secernentea). *International Journal for Parasitology*. Vol.32, No.2, (February 2002), pp. 145-58, ISSN 0020-7519.

Huang, J. & Lemire, B.D. (2009). Mutations in the *C. elegans* Succinate Dehydrogenase Iron-Sulfur Subunit Promote Superoxide Generation and Premature Aging. *Journal of Molecular Biology*, Vol.387, No.3, (April 2009), pp. 559-569, ISSN 0022-2836.

Itoi, S.; Kinoshita, S.; Kikuchi, K. & Watabe, S. (2003). Changes of Carp FoF$_1$-ATPase in Association with Temperature Acclimation. *American Journal of Physiology - Regulatory, Integrative and Comparative Physiology*, Vol.284, No.1, (September 2002), pp. R153–R163, ISSN 0363-6119.

Jacobs, H.T.; Elliott, D.J.; Math, V.B. & Farquharson, A. (1988). Nucleotide Sequence and Gene Organization of Sea Urchin Mitochondrial DNA. *Journal of Molecular Evolution*. Vol.202, No.2, (July 1988), pp.185-217, ISSN 0022-2844.

Jiang, W. & Wang, X. (2004). Cytochrome C-Mediated Apoptosis. *Annual Review of Biochemistry*, Vol.73, (July 2004), pp. 87-106, ISSN 0066-4154.

Jiang, S.T.; Hong, G.Y.; Yu, M.; Li, N.; Yang, Y.; Liu, Y.Q. & Wei, Z.J. (2009). Characterization of the Complete Mitochondrial Genome of the Giant Silkworm Moth, *Eriogyna pyretorum* (Lepidoptera: Saturniidae). *International Journal of Biological Sciences*, Vol.5, No.4, (May 2009), pp. 351-365, ISSN 1449-2288.

Jubrias, S.A.; Esselman, P.C.; Price, L.B.; Cree, M.E. & Conley, K.E. (2001). Large energetic Adaptations of Elderly Muscle to Resistance and Endurance Training. *Journal of Applied Physiology*, Vol.90, No.5, (May 2001), pp. 1663-1770, ISSN 0363-6143.

Julian, D.; Dalia, W.E. & Arp, A. (1998). Neuromuscular Sensitivity to Hydrogen Sulfide in the Marine Invertebrate *Urechis caupo*. *The Journal of Experimental Biology*, Vol.201, No.9, (May 1998), pp. 1393-1403, ISSN 0022-0949.

Kayal, E. & Lavrov, D.V. (2008). The Mitochondrial Genome of *Hydra oligactis* (Cnidaria, Hydrozoa) Sheds New Light on Animal mtDNA Evolution and Cnidarian Phylogeny. *Gene*, Vol.410, No.1, (February 2008), pp. 177-186, ISSN 0378-1119.

Keddie, E.M.; Higazi, T. & Unnasch, T.R. (1998). The Mitochondrial Genome of *Onchocerca volvulus*: Sequence, Structure and Phylogenetic Analysis. *Molecular and Biochemical Parasitology*, Vol.95, No.1, (September 1998), pp. 111–127, ISSN 0166-6851.

Kern, B.; Ivanina, A.V.; Piontkivska, H.; Sokolov, E.P. & Sokolova, I.M. (2009). Molecular Characterization and Expression of a Novel Homolog of Uncoupling Protein 5 (UCP5) from the Eastern Oyster *Crassostrea virginica* (Bivalvia: Ostreidae). *Comparative Biochemistry and Physiology Part D*, Vol.4, No.2, (June 2009), pp. 121-127, ISSN 744-117X.

Khalimonchuk, O. & Rödel, G. (2005). Biogenesis of Cytochrome c Oxidase. *Mitochondrion*, Vol.5, No.6, (December 2005), pp. 363-388, ISSN 1567-7249.

Kidd, T.; Abu-Shumays, R.; Katzen, A.; Sisson, J.C.; Jimenez, G.; Pinchin, S.; Sullivan, W. & Ish-Horowicz, D. (2005). The Epsilon-subunit of Mitochondrial ATP synthase is Required for Normal Spindle Orientation during the Drosophila Embryonic Divisions. *Genetics*, Vol. 170, No.2, (June 2005), pp. 697-708, ISSN 0016-6731.

Klimova, T. & Chandel, N.S. (2008). Mitochondrial Complex III Regulates Hypoxic Activation of HIF. Cell Death and Differentiation. Vol. 15, (January 2008), pp 660-666, ISSN 1350-9047.

Kotlyar, S.; Weihrauch, D.; Paulsen, R. & Towle, D.W. (2000). Expression of Arginine Kinase Enzymatic Activity and mRNA in Gills of the Euryhaline Crabs *Carcinus maenas* and *Callinectes sapidus*. *The Journal of Experimental Biology*. Vol.203, (July 2000), pp. 2395-2404, ISSN 0022-0949.

Kroemer, G. (1997). Mitochondrial Implication in Apoptosis. Towards an Endosymbiont Hypothesis of Apoptosis Evolution. *Cell Death & Differentiation*, Vol. 4, No. 6, (August 1997), pp. 443-456, ISSN 1350-9047.

Lai-Zhang, J. & Mueller, D. (2000). Complementation of Deletion Mutants in the Genes Encoding the F1- ATPase by Expression of the Corresponding Bovine Subunits in Yeast *S. cerevisiae*. *European Journal of Biochemistry*, Vol.267, No.8, (April 2000), pp. 2409-2418, ISSN 0014-2956.

Lavrov, D.V.; Wang, X. & Kelly, M. (2008). Reconstructing Ordinal Relationships in the Demospongiae Using Mitochondrial Genomic Data. *Molecular Phylogenetics and Evolution*. Vol.49, No.1, (October 2008), pp. 111-24, ISSN 1055-7903.

Lee, S.S.; Lee, R.Y.N.; Fraser, A.G.; Kamath, R.S.; Ahringer, J. & Ruvkun, G. (2003). A Systematic RNAi Screen Identifies a Critical Role for Mitochondria in *C. elegans* Longevity. *Nature Genetics*, Vol.33, No.1, (January 2003), pp. 40–48, ISSN 1061-4036.

Lee, S.J.; Hwang, A.B. & Kenyon, C. (2010). Inhibition of Respiration Extends *C. elegans* Life Span via Reactive Oxygen Species that Increase HIF-1 Activity. *Current Biology*, Vol.20, No.23, pp.2131-2136, ISSN 0960-9822.

Levy, O.; Kaniewska, P.; Alon, S.; Eisenberg, E.; Karako-Lampert, S.; Bay, L.K.; Reef, R.; Rodriguez-Lanetty, M.; Miller, D.J. & Hoegh-Guldberg, O. (2011). Complex Diel

Cycles of Gene Expression in Coral-Algal Symbiosis. *Science*, Vol.331, No.6014, (January 2011), pp. 175, ISSN 0036-8075.

Li, T. & Brouwer, M. (2009). Gene Expression Profile of the Grass Shrimp *Palaemonetes pugio* Exposed to Chronic Hypoxia. *Comparative Biochemistry and Physiology Part D*, Vol.4, No.3, (September 2009), pp. 196-208, ISSN 1744-117X.

Li, Z. & Neufeld, G.J. (2001a). Isolation and Characterization of Mitochondrial F(1)-ATPase from Crayfish (*Orconectes virilis*) Gills. *Comparative Biochemistry and Physiology. Part B*, Vol.128, No.2, (February 2001), pp. 325-338, ISSN 1096-4959.

Li, Z. & Neufeld, G.J. (2001b). Kinetic Studies on Mitochondrial F(1)-ATPase from Crayfish (*Orconectes virilis*) Gills. *Comparative Biochemistry and Physiology. Part B*, Vol.128, No.2, (February 2001), pp. 339-350, ISSN 1096-4959.

Liang, Y.; Cheng, J.J.; Yang, B. & Huang, J. (2010). The Role of F1 ATP synthase Beta Subunit in WSSV Infection in the Shrimp, *Litopenaeus vannamei*. *Virology Journal*, Vol.7, (June 2010), pp. 144, ISSN 1098-5514.

Liu, G.H.; Lin, R.Q.; Li, M.W.; Liu, W.; Liu, Y.; Yuan, Z.G.; Song, H.Q.; Zhao, G.H.; Zhang, K.X. & Zhu, X.Q. (2011). The Complete Mitochondrial Genomes of Three Cestode Species of *Taenia* Infecting Animals and Humans. *Molecular Biology Reports*, Vol.38, No.4, (April 2011), pp. 2249-56, ISSN 0301-4851.

Mao, X. R. & Crowder, C.M. (2010). Protein Misfolding Induces Hypoxic Preconditioning via a subset of the Unfolded Protein Response Machinery. *Molecular and Cellular Biology*, Vol.30, No.21, (November 2010), pp. 5033-42, ISSN 0270-7306.

Martinez-Cruz, O. (2007). Expresion Genica de las Subunidades atp6 Mitocondrial y atpc Nuclear del Complejo ATP-Sintasa en el Camaron Blanco (*Litopenaeus vannamei*) en Condiciones de Hipoxia. Master of Science Thesis. Centro de Investigacion en Alimentacion y Desarrollo, A.C. Hermosillo, Sonora. pp. 1-55.Mexico.

Martinez-Cruz, O.; Garcia-Carreño, F.; Robles-Romo, A.; Varela-Romero, A. & Muhlia-Almazan, A. (2011). Catalytic Subunits $atp\alpha$ and $atp\beta$ from the Pacific White shrimp *Litopenaeus vannamei* FoF$_1$ ATP-synthase complex: cDNA Sequences, Phylogenies, and mRNA Quantification During Hypoxia. *Journal of Bioenergetics and Biomembranes*, Vol.43, (March 2011), pp. 119-133, ISSN 1573-6881.

Martinez–Cruz, O.; Arvizu-Flores, A.; Sotelo-Mundo, R.; Garcia-Carreño, F.; Yepiz-Plascencia, G. & Muhlia-Almazan, A. (2011). Molecular Characterization and Relative Expression of the F1 Subunits from the Mitochondrial ATP-synthase Complex in the Tail Muscle of the White Shrimp *Litopenaeus vannamei* During Hypoxia. In Preparation.

Mattisson, A.G.M. (1965). The Localization of Succinic Dehydrogenase in the Muscles of *Nereis virens* and *Homarus gammarus*. *Histochemistry and Cell Biology*, Vol.5, No.5, (June 1965), pp. 97-115, ISSN 0948-6143.

Mayevsky A. & Rogatsky, G. (2007). Mitochondrial Function *in vivo* Evaluated by NADH Fluorescence: from Animal Models to Human Studies. *American Journal of Physiology - Cell Physiology*, Vol.292, No.2, (February 2007), pp. C615-C640, ISSN 0363-6143.

McDonald, A.E.; Vanlerberghe, G.C. & Staples, J.F. (2009). Alternative Oxidase in Animal: Unique Characteristics and Taxonomic Distribution. *The Journal of Experimental Biology*, Vol.212, (August 2009), pp. 2627-2634, ISSN 0022-0949.

Menze, M.A.; Fortner, G.; Nag, S. & Hand, S.C. (2010). Mechanisms of Apoptosis in Crustacea: What Conditions Induce Versus Suppress Cell Death? *Apoptosis: an International Journal on Programmed Cell Death,* Vol.15, No.3, (March 2010), pp. 293-312, ISSN 1360-8185.

Mitchell, P. (1976). Possible Molecular Mechanisms of the Protonmotive Function of Cytochrome Systems. *The Journal of Theoretical Biology,* Vol.62, No.2, (October 1976), pp. 327-367, ISSN 0022-5193.

Moraes, C.T. (2001). What Regulates Mitochondrial DNA Copy Number in Animal Cells? *TRENDS in Genetics,* Vol.17, No.4, (April 2001), pp. 199- 205, ISSN 0168-9525.

Morris, S.; Aardt, W. & Ahern, M. (2005). The Effect of Lead on the Metabolic and Energetic Status of the Yabby, *Cherax destructor* During Environmental Hypoxia. *Aquatic Toxicology,* Vol.75, No.1, (October 2005), pp. 16-31, ISSN 0166-445X.

Mueller, D.M.; Puri, N.; Kabaleeswaran, V.; Terry, C.; Leslie, A.G.W. & Walker J.E. (2004). Ni-chelate-Affinity Purification and Crystallization of the Yeast Mitochondrial F1-ATPase. *Protein Expression and Purification,* Vol.37, No.2, (October 2004), pp. 479-485, ISSN 1046-5928.

Muhlia-Almazan, A.; Martinez-Cruz, O.; Navarrete del Toro, M. A.; Garcia-Carreño, F.; Arreola, R.; Sotelo-Mundo, R. & Yepiz-Plascencia, G. (2008). Nuclear and Mitochondrial Subunits from the White Shrimp *Litopenaeus vannamei* FoF$_1$ ATP-synthase complex: cDNA Sequence, Molecular Modeling, and mRNA Quantification of *atp9* and *atp6*. *Journal of Bioenergetics and Biomembranes,* Vol.40, No.4, (September 2008), pp. 359–369, ISSN 0145-479X.

Mwinyi, A.; Bailly, X.; Bourlat, S.J.; Jondelius, U.; Littlewood, D.T. & Podsiadlowski, L. (2010). The Phylogenetic position of Acoela as Revealed by the Complete Mitochondrial Genome of *Symsagittifera roscoffensis*. *BMC Evolutionary Biology,* Vol.13, (October 2010), pp. 309, ISSN 1471-2148.

Negrisolo, E.; Minelli, A. & Valle, G. (2004). The Mitochondrial Genome of the House Centipede *Scutigera* and the Monophyly Versus Paraphyly of Myriapods. *Molecular Phylogenetics and Evolution,* Vol.21, No.4, (April 2004), pp. 770-780, ISSN 1055-7903.

Nestel, D.; Tomalsky, D.; Rabossi, A. & Quesada-Alllue, L.A. (2003). Lipid, Carbohydrates and Protein Patterns During Metamorphosis of the Mediterranean Fruit Fly, *Ceratitis capitata,* (Diptera: Tephritidae). *Annals of the Entomological Society of America,* Vol.96, No.3, (May 2003), pp. 237-244, ISSN 0013-8746.

Noguchi, Y.; Endo, K.; Tajima, F. & Ueshima, R. (2000). The Mitochondrial Genome of the Brachiopod *Laqueus rubellus*. *Genetics,* Vol.155, No.1, (May 2000), pp. 245-259, ISSN 0016-6731.

Nübel, E.; Wittig, I.; Kerscher, S.; Brandt, U. & Schägger, H. (2009). Two-Dimensional Native Electrophoresis Analysis of Respiratory Supercomplexes from *Yarrowia lipolytica*. *Proteomics,* Vol.9, No.9, (May 2009), pp. 2408-2418, ISSN 1615-9853.

Okimoto, R., Macfarlane, J.L., Clary, D.O. & Wolstenholme, D.R. (1992). The Mitochondrial Genomes of Two Nematodes, *Caenorhabditis elegans* and *Ascaris suum*. *Genetics,* Vol.130, No.3, (March 1992), pp. 471-498, ISSN 0016-6731.

Oleaga, A.; Escudero-Poblacion, A.; Camafeita, E. & Perez-Sanchez, R. (2007). A Proteomic Approach to the Identification of Salivary Proteins from the Argasid Ticks *Ornithodoros moubata* and *Ornithodoros erraticus*. *Insect Biochemistry and Molecular Biology,* Vol.37, No.11, (November 2007), pp. 1149-1159, ISSN 0965-1748.

Peck, L.S. (2002). Ecophysiology of Antarctic Marine Ectotherms: Limits to Life. *Polar Biology*, Vol.25, No.1, (September 2001), pp. 31-40, ISSN 0722-4060.

Pedler, S.; Fuery, C.J.; Withers, P.C.; Flanigan, J. & Guppy, M. (1996). Effectors of Metabolic Depression in an Estivating Pulmonate Snail (*Helix aspersa*): Whole Animal and *in vitro* Tissue Studies. *Journal of Comparative Physiology Part B*, Vol.166, No.6, pp. 375–381, ISSN 0174-1578.

Peregrino-Uriarte, A., Varela-Romero, A., Muhlia-Almazan, A., Anduro-Corona, I., Vega-Heredia, S., Gutierrez-Millan, L., De la Rosa-Velez, J., Yepiz-Plascencia, G. (2009). The Complete Mitochondrial Genomes of the Yellowleg Shrimp *Farfantepenaeus californiensis* and the Blue Shrimp *Litopenaeus stylirostris* (Crustacea: Decapoda). *Comparative Biochemistry and Physiology Part D*, Vol.4, No.1, (March 2009), pp. 45-53, ISSN 1744-117X.

Popova-Butler, A. & Dean, D.H. (2009). Proteomic Analysis of the Mosquito *Aedes aegypti* Midgut Brush Border Membrane Vesicles. *Journal of Insect Physiology*, Vol.55, No.3, (March 2009), pp. 264–272, ISSN: 0022-1910.

Qureshi, S.A. & Jacobs, H.T. (1993). Two Distinct, Sequence-Specific DNA-Binding Proteins Interact Independently with the Major Replication Pause Region of Sea Urchin mtDNA. *Nucleic Acids Research*, Vol.21, No.12, (January 1993), pp. 2801-8, ISSN 0305-1048.

Racotta I.; Palacios, E. & Mendez, L. (2002). Metabolic Responses to Short and Long-Term Exposure to Hypoxia in White Shrimp (*Penaeus vannamei*). *Marine and Freshwater Behaviour and Physiology*, Vol.35, pp. 269-275, ISSN 023-6244.

Ren, J.; Liu, X.; Jiang, F.; Guo, X. & Liu, B. (2010). Unusual Conservation of Mitochondrial Gene Order in *Crassostrea* Oysters: Evidence for Recent Speciation in Asia. *BMC Evolutionary Biology*. Vol.10, (December 2010), pp.394, ISSN 1471-2148.

Rich, P.R. & Marechal, A. (2010). The Mitochondrial Respiratory Chain. *Essays in Biochemistry*. Vol.47, pp.1-27. ISSN 0071-1365.

Ripamonti, M.; Vigano, A.; Moriggi, M.; Milano, G.; von Segesser, L.K.; Samaja, M. & Gelfi, C. (2006). Cytochrome c Oxidase Expression in Chronic and Intermittent Hypoxia Rat Gastrocnemius Muscle Quantitated by CE. *Electrophoresis*, Vol.27, No.19, (October 2006), pp. 3897-3903, ISSN 0173-0835.

Robin, E.D. & Wong, R. (1988). Mitochondrial DNA Molecules and Virtual Number of Mitochondria per Cell in Mammalian Cells. *Journal of Cellular Physiology*, Vol.136, No.3, (September 1988), pp. 507-13, ISSN 0021-9541.

Rolland, S. & Conradt, B. (2006). The Role of Mitochondria in Apoptosis Induction in *Caenorhabditis elegans*: More than Just Innocent Bystanders? *Cell Death and Differentiation*, Vol.13, No.8, (August 2006), pp. 1281–1286, ISSN 1350-9047.

Rosa, R.D. & Barracco, M.A. (2008). Shrimp Interferon is Rather a Portion of the Mitochondrial Fo-ATP Synthase than a True α-interferon. *Molecular Immunology*, Vol.45, No.12, (July 2008), pp. 3490-3493, ISSN 0161-5890.

Rosengarten, R.D.; Sperling, E.A.; Moreno, M.A.; Leys, S.P. & Dellaporta, S.L. (2008). The Mitochondrial Genome of the Hexactinellid Sponge *Aphrocallistes vastus*: Evidence for Programmed Translational Frameshifting. *BMC Genomics*, Vol.9, (January 2008), pp.33, ISSN 1471-2164.

Ryan, M.T. & Hoogenraad, N.J. (2007). Mitochondrial-Nuclear Communications. *Annual Review of Biochemistry*, Vol.76, pp. 701-722, ISSN 0066-4154.

Sanchez-Paz, A.; Garcia-Carreño, F.; Hernandez-Lopez J.; Muhlia-Almazan, A. & Yepiz-Plascencia, G. (2007). Effect of Short-term Starvation on Hepatopancreas and Plasma Energy Reserves of the Pacific White Shrimp (*Litopenaeus vannamei*). *Journal of Experimental Marine Biology and Ecology*, Vol.340, pp. 184-193, ISSN 0022-0981.

Sanchez-Paz, A. (2010). White Spot Syndrome Virus: An Overview on an Emergent Concern. *Veterinary Research*, Vol.41, No.6, (November-December 2010), pp. 43, ISSN 1573-7446.

Sanz, A.; Soikkeli, M.; Portero-Otin, M.; Wilson, A.; Kemppainen, E.; McIlroy, G.; Ellilä, S.; Kemppainen, K.K.; Tuomela, T.; Lakanmaa, M.; Kiviranta, E.; Stefanatos, R.; Dufour, E.; Hutz, B.; Naudi, A.; Jove, M.; Zeb, A.; Vartiainen, S.; Matsuno-Yagi, A.; Yagi, T.; Rustin, P.; Pamplona, R. & Jacobs, H.T. (2010). Expression of the Yeast NADH Dehydrogenase Ndi1 in Drosophila Confers Increased Lifespan Independently of Dietary Restriction. *Proceedings of the National Academy of Sciences, USA*, Vol.107, No.20, (May 2010), pp. 9105–9110, ISSN 1091-6490.

Segovia, R.; Pett, W.; Trewick, S. & Lavrov, D.V. (2011). Extensive and Evolutionarily Persistent Mitochondrial tRNA Editing in Velvet Worms (phylum Onychophora). *Molecular Phylogenetics and Evolution*, (May 2011), ISSN (electronic) 1537-1719.

Seibel, B.A. (2011). Critical Oxygen Levels and Metabolic Suppression in Oceanic Oxygen Minimum Zones. *The Journal of Experimental Biology*, Vol.214, (January 2011), pp. 326-336, ISSN 0022-0949.

Semenza, G.L. (2007). Oxygen-dependent Regulation of Mitochondrial Respiration by Hypoxia-Inducible Factor. *Biochemical Journal*, Vol.405, No.1, (July 2007), pp. 1-9, ISSN 0264-6021.

Shao, Z.; Graf, S.; Chaga, O.Y. & Lavrov, D.V. (2006). Mitochondrial Genome of the Moon Jelly *Aurelia aurita* (Cnidaria, Scyphozoa): A linear DNA Molecule Encoding a Putative DNA-Dependent DNA polymerase. *Gene*, Vol.381, (October 2006), pp. 92-101, ISSN 0378-1119.

Shen, X.; Ren, J.; Cui, Z.; Sha, Z.; Wang, B.; Xiang, J., & Liu, B. (2007). The Complete Mitochondrial Genomes of Two Common Shrimps (*Litopenaeus vannamei* and *Fenneropenaeus chinensis*) and their Phylogenomic Considerations. *Gene*. Vol.403, No.1-2, (November 2007), pp. 98–109, ISSN 0378-1119.

Sherwood, L.; Klandorf, H. & Yancey, P.H. (2005). *Animal Physiology*, Brooks/Cole, ISBN 978-0-534-55404-0, Velmont, California. USA.

Shin, D.S.; Didonato, M.; Barondeau, D.P.; Hura, G.L.; Hitomi, C.; Berglund, J.A.; Getzoff, E.D.; Cary, S.C. & Tainer, J.A. (2009). Superoxide Dismutase from the Eukaryotic Thermophile *Alvinella pompejana*: Structures, Stability, Mechanism, and Insights into Amyotrophic Lateral Sclerosis. *Journal of Molecular Biology*, Vol.385, No.5, (February 2009), pp. 1534-1555, ISSN 0022-2836.

Silvestre, F.; Dierick, J.F.; Dumont, V.; Dieu, M.; Raes, M. & Devos, P. (2006). Differential Protein Expression Profiles in Anterior Gills of *Eriocheir sinensis* During Acclimation to Cadmium. *Aquatic Toxicology*, Vol.76, No.1, (January 2006), pp. 46-58, ISSN 0166-445X.

Sokolova, I.M. (2004). Cadmium Effects on Mitochondrial Function are Enhanced by Elevated Temperatures in a Marine Poikilotherm, *Crassostrea virginica* Gmelin (Bivalvia: Ostreidae). *The Journal of Experimental Biology*, Vol.207, No.15, (July 2004), pp. 2639-2648, ISSN 0022-0949.

Sokolova, I.M. & Sokolov, E.P. (2005). Evolution of Mitochondrial Uncoupling Proteins: Novel Invertebrate UCP Homologues Suggest Early Evolutionary Divergence of the UCP Family. *FEBS Letters*, Vol. 579, No. (2): 313-317, ISSN 0014-5793.

Soldatov, A.A.; Andreenko, T.I.; Golovina, I.V. & Stolbov, A. (2010). Peculiarities of Organization of Tissue Metabolism in Molluscs with Different Tolerance to External Hypoxia. *Zhurnal evoliutsionnoi biokhimii i fiziologii*, Vol.46, No.4, (July-August 2010), pp. 284-90, ISSN 0044-4529.

Sommer, A. & Portner, H.O. (1999). Exposure of *Arenicola marina* to Extreme Temperatures: Adaptive Flexibility of a Boreal and Subpolar Subpopulation. *Marine Ecology Progress Series*, Vol.181, (May 1999), pp. 215-226, ISSN 0171-8630.

Soñanez-Organis, J.G.; Peregrino-Uriarte, A.B.; Gomez-Jimenez, S.; Lopez-Zavala, A.; Forman, H.J. & Yepiz-Plascencia, G. (2009). Molecular Characterization of Hypoxia Inducible Factor-1 (HIF-1) from the White Shrimp *Litopenaeus vannamei* and Tissue-Specific Expression under Hypoxia. *Comparative Biochemistry and Physiology Part C*, Vol.150, No.3, (September 2009), pp. 395-405, ISSN 1532-0456.

Staton, J.L.; Daehler, L.L. & Brown, W.M. (1997). Mitochondrial Gene Arrangement of the Horseshoe Crab *Limulus polyphemus* L.: Conservation of Major Features Among Arthropod Classes. *Molecular and Biological Evolution*, Vol.14, No.8, (August 1997), pp. 867-874, ISSN 0737-4038.

Steinauer, M.L; Nickol, B.B.; Broughton, R. & Orti, G. (2005). First Sequenced Mitochondrial Genome from the Phylum Acanthocephala (*Leptorhynchoides thecatus*) and its Phylogenetic Position within Metazoa. *Journal of Molecular Evolution*, Vol.60, No.6, (January 2005), pp. 706-15, ISSN 0022-2844.

St. John, J.C.; Jokhi, R.P. & Barrat, C.L.R. (2005). The Impact of Mitochondrial Genetics on Male Infertility. *International Journal of Andrology*, Vol.28, No.2, (April 2005), pp.65-73, ISSN 0105-6263.

Storey, K.B. (1993). *Molecular Mechanisms of Metabolic Arrest in Mollusks*. In: *Surviving Hypoxia: Mechanisms of Control and Adaptation* (eds. Hochachka, P.W., Lutz, P.L., Sick, T.J., Rosenthal, M., and Thillart, G. van den), pp. 253-269. CRC Press, Boca Raton.

Storz, P. (2006). Reactive Oxygen Species-Mediated Mitochondria-to-Nucleus Signaling: A Key to Aging and Radical-caused Diseases. *Science Signaling*, Vol.2006, No.332, (April 2006), pp. re3, ISSN 1937-9145.

Stuart, J.A. & Brown, M.F. (2006). Energy, Quiescence and the Cellular Basis of Animal Life Spans. *Comparative Biochemistry and Physiology Part A*. Vol.143, No.1, (January 2006), pp. 12–23, ISSN 1095-6433.

Suarez, R.K.; Staples, J.F.; Lighton, J.R. & Mathieu-Costello, O. (2000). Mitochondrial Function in Flying Honeybees (*Apis mellifera*): Respiratory Chain Enzymes and Electron Flow from Complex III to Oxygen. *The Journal of Experimental Biology*, Vol.203, (March 2000), pp. 905–911, ISSN 0022-0949.

Suga, K.; Mark-Welch, D.B.; Tanaka, Y.; Sakakura, Y. & Hagiwara, A. (2008). Two Circular Chromosomes of Unequal Copy Number Make Up the Mitochondrial Genome of the Rotifer *Brachionus plicatilis*. *Molecular Phylogenetics and Evolution*, Vol.25, No.6, (January 2008), pp. 1129-37, ISSN 1055-7903.

Talamillo, A.; Chisholm, A.A.; Garesse, R.; & Jacobs, H.T. (1998). Expression of the Nuclear Gene Encoding Mitochondrial ATP synthase Subunit Alpha in Early Development of Drosophila and Sea Urchin. *Molecular Biology Reports*. Vol.25, No.2, (March 1998), pp. 87-94, ISSN 0301-4851.

Terrett, J.A.; Miles, S. & Thomas, R.H. (1996). Complete DNA Sequence of the Mitochondrial Genome of *Cepaea nemoralis* (Gastropoda: Pulmonata). *Journal of Molecular Evolution*, Vol.42, No.2, (February 1996), pp. 160-8, ISSN 0022-2844.

Toivonen, J.M.; O'Dell, K.M.C.; Petit, N.; Irvine, S.C.; Knight, G.K.; Lehtonen, M.; Longmuir, M.; Luoto, K.; Touraille, S.; Wang, Z.; Alziari, S.; Shah, Z.H. & Jacobs, H.T. (2001). Technical knockout, a Drosophila Model of Mitochondrial Deafness. *Genetics*, Vol.159, No.1, (September 2001), pp. 241–254, ISSN 0016-6731.

Tuena de Gomez-Poyou, M.; Perez-Hernandez, G. & Gomez-Poyou A. (1999). Synthesis and Hydrolysis of ATP and the Phosphate-ATP Exchange Reaction in Soluble Mitochondrial F_1 in the Presence of Dimethylsulfoxide, *European Journal of Biochemistry*, Vol.266, No.2, (December 1999), pp. 691-696, ISSN 0014-2956.

Turrens, J.F. (2003). Mitochondrial Formation of Reactive Oxygen Species. *The Journal of Physiology*, Vol.552, No.2, (October 2003), pp. 335-344, ISSN 0022-3751.

Uno, T.; Nakasuji, A.; Shimoda, M. & Aizono, Y. (2004). Expression of Cytochrome c Oxidase Subunit 1 Gene in the Brain at Early Stage in the Termination of Pupal Diapauses in the Sweet Potato Hornworm *Agrius convolvuli*. *Journal of Insect Physiology*, Vol.50, No.1, (January 2004), pp. 35-40, ISSN 0022-1910.

Van der Giezen, M. (2009). Eukaryotic Life Without Mitochondria?. *Comparative Biochemistry and Physiology Part A*, Vol.153, (June 2009), pp. S165–S167, ISSN 0300-9629.

Vendemiale, G.; Grattagliano, I.; Caraceni, P.; Caraccio, G.; Domenicali, M.; Dall'Agata, M.; Trevisani, F.; Guerrieri, F.; Bernardi, M. & Altomare, E. (2001). Mitochondrial Oxidative Injury and Energy Metabolism Alteration in Rat Fatty Liver: Effect of the Nutritional Status. *Hepatology*, Vol.33, No.4, (April 2011), pp. 808-815, ISSN 0270-9139.

Vishnudas, V. & Vigoreaux, J.O. (2006). Sustained High Power Performance Possible Strategies for Integrating Energy Supply and Demand in Flight Muscle, In: *Nature's Versatile Engine: Insect Flight Muscle Inside and Out*, J.O. Vigoreaux, (Ed.), 188-196, Springer, ISBN 978-0387257983, NY, USA.

Voigt, O.; Erpenbeck, D. & Worheide, G. (2008). A Fragmented Metazoan Organellar Genome: Two Mitochondrial Chromosomes of *Hydra magnipapillata*. *BMC Genomics*, Vol.9, (July 2008), pp. 350, ISSN 1471-2164.

Wang, B.; Li, F.; Dong, B.; Zhang, X.; Zhang, C. & Xiang, J. (2006). Discovery of the Genes in Response to White Spot Syndrome Virus (WSSV) Infection in *Fenneropenaeus chinensis* through cDNA Microarray. *Marine Biotechnology*, Vol.8, No.5, (September-October 2006), pp. 491-500, ISSN 1436-2228.

Wang, C. & Youle, R. (2009). The Role of Mitochondria in Apoptosis. *Annual Review of Genetics*, Vol.43, pp. 95-118, ISSN 0066-4197.

Wang, T.; Lefevre, S.; Thanh-Huong D.T.; van Cong N. & Bayley M. (2009). The Effects of Hypoxia on Growth and Digestion, *Fish Physiology*, Vol. 27, (February 2009), pp. 361-396, ISSN 0920-1742.

Watanabe, K. (2010). Unique Features of Animal Mitochondrial Translation Systems. The Non-Universal Genetic Code, Unusual Features of the Translational Apparatus and their Relevance to Human Mitochondrial Diseases. *Proceedings of the Japan Academy. Series B*, Vol.86, No.1, pp. 11-39, ISSN 0386-2208.

Wegener, G. (1996). Flying Insects: Model Systems in Exercise Physiology. *Experientia*, Vol.52, No.5, (May 1996), pp. 404-12, ISSN 0014-4754.

Wei, Y.H.; Lu, C.Y.; Lee, H.C.; Pang, C.Y. & Ma, Y.S. (1998). Oxidative Damage and Mutation to Mitochondrial DNA and Age-dependent Decline of Mitochondrial Respiratory Function. *Annals of the New York Academy of Sciences*, Vol.854, (November 1998), pp. 155-170, ISSN 0077-8923.

Wei, L.Z.; Zhang, X.M.; Li, J. & Huang, G.Q. (2008). Compensatory Growth of Chinese Shrimp, *Fenneropenaeus chinensis* Following Hypoxic Exposure. *Aquaculture International*, Vol.16, No.5, pp. 455-470, ISSN 0967-6120

Williams, T.; Chitnis, N.; & Bilimoria, S. (2009). Invertebrate Iridovirus Modulation of Apoptosis. *Virologica Sinica*, Vol.24, No.4, (August 2009), pp. 295-304, ISSN 1674-0769.

Wilson, D.F.; Rumsey, W.L.; Green, T.J. & Vanderkooi, J.M. (1988). The Oxygen Dependence of Mitochondrial Oxidative Phosphorylation Measured by a New Optical Method for Measuring Oxygen Concentration. *The Journal of Biological Chemistry*. Vol.263, No.6, (February 1988), pp. 2712-2718, ISSN 0021-9258.

Wittig, I.; Carrozzo, R.; Santorelli, F.M. & Schägger, H. (2006). Supercomplexes and Subcomplexes of Mitochondrial Oxidative Phosphorylation. *Biochimica et Biophysica Acta*, Vol. 1757, No.9-10, (September-October 2006), pp.1066-1072, ISSN 0006-3002.

Wittig, I. & Schägger, H. (2009). Supramolecular Organization of ATP synthase and Respiratory Chain in Mitochondrial Membranes. *Biochimica et Biophysica Acta*. Vol.1787, No.6, (June 2009), pp. 672-680, ISSN 0006-3002.

Wolstenholme, D.R.; Okimoto, R. & Macfarlane, J.L. (1994). Nucleotide Correlations that Suggest Tertiary Interactions in the TV-Replacement Loop-Containing Mitochondrial tRNAs of the Nematodes, *Caenorhabditis elegans* and *Ascaris suum*. *Nucleic Acids Research*, Vol.22, No.20, (October 1994), pp. 4300-6, ISSN 0305-1048.

Wu, R.S. (2002). Hypoxia: from Molecular Responses to Ecosystem Responses. *Marine Pollution Bulletin*, Vol.14, No.45, pp. 35-45, ISSN 0025-326X.

Xia, D.; Yu. C.A.; Kim, H.; Xia, J.Z.; Kachurin, A.M.; Zhang, L.; Yu, L. & Deisenhofer, J. (1997). Crystal Structure of the Cytochrome bc1 Complex from Bovine Heart Mitochondria. *Science*, Vol.277, No.5322, (July 1997), pp. 60-66, ISSN 0036-8075.

Yang, J.; Zhu, J. & Xu, W.H. (2010). Differential Expression Phosphorylation of COX Subunit 1 and COX Activity During Diapause Phase in the Cotton Bollworm *Helicoverpa armigera. Journal of Insect Physiology*, Vol.56, No.12, (December 2010), pp. 1992-1998, ISSN 0022-1910.

Yang, W. & Hekimi, S. (2010). A Mitochondrial Superoxide Signal Triggers increased Longevity in *Caenorhabditis elegans. PLos Biology*, Vol. 8, No. 12, (December 2010), pp. e1000556, ISSN 1544-9173.

Yasuda, N.; Hamaguchi, M.; Sasaki, M.; Nagai, S.; Saba, M. & Nadaoka, K. (2006). Complete Mitochondrial Genome Sequences for Crown-of-thorns Starfish *Acanthaster planci* and *Acanthaster brevispinus. BMC Genomics*. Vol.7, (January 2006), pp. 17, ISSN 1471-2164.

Yokobori, S.; Ueda, T.; Feldmaier-Fuchs, G.; Pääbo, S.; Ueshima, R.; Kondow, A.; Nishikawa, K. & Watanabe, K. (1999). Complete DNA Sequence of the Mitochondrial Genome of the Ascidian *Halocynthia roretzi* (Chordata, Urochordata). *Genetics*, Vol.153, No.4, (December 1999), pp. 1851-1862, ISSN 0016-6731.

Yu, X.X.; Mao, W.; Zhong, A.; Schow, P.; Brush, J.; Sherwood, S.W.; Adams, S.H. & Pan, G. (2000). Characterization of Novel UCP5/BMCP1 isoforms and Differential Regulation of UCP4 and UCP5 Expression through Dietary or Temperature Manipulation. *The FASEB Journal*, Vol. 14, No. 11. (August 2000), pp. 1611-1618, ISSN 892-6638.

Yu, H. & Li, Q. (2011). Complete Mitochondrial DNA Sequence of *Crassostrea nippona*: Comparative and Phylogenomic Studies on Seven Commercial *Crassostrea* species. *Molecular Biology Reports*, (May 2011), ISSN 1573-4978.

Zickermann, V.; Dröse, S.; Tocilescu, M.A.; Zwicker, K.; Kerscher, S. & Brandt, U. (2008). Challenges in Elucidating Structure and Mechanism of Proton Pumping NADH: Ubiquinone Oxidoreductase (Complex I). *Journal of Bioenergetics and Biomembranes*. Vol.40, No.5, (October 2008), pp. 475-483, ISSN 0145-479X.

Part 3

New Techniques and Findings
in Bioenergetics Research

10

Targeting the Mitochondria by Novel Adamantane-Containing 1,4-Dihydropyridine Compounds

Linda Klimaviciusa[1], Maria A. S. Fernandes[2], Nelda Lencberga[1],
Marta Pavasare[1], Joaquim A. F. Vicente[2], António J. M. Moreno[2],
Maria S. Santos[3], Catarina R. Oliveira[4], Imanta Bruvere[5],
Egils Bisenieks[5], Brigita Vigante[5] and Vija Klusa[1]

[1]*Department of Pharmacology, Faculty of Medicine, University of Latvia, Riga*
[2]*IMAR-CMA, Department of Life Sciences, University of Coimbra, Coimbra*
[3]*CNC, Department of Life Sciences, University of Coimbra, Coimbra*
[4]*CNC, Faculty of Medicine, University of Coimbra, Coimbra*
[5]*Laboratory of Membrane Active and beta-Diketone Compounds,*
Latvian Institute of Organic Synthesis, Riga
[2,3,4]*Portugal*
[1,5]*Latvia*

1. Introduction

Mitochondria are important regulators of cellular functions and energy metabolism, therefore mitochondrial dysfunction leads to a compromised energy-generating system, deteriorated cellular homeostasis and neurodegenerative disorders, such as Parkinson's disease and Alzheimer's disease (Shapira, 1999; 2009). Hence, the protection of mitochondria, even their repair mechanisms at the level of complex I, may be a key strategy in limiting mitochondrial damage and ensuring cellular integrity (Dawson & Dawson, 2003). Thus, in addition to traditionally used antiparkinsonian drugs, which are focused on the activation of the dopaminergic system, different mitochondria-protecting agents are being used in clinics for the treatment of Parkinson's disease. For instance, agents with antioxidant properties, such as melatonin (Esposito & Cuzzocrea, 2010), coenzyme Q10 and creatine (Kones, 2010), lipoic acid (De Araújo et al., 2011), and the extract of Hyoscyamus niger seeds (Sengupta et al., 2011), are currently used to treat Parkinson's disease.

Recently, antihypertensive drugs of the calcium antagonistic series, which belong to 1,4-dihydropyridine (DHP) class and are capable of penetrating the blood-brain barrier (e.g., nifedipine, nimodipine), were shown to significantly reduce the risk of developing Parkinson's disease (Becker et al., 2008; Ritz et al., 2010). This was explained by blocking L-type calcium channels in the dopaminergic neurons of the substantia nigra, where elevated calcium ion concentrations initiate cell death (Sulzeret & Schmitz, 2007). However, the mechanism of the antiparkinsonian action of DHPs is not yet understood.

Our investigation of DHP compounds showed that many of them are capable of protecting mitochondrial processes (Fernandes et al., 2003, 2005, 2008, 2009). For instance, the most active compound cerebrocrast {4-[2-(difluoromethoxy)phenyl-2,6-dimethyl-1,4-dihydropiridine-3,5-dicarboxylic acid di(2-propoxyethyl)diester}, which has shown neuroprotective effects in different neurodeficiency models (Klusa, 1995), decreased mitochondrial toxin 1-methyl-4-phenylpyridinium (MPP+)-induced cell death in rat cerebellar granule cells (Klimaviciusa et al., 2007). In isolated mitochondria of rat liver, cerebrocrast inhibited the inner mitochondrial anion channel, Ca^{2+}-induced opening of the mitochondrial membrane permeability transition pore and permeabilization of the mitochondrial inner membrane (Vicente et al., 2006). In addition, it normalized oxidative phosphorylation and increased adenosine triphosphate (ATP)-induced contraction in swollen mitochondria of isolated rat skeletal muscle (Velena et al., 1997). Cerebrocrast and its congeners also protected against histopathological changes caused by azidothymidine, known to be a mitochondrial toxin (Pupure et al., 2008).

The present study investigates two novel DHP compounds, cerebrocrast analogues containing structure elements that may enhance the delivery of molecules through the blood-brain barrier and improve their access to mitochondria. The compounds are composed of either one adamantane moiety in position 3 (AV-6-93) or two adamantane moieties in positions 3 and 5 (diflurone) of the DHP ring. We suggest that these DHP structures may possess mitochondria-protecting and antiparkinsonian activity due to both the adamantane moiety, which can be considered to be an important functional unit, and the DHP structure, which may serve as the carrier molecule. Adamantane molecules were previously used in the design of neuroprotective drugs. For example, amantadine (1-amino-adamantane) is used in antiparkinsonian drugs with mechanisms focused on NMDA-receptor gated ion channels (Kornhuber et al., 1991). Adamantane derivatives, particularly memantine, are reported as neuroprotective agents against mitochondrial toxicity *in vivo* (Rojas et al., 2008) and *in vitro* (McAllister et al., 2008). Memantine may act directly on dopamine D2High receptors (Seeman et al., 2008), whereas amantadine may stimulate the synthesis and release of dopamine in the rat striatum (Spilker & Dhasmana, 1973), which is beneficial in the treatment of Parkinson's disease. Aminoadamantane derivatives 4-(1-adamantylamino)-2,2,6,6-tetramethylpiperidine-1-oxyl and 4-(1-adamantylammonio)-1-hydroxy-2,2,6,6-tetramethylpiperidinium dihydrochloride were also synthesised as antiparkinsonian drugs (Skolimowski et al., 2003). However, compounds with adamantane moieties attached to the DHP structure have not yet been synthesised.

In this study, we tested novel compounds *in vitro* to assess their influence on mitochondrial processes in primary cultures of rat cortical neurons, using mitochondrial toxin MPP+, and on isolated rat liver mitochondria.

2. Materials and methods

2.1 Animals
Male Wistar rats (250-350 g), housed at 22 ± 2 °C under artificial light for a 12-h light/dark cycle and with access to water and food *ad libitum*, were used for these experiments. All of the experimental procedures were performed in accordance with the guidelines of Directive 86/609/EEC "European Convention for the Protection of Vertebrate Animals Used for Experimental and Other Scientific Purposes" (1986) and were approved by the National Ethics Committee.

2.2 Chemicals
AV-6-93 [2,6- dimethyl-3-(1-adamantyloxycarbonyl)-4-(2-difluoromethoxyphenyl)-5-[(2-propoxy)ethoxycarbonyl]-1,4-dihydropyridine] (Fig. 1A) and diflurone [2,6- dimethyl-3,5-

bis(1-adamantyloxycarbonyl)-4-(2-difluoromethoxyphenyl)-1,4-dihydropyridine] (Fig. 1B) were synthesised at the Latvian Institute of Organic Synthesis, 21 Aizkraukles Street, Riga, LV-1006. AV-6-93 and diflurone were dissolved in 100% DMSO and further diluted to concentrations of 0.1% (v/v) and less.

Chemicals for the mitochondrial studies were obtained from Sigma Chemical Company (St Louis, MO, USA); chemicals for cytotoxicity studies mentioned in 2.3. and 2.4.

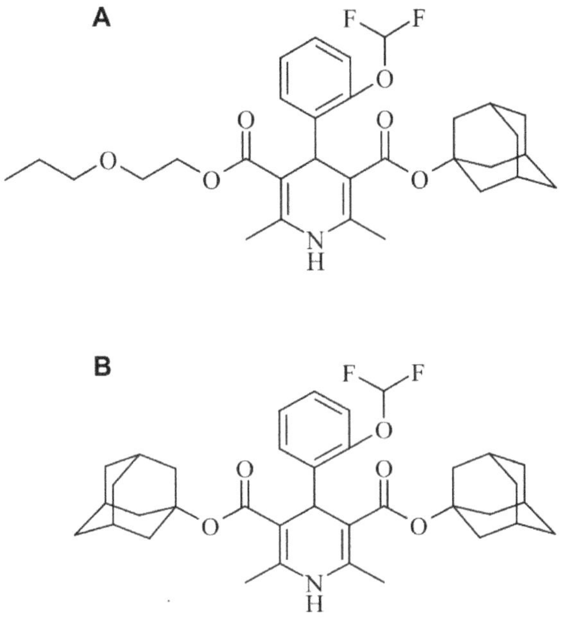

Fig. 1. The structures of AV-6-93 (A) and diflurone (B).

2.3 Primary culture of rat cortical neurons

Primary cultures were prepared from 1-day-old Wistar rat pups, according to the method of Alho et al., 1988, with minor modifications. Briefly, cortices were dissected in ice-cold Krebs-Ringer solution (135 mM NaCl, 5 mM KCl, 1 mM MgSO$_4$, 0.4 mM KH$_2$PO$_4$, 15 mM glucose, 20 mM HEPES, pH 7.4, containing 0.3% bovine serum albumin) and trypsinised in 0.8% trypsin-EDTA (Invitrogen, U.K.) for 10 min at 37 °C, followed by trituration in 0.008% DNAse I solution containing 0.05% soybean trypsin inhibitor (both obtained from Surgitech AS, Estonia). Cells were resuspended in Eagle's basal medium with Earle's salts (BME, Invitrogen, U.K.), containing 10% heat-inactivated foetal bovine serum (FBS, Invitrogen, U.K.), 25 mM KCl, 2 mM GlutaMAX™-I (Invitrogen, U.K.) and 100 µg/mL gentamycin. Cells were plated onto poly-L-lysine- (Sigma Chemical Co., MO, USA) coated 48-well plates at a density of 1.8 x 10^5 cells/cm^2. The medium was changed to Neurobasal™-A medium containing 2 mM GlutaMAX™-I with B-27 supplement and 100 µg/mL gentamycin 2.5 hr later. Cultures were incubated for 6 days in a 5% CO$_2$/95% air atmosphere at 37 °C, and one-fifth of the culture medium was changed on DIV 3 (day 3 in vitro).

2.4 Measurement of cell death in cytotoxicity assay

Primary rat cortical neurons were cultured for 5 days as described above. On DIV 5, cultures were incubated with 1-methyl, 4-phenylpyridinium (MPP$^+$, Sigma Chemical Co., MO, USA) for the following 24 hr at a concentration of 300 µM. Cells were pre-incubated with the tested compounds AV-6-93 and diflurone for 90 min followed by the addition of MPP$^+$ and further incubation with MPP$^+$ plus the tested compounds or a solvent (control) for the next 24 hours. Cell death was measured with a Trypan blue assay (Tymianski et al., 1993). Cells were incubated with 0.4% Trypan blue solution in phosphate buffered saline (PBS, 145 mM NaCl, 3 mM KCl, 0.42 mM Na$_2$HPO$_4$, 2.4 mM KH$_2$PO$_4$, pH = 7.4) at 37 °C for 7 min and then washed twice with PBS and fixed with 4% paraformaldehyde in PBS. Only dead neurons were stained with Trypan blue (Tymianski et al., 1993). The fixed cultures were rinsed with PBS for microscopic observation, and approximately 150 cells per 5 fields in each well were counted to determine the number of dead cells and the total number of cells. Neuronal death was calculated as the percentage of dead cells from the total (viable plus dead) number of cells, and the obtained data were averaged for each well.

2.5 Isolation of rat liver mitochondria

Rat liver mitochondria were isolated from male Wistar rats by differential centrifugation according to conventional methods (Gazotti et al., 1979). After washing, the pellet was gently resuspended in the washing medium at a protein concentration of about 50 mg/ml. Protein content was determined by the biuret method (Gornall et al., 1949), using bovine serum albumin as a standard.

2.6 Measurement of respiratory activities

Oxygen consumption was monitored polarographically with a Clark-type electrode at 30 °C in a closed glass chamber equipped with magnetic stirring. Mitochondria (1 mg/ml) were incubated in a respiratory medium containing 130 mM sucrose, 5 mM HEPES (pH 7.2), 50 mM KCl, 2.5 mM K$_2$HPO$_4$, and 2.5 mM MgCl$_2$ (in the presence and absence of AV-6-93 or diflurone) for 3 min before energisation with 10 mM glutamate/5 mM malate. When 10 mM succinate was used as the respiratory substrate, the reaction medium was supplemented with 2 µM rotenone. To induce state 3 respiration, adenosine diphosphate (ADP, 150 µM) was added. FCCP (p-trifluoromethoxyphenylhydrazone)-stimulated respiration was initiated by the addition of 1µM FCCP. The respiratory control ratio (RCR), which is calculated by the ratio between state 3 (consumption of oxygen in the presence of substrate and ADP) and state 4 (consumption of oxygen after ADP phosphorylation), is an indicator of mitochondrial membrane integrity. The ADP/O ratio, which is expressed by the ratio between the amounts of ADP added and the oxygen consumed during state 3 respiration, is an index of oxidative phosphorylation efficiency. Respiration rates were calculated assuming that the saturation of oxygen concentration was 250 µM at 30 °C (Chance & Williams, 1956), and the values are expressed in percentage of control (% of control).

2.7 Measurement of mitochondrial transmembrane potential

The mitochondrial transmembrane potential ($\Delta\psi$) was measured indirectly based on the detection of lipophilic cation tetraphenylphosphonium (TPP$^+$) using a TPP$^+$-selective electrode, as previously described (Kamo et al., 1979). The $\Delta\psi$ was estimated from the following equation [1]:

$$\Delta\psi = 59 \times \log (v/V) - 59 \times \log (10^{\Delta E/59} - 1) \tag{1}$$

where v, V, and ΔE stand for inner mitochondrial volume, incubation medium volume, and deflection of the electrode potential from the baseline, respectively. A mitochondrial matrix volume of 1.1 $\mu l/mg$ protein was assumed. No correction was made for the "passive" binding of TPP^+ to the mitochondrial membranes because the purpose of the experiments was to show relative changes in potential rather than absolute values. As a consequence, we anticipate some overestimation for the $\Delta\psi$ values. To monitor $\Delta\psi$ associated with mitochondrial respiration, liver mitochondria (1 mg/ml) were incubated for 3 min in the respiratory medium described above, supplemented with 3 μM TPP^+, at 30 °C in the absence or presence of different concentrations of AV-6-93 or diflurone before energisation with 10 mM glutamate/5 mM malate or 10 mM succinate. When succinate was used as the respiratory substrate, the medium was supplemented with 2 μM rotenone. AV-6-93 or diflurone did not affect TPP^+ binding to mitochondrial membranes or the electrode response.

2.8 Ca^{2+}-induced mitochondrial membrane transition pore (MPT)

Ca^{2+}-induced MPT was evaluated by measuring changes in mitochondrial transmembrane potential ($\Delta\psi$) using a TPP^+ electrode, changes in oxygen consumption using a Clark-type electrode, and changes in Ca^{2+} fluxes using a Ca^{2+}-selective electrode. The reactions were conducted in a medium containing 200 mM sucrose, 10 mM Mops-Tris (pH 7.4), 1 mM KH_2PO_4, and 10 μM EGTA, supplemented with 2 μM rotenone, as previously described (Custódio et al., 1998a, 1998b). Mitochondria (1mg/ml) that were incubated at 30 °C for 3 min (in the absence and presence of AV-6-93 or diflurone) were energised with 10 mM succinate, and the single addition of Ca^{2+} (100 nmol/mg protein) was used to induce MPT. Control assays, in both the absence and presence of Ca^{2+} plus 0.75 nmol/mg protein cyclosporin A (CsA) and compound (when necessary) were also performed.

2.9 Lipid peroxidation

The extent of lipid peroxidation was evaluated by oxygen consumption using a Clark-type electrode at 30 °C in an open glass chamber equipped with magnetic stirring. Mitochondria (1 mg/ml) were pre-incubated for 3 min in a medium containing 175 mM KCl, 10 mM Tris–Cl (pH 7.4), supplemented with 3 μM rotenone (in the presence or absence of tested compounds) to avoid mitochondrial respiration induced by endogenous respiratory substrates. The iron solution was prepared immediately before use and was protected from light. The changes in O_2 tension were recorded in a potentiometric chart record and oxygen consumption was calculated assuming an oxygen concentration of 230 nmol/ml. Membrane lipid peroxidation was initiated by adding 1 mM $ADP/0.1$ mM Fe^{2+} as oxidizing agents. Controls, in the absence of ADP/Fe^{2+}, were performed under the same conditions.

Lipid peroxidation was also determined by measuring thiobarbituric acid reactive substances (TBARs), using the thiobarbituric acid assay (Ernster & Nordenbrand, 1967). Aliquots of mitochondrial suspensions (0.5 ml each), removed 10 min after the addition of ADP/Fe^{2+}, were added to 0.5 ml of ice cold 40% trichloroacetic acid. Then, 2 ml of 0.67% of aqueous thiobarbituric acid containing 0.01% of 2,6-di-*tert*-butyl-*p*-cresol was added. The mixtures were heated at 90 °C for 15 min, then cooled on ice for 10 min, and centrifuged at 850 g for 10 min. Controls, in the absence of ADP/Fe^{2+}, were performed under the same conditions. The supernatant fractions were collected and lipid peroxidation was estimated spectrophotometrically at 530 nm. As blanks, we used control reactions performed in the absence of mitochondria and ADP/Fe^{2+}. The amount of TBARs formed was calculated using a molar extinction coefficient of 1.56×10^5 mol^{-1} cm^{-1} and expressed as nmol TBARs/mg protein (Buege & Aust, 1978).

2.10 Statistical analysis

The cytotoxicity data were calculated as a mean ± S.E. Statistical analysis was performed using Student's t-test or one-way analysis of variance (ANOVA), followed by a Bonferroni multiple comparisons test.

The mitochondrial experiments were performed using three independent experiments with different mitochondrial preparations. The values are expressed as means ± S.E. Means were compared using one-way ANOVA for multiple comparisons, followed by Tukey's test. Statistical significance was set at $p < 0.05$.

3. Results

3.1 Protection against the cell death induced by MPP⁺

In primary rat cortical neurones, AV-6-93 at concentrations of 1 and 10 µM decreased MPP⁺-induced cell death by 75% and 56%, respectively (Fig. 2A). Diflurone exerted the protective ability only at the highest tested concentration, 10 µM, and decreased the MPP⁺-induced cell death by 35% (Fig. 2B). Neither AV-6-93 nor diflurone, added without MPP⁺, changed cell viability at the highest tested concentrations (Fig. 2).

3.2 Effects of AV-6-93 and diflurone on rat liver mitochondrial bioenergetics

AV-6-93 and diflurone (up to 100 µM) were studied for their effects on mitochondrial bioenergetics by evaluating several mitochondrial respiratory chain parameters (state 2, state 3, state 4, FCCP-stimulated respiration, RCR, ADP/O ratio, Δψ, and phosphorylation rate) using glutamate/malate as the respiratory substrate.

The effects of AV-6-93 on glutamate/malate-supported respiratory rates (state 2, state 3, state 4 and FCCP-stimulated respiration), respiratory indices RCR and ADP/O of rat liver mitochondria were almost non-existent and insignificant at concentrations of up to 100 µM (Table 1), indicating that the compounds did not significantly affect mitochondrial bioenergetics.

These results are demonstrated in Table 2, where AV-6-93 and diflurone, at concentrations of up to 100 µM, did not significantly affect either the Δψ induced by glutamate/malate-dependent respiration or the phosphorylation time.

As for glutamate/malate-supported respiration, the effects of AV-6-93 and diflurone on succinate-supported respiratory rates (state 2, state 3, state 4 and FCCP) and respiratory indices RCR and ADP/O of rat liver mitochondria were not significantly affected (results not shown), further supporting the finding that these compounds did not affect mitochondrial bioenergetics.

3.3 Effects of AV-6-93 and diflurone on Ca²⁺-induced MPT

The effect of AV-6-93 and diflurone on Ca^{2+}-induced MPT was studied in order to evaluate their capacity to protect mitochondria against MPT opening by measuring the decrease in Δψ, the increase in oxygen consumption, and the Ca^{2+}-induced release of mitochondrial Ca^{2+}, which are typical phenomena that follow the induction of MPT. The amount of Ca^{2+} used to induce MPT was 100 nmol/mg protein.

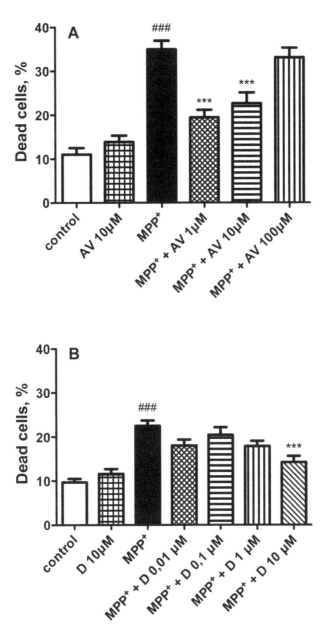

Fig. 2. Influence of AV-6-93 (AV) and diflurone (D) on MPP+-induced cell death in primary rat cortical neurons (A and B, respectively). Cell death measured by Trypan blue method. Data are presented as a mean ± S.E. ### p < 0.001 vs control, t-test, *** p < 0.001 vs MPP+, one-way ANOVA followed by Bonferroni multiple comparison's test.

Compounds (µM)	Oxygen consumption (% of control)				RCR	ADP/O
	State 2	State 3	State 4	State FCCP		
AV-6-93						
0.0	100.0 ± 0.0	100 ± 0.0	100.0 ± 0.0	100.0 ± 0.0	6.8 ± 1.3	3.01 ± 0.1
1.0	100.0 ± 0.0	113.4 ± 6.7	94.4 ± 5.6	100.2 ± 8.7	6.5 ± 1.5	2.9 ± 0.1
10.0	105.0 ± 18.9	104.0 ±7.3	119.4 ± 10.0	101.9 ± 7.3	6.4 ± 1.0	2.8 ± 0.4
100.0	127.2 ± 31.9	109.9 ± 6.6	145.8 ± 44.5	91.2 ± 8.5	6.3 ± 2.7	2.8 ± 0.1
Diflurone						
0	100.0± 0.0	100.0± 0.0	100.0± 0.0	100.0± 0.0	6.8 ± 1.3	3.01 ± 0.1
100	107.3 ± 4.8	95.9 ± 2.9	104.4 ± 5.27	94.4 ± 4.03	6.4 ± 1.2	2.9 ± 0.09

Table 1. Effects of AV-6-93 and diflurone on the respiratory parameters (state 2, state 3, state 4, FCCP-stimulated respiration) and respiratory indices (RCR and ADP/O ratio) of rat liver mitochondria using glutamate/malate as respiratory substrate.

The values, which are given in percentage of control (% of control), correspond to the mean ± S.E. of the respiratory parameters, evaluated in three different mitochondrial preparations, at the different indicated situations. Control values are expressed in nmol O_2. mg^{-1} protein min^{-1}: state 2 = 7.1 ± 0.7; state 3 = 39.4 ± 4.0; state 4 = 5.62 ± 0.8; FCCP-stimulated respiration = 57.14 ± 10.4.

Compounds (µM)	$\Delta\psi$ (mV)			Phosphorylation time (s)
	Glu/Mal energisation	ADP depolarisation	Repolarisation	
AV-6-93				
0	-220.5 ± 5.1	21.7 ± 2.2	-216.8 ± 2.6	33.0 ± 3.0
1	-219.0 ± 4.0	21.3 ± 1.9	-217.0 ± 4.3	32.7 ± 2.9
10	-220.6 ± 4.1	23.5 ± 2.1	-218.7 ± 4.2	32.5 ± 4.5
100	-216.0 ± 5.6	23.4 ± 1.1	-212.5 ± 4.7	39.0 ± 1.7
Diflurone				
0	-220.5 ± 5.1	21.7 ± 2.2	-216.8 ± 2.6	33.0 ± 3.0
100	-218.0 ± 1.7	21.0± 0.3	-216.0 ± 1.4	32.8 ± 2.3

Table 2. Effects of AV-6-93 and diflurone on glutamate/malate-dependent transmembrane potential ($\Delta\psi$) and phosphorylation time of rat liver mitochondria.

The values correspond to the mean ± S.E. of the $\Delta\psi$ and the phosphorylation time, evaluated in three different mitochondrial preparations, at the different indicated situations.

The results of the effect of AV-6-93 on MTP protection are depicted in Fig. 3. Under control conditions, the addition of 10 mM succinate to mitochondrial suspensions produced a $\Delta\psi$ of about -216 mV (negative inside mitochondria) (Fig. 3A), corresponding to respiratory state 4 (Fig. 3B). The addition of Ca^{2+} led to a rapid depolarisation (decrease of $\Delta\psi$), followed by a partial repolarisation (recover of $\Delta\psi$), the subsequent total depolarisation of mitochondria (Fig. 3A), and an increase in respiratory state 4 (Fig. 3B).

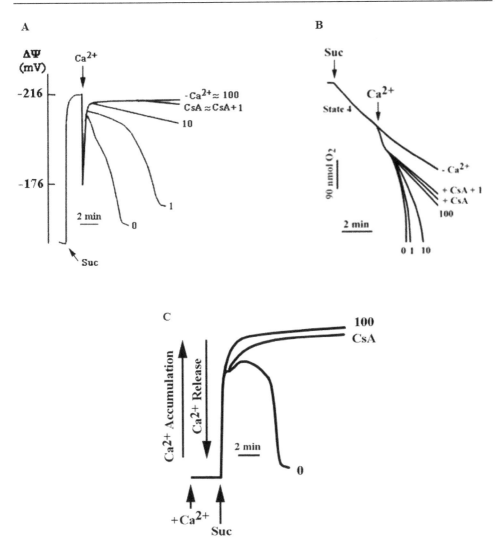

Fig. 3. Effect of AV-6-93 on rat liver MPT induced by Ca²⁺.

Evaluation was performed by measuring succinate-supported transmembrane potential ($\Delta\psi$) (A), oxygen consumption (B), and mitochondrial Ca²⁺ fluxes (C). Additions of 100 nmol calcium/mg protein (Ca²⁺) and 10 mM succinate (Suc); additions of AV-6-93 at the concentrations of 1, 10, and 100 µM (1, 10, 100) are indicated. Assays in the absence of Ca²⁺ (-Ca²⁺); assays in the presence of Ca²⁺ plus CsA (0.75 nmol/mg protein (CsA); assays in the presence of Ca²⁺ plus CsA + 1µM AV-6-93 (CsA+ 1). The traces are representative of assays with three different mitochondrial preparations.

These effects were due to the entry of Ca²⁺ into the electronegative mitochondrial matrix (Fig. 3C), followed by the efflux of H⁺ for restoring the $\Delta\psi$. Incubation of mitochondria with

AV-6-93 concentrations of up to 100 µM for 3 min before energisation with succinate prevented total depolarisation of mitochondria (Fig. 3A), the increase in respiratory state 4 (Fig. 3B), and the release of mitochondrial Ca^{2+} (Fig. 3C), suggesting that this compound has a high ability to protect mitochondria against MPT induction. Incubation of mitochondria with 0.75 nmol/mg protein, CsA, a specific inhibitor of MPT (Broekemeier et al., 1989), for 2 min before energising with succinate, either in the absence or presence of 1 µM AV-6-93, completely blocked mitochondrial depolarisation (Fig. 3A), the increase in respiratory state 4 (Fig. 3B), and the Ca^{2+}-induced release of mitochondrial Ca^{2+} (Fig. 3C). These data show that these effects had been induced by MPT. In contrast to AV-6-93, diflurone, in the same concentration range, did not prevent either the depolarisation of mitochondria or the release of mitochondrial Ca^{2+} (results not shown), indicating that this compound did not protect mitochondria against MPT.

3.4 Effects of AV-6-93 and diflurone on mitochondrial oxidative stress

The effects of AV-6-93 and diflurone on mitochondrial oxidative damage were assessed by detecting the mitochondrial membrane lipid peroxidation induced by the pro-oxidant pair ADP/Fe^{2+}. Lipid peroxidation was evaluated by measuring oxygen consumption (Fig. 4) and TBARs formation (Table 3). In the absence of AV-6-93 and after the addition of the pro-oxidant pair, it is possible to distinguish two-phase kinetics in oxygen consumption: an initial lag phase, characterized by slow oxygen consumption lasting about 2 min, is followed by a rapid oxygen consumption phase. The lag phase is probably related with the time required for the generation of a sufficient amount of the perferryl ion complex ($ADP\text{-}Fe^{2+}\text{-}O_2 \Rightarrow ADP\text{-}Fe^{3+}\text{-}O_2.$), which has been suggested to be responsible for the initiation of lipid peroxidation. The rapid oxygen consumption phase is probably due to the oxidation of the polyunsaturated fatty acid acyl chain of membrane phospholipids by reactive oxygen species (ROS) and, consequently, due to the propagation phase of lipid peroxidation (Sassa et al., 1990). AV-6-93 concentrations up to 100 µM enlarged the lag phase of slow oxygen consumption before the oxygen uptake burst induced by the ADP/Fe^{2+} complex and increased the rate of the rapid oxygen consumption phase (Fig. 4), suggesting that the compounds affected both the initiation and the propagation of lipid peroxidation of mitochondrial membranes.

These results agree with the quantitative evaluation of TBARs formation performed to confirm the protective effects of AV-6-93. The data in Table 3 show that the kinetics of TBARs formation induced by ADP/Fe^{2+} are similar to that observed for oxygen consumption. The same range of AV-6-93 concentrations used in the oxygen consumption assays also affected TBARs formation. TBARs formation in the absence of ADP/Fe^{2+} was negligible (0.44 ± 0.25 nmol/mg of protein). In contrast to AV-6-93, diflurone, in the same concentration range, did not affect oxygen consumption induced by the ADP/Fe^{2+} complex or TBARs formation (results not shown), indicating that this compound has no capacity to protect mitochondria against the lipid peroxidation induced by the pro-oxidant pair ADP/Fe^{2+}.

Lipid peroxidation was evaluated by oxygen consumption and initiated by adding 1 mM $ADP/0.1$ mM Fe^{2+} to mitochondrial suspensions (Fig. 4). The traces represent typical direct oxygen consumption recordings of three experiments obtained from different mitochondrial preparations; controls in the absence of ADP/Fe^{2+} ($-ADP/Fe^{2+}$); assays in the presence of AV-6-93 at the concentrations 1, 10, 20, 50, 100 µM (1, 10, 20, 50, 100).

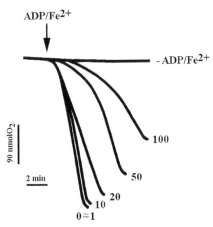

Fig. 4. Effect of AV-6-93 on membrane lipid peroxidation of rat liver mitochondria induced by the pro-oxidant pair ADP/Fe^{2+} evaluated by oxygen consumption.

AV-6-93 (µM)	TBARs (nmol/mg protein/10 min)
0	10.2 ± 1.0
1	9.6 ± 1.0
10	8.5 ± 0.5
20	7.6 ± 1.0
50	5.2 ± 2.1
100	3.1± 1.9*

Table 3. Effect of AV-6-93 on membrane lipid peroxidation of rat liver mitochondria induced by the pro-oxidant pair ADP/Fe^{2+} evaluated by TBARs assay. The data correspond to the mean ± S.E. of three independent experiments. *$p< 0.05$ vs control (in the absence of AV-6-93).

4. Discussion

Studies examining the importance of mitochondrial pathophysiology in neurodegeneration provide a target for additional treatments with agents that improve mitochondrial function, protect MPT, and/or exert antioxidant activity (Petrozzi et al., 2007). These studies lead to novel approaches in the treatment of neurodegenerative diseases, such as Parkinson's disease, with disease-modifying drugs.

The aim of the present study was to examine the abilities of two novel adamantane-containing DHP analogues, AV-6-93 and diflurone, to protect against cell death induced by mitochondrial toxin MPP$^+$ and beneficially influence mitochondrial processes in an attempt to identify putative antiparkinsonian drugs.

First, we examined how both compounds acted in primary cortical cultures in response to MPP$^+$. AV-6-93, at concentrations of 1 and 10 µM, significantly protected against MPP$^+$-induced cell death by 75% and 56%, respectively, whereas diflurone protected against cell death by 35% at a concentration of 10 µM. Neither AV-6-93 nor diflurone, added without MPP$^+$, changed cell viability.

A larger difference between the compounds' activities was observed in isolated rat liver mitochondria by the assessment of their ability to affect both the Ca^{2+}-induced mitochondrial permeability transition (MPT) and lipid peroxidation. To assess the Ca^{2+}-induced MPT, the evaluation of the drop of $\Delta\psi$, the increase in mitochondrial respiration associated with Ca^{2+} accumulation in the mitochondrial matrix, and the mitochondrial Ca^{2+} fluxes were carried out. Changes in these parameters help us to conclude whether the compound protects mitochondria against MPT induction and, consequently, to discern whether the compound alters mitochondrial Ca^{2+} homeostasis. AV-6-93, at a concentration of 10 μM, significantly protected mitochondria against MTP induction and provided complete protection at 100 μM, as revealed by its ability to prevent the depolarisation of mitochondria, the increase in mitochondrial respiration and mitochondrial Ca^{2+} release. These effects were comparable with that of CsA (0.75 nmol/mg protein), a specific inhibitor of the mitochondrial permeability transition pore. Diflurone was ineffective in these tests. The effectiveness of AV-6-93 can be considered to be very promising because it indicates the ability of this compound to halt mitochondrial swelling and cell death, both consequences of the induction of the permeability transition pore.

A critical factor for induction of MPT is the oxidation of thiol groups of the MPT complex, creating diethyl cross-links (Costantini et al., 1996; 1998, Halestrap et al., 1997; McStay et al., 2002). Therefore, the most plausible hypothesis to explain the partial MPT protection induced by AV-6-93 is that changes in the redox-state of thiol groups of the MPT complex is provided via avoiding of diethyl cross-links. This hypothesis is supported by the observation that AV-6-93 protected mitochondria against oxidative stress. Oxidative stress was assessed by evaluating the extent of lipid peroxidation by measuring oxygen consumption and TBARs formation. Alterations of these parameters may reveal whether the compound protects mitochondria against oxidative stress, i.e., whether the compound acts as an antioxidant. AV-6-93, at concentrations up to 100 μM, protected (by about a half) mitochondria against membrane lipid peroxidation, as inferred by its ability to inhibit both oxygen consumption and TBARs formation induced by the pro-oxidant pair ADP/Fe^{2+}. These data suggest that this compound may act as antioxidant because it can avoid both the initiation and the propagation of the oxidation of polyunsaturated fatty acid acyl chains of membrane phospholipids induced by the perferryl ion complex $ADP-Fe^{3+}-O_2^-$, a mechanism suggested to be responsible for lipid peroxidation (Sassa et al., 1990). In contrast to AV-6-93, diflurone, under the same conditions, had no capacity to protect mitochondria against oxidative damage induced by the pro-oxidant pair ADP/Fe^{2+}.

The only common feature of both compounds was a lack of influence on mitochondrial bioenergetics, which was assessed by analysing several mitochondrial functioning parameters of the respiratory chain (respiration states 2, 3, 4, FCCP-stimulated respiration, the RCR, and the ADP/O ratio) and the oxidative phosphorylation system ($\Delta\psi$ and phosphorylation time), using both glutamate/malate and succinate as respiratory substrates. According to the mitochondrial parameters affected, it is possible to assess how the compound interferes with mitochondrial bioenergetics: by perturbing the permeability (integrity) of the inner mitochondrial membrane (stimulation of respiration states 2 and 4), by impairing the respiratory chain (inhibition of FCCP-stimulated respiration), and/or by acting at the level of the phosphorylation system (affecting respiration state 3). Both AV-6-93 and diflurone, at concentrations of up to 100 μM, failed to significantly affect liver mitochondrial bioenergetics, as shown by the lack of effects on both glutamate/malate- and succinate-supported respiration in state 2, state 3, state 4, FCCP-stimulated respiration, RCR and ADP/O ratios, $\Delta\psi$ and phosphorylation time.

To address why both adamantane-containing compounds showed very distinct effects on mitochondrial damage induced by both Ca^{2+} and ADP/Fe^{2+}, one may suggest that the molecular "volume" of AV-6-93 (one adamantane ring-containing DHP) is more optimal than that of diflurone (two adamantane ring-containing DHP) for mitochondrial protection. The two adamantane rings in the diflurone molecule probably generate a steric hindrance that prevents or delays the chemical reaction, which can easily occur in the case of AV-6-93, a one adamantine ring-containing DHP.

Based on the results obtained in primary cortical cultures, the two-adamantane DHP structure is not as crucial as it is in isolated rat liver mitochondria because diflurone has not lost its activity to prevent cell death caused by MPP^+ (a toxin focused on mitochondrial complex I). However, the activity of diflurone was lower than that of AV-6-93. One could suggest that, in addition to the protection of complex I, other cellular signalling mechanisms may be initiated by DHP compounds to increase cell survival.

5. Conclusion

The novel one-adamantane 1,4-dihydropyridine compound AV-6-93 is capable of regulating cell survival processes with regards to mitochondrial processes, such as inhibition of the induction of the permeability transition pore and prevention of oxidative stress. The effectiveness of AV-6-93 can be considered to be very promising in the treatment of neurodegenerative diseases associated with compromised mitochondrial processes, e.g., Parkinson's disease.

6. Acknowledgment

ESF project No. 2009/0217/1DP/1.1.1.2.0/09/APIA/VIAA/031; Latvian Science Council grant: No.10.0030. Center for Neuroscience and Cell Biology (CNC), and Center for Marine and Environmental Research (IMAR-CMA) of the University of Coimbra, Portugal.

7. References

Alho, H., Ferrarese, C., Vicini, S. & Vaccarino, F. (1988) Subsets of GABAergic neurons in dissociated cell cultures of neonatal rat cerebral cortex show co-localization with specific modulator peptides. *Brain Research*, Vol.467, No.2, (April 1988), pp. 193-204, ISSN 0006-8993

Becker, C., Jick, S.S. & Meier, C.R. (2008) Use of antihypertensives and the risk of Parkinson disease. *Neurology*, Vol.15, No.70, (April 2008), pp. 1438-1444, ISSN 0028-3878

Broekemeier, K. M., Dempsey, M. E. & Pfeiffer, D. R. (1989) Cyclosporin A is a potent inhibitor of the inner membrane permeability transition in heart mitochondria. *The Journal of Biological Chemistry*, Vol.264, No.14, (May 1989), pp. 7826-7830, ISSN 0021-9258

Buege, J.A. & Aust, S.D. (1978) Microsomal lipid peroxidation. *Methods in Enzymology*, Vol.52, pp. 302-310, ISSN 0076-6879

Chance, B. & Williams, G. R. (1956) The respiratory chain and oxidative phosphorylation. *Advances in Enzymology and Related Subjects of Biochemistry*, Vol.17, pp. 65-134, ISSN 0096-5316

Costantini, P., Chernyak, B.V., Petronilli, V., Bernardi, P. (1996) Modulation of the mitochondrial permeability transition pore by pyridine nucleotides and dithiol oxidation at two separate sites. *The Journal of Biological Chemistry*, Vol.271, No.12, (March 1996), pp. 6746-6751, ISSN 0021-9258

Costantini, P., Colonna, R., Bernardi, P. (1998) Induction of the mitochondrial permeability transition by N-ethylmaleimide depends on secondary oxidation of critical thiol groups. Potentiation by copper-ortho-phenanthroline without dimerization of the adenine nucleotide translocase. *Biochimica et Biophysica Acta*, Vol.1365, No.3, (July 1998), pp. 385-392, ISSN 0006-3002

Custódio, J. B. A., Palmeira, C. M., Moreno, A. J. M. & Wallace, K. B. (1998a) Acrylic acid induces the glutathione-independent mitochondrial permeability transition *in vitro*. *Toxicological Sciences : an Official Journal of the Society of Toxicology*, Vol.43, No.1 (May 1998), pp. 19-27, ISSN 1096-6080

Custódio, J. B. A., Moreno, A. J. M. & Wallace, K. B. (1998b) Tamoxifen inhibits induction of the mitochondrial permeability transition by Ca^{2+} and inorganic phosphate. *Toxicology and Applied Pharmacology*, Vol.152, No.1, (September 1998), pp. 10-17, ISSN 0041-008X

Dawson, T.M. & Dawson, V.L. (2003) Molecular pathways of neurodegeneration in Parkinson's disease. *Science (New York, N.Y.)*, Vol.302, No.5646, (October 2003), pp. 819-822, ISSN 0036-8075

De Araújo, D.P., Lobato, Rde. F., Cavalcanti, J.R., Sampaio, L.R., Araújo, P.V., Silva, M.C., Neves, K.R., Fonteles, M.M., Sousa, F.C. & Vasconcelos, S.M. (2011) The contributions of antioxidant activity of lipoic acid in reducing neurogenerative progression of Parkinson's disease: a review. *The International Journal of Neuroscience*, Vol.121, No.2, (February 2011), pp. 51-57, ISSN 0020-7454

Ernster, L. & Nordenbrand, K. (1967) Microsomal lipid peroxidation. In: *Methods in Enzymology*, S.P. Colowick & N.O. Kaplan, (Eds.), 574-580, Academic Press, ISBN 0121820181, New York, USA

Esposito, E. & Cuzzocrea S. (2010) Antiinflammatory activity of melatonin in central nervous system. *Current Neuropharmacology*, Vol.8, No.3, (September 2008), pp. 228-242, ISSN 1570-159X

Fernandes, M.A.S., Santos, M.S., Vicente, J.A.F., Moreno, A.J.M., Velena, A., Duburs, G. & Oliveira, C.R. (2003) Effects of 1,4-dihydropyridine derivatives (cerebrocrast, gammapyrone, glutapyrone, and diethone) on mitochondrial bioenergetics and oxidative stress: a comparative study. *Mitochondrion*, Vol.3, (August 2003), pp. 47-59, ISSN 1567-7249

Fernandes, M.A.S., Jurado, A.S., Videira, R.A., Santos, M.S., Moreno, A.J.M., Velena, A., Duburs, G., Oliveira, C.R. & Vicente, J.A.F. (2005) Cerebrocrast promotes the cotransport of H+ and Cl- in rat liver mitochondria. *Mitochondrion*, Vol.5, (October 2005), pp. 341-351, ISSN 1567-7249

Fernandes, M.A.S., Pereira, S.P.S., Jurado, A.S., Custódio, J.B.A., Santos, M.S., Moreno, A.J.M., Duburs, G. & Vicente, J.A.F. (2008) Comparative effects of three 1,4-dihydropyridine derivatives [OSI-1210, OSI-1211 (etaftoron), and OSI-3802] on rat liver mitochondrial bioenergetics and on the physical properties of membrane lipid bilayers: relevance to the length of the alkoxyl chain in positions 3 and 5 of the DHP ring. *Chemico-Biological Interactions*, Vol.173, No.3, (June 2008), pp. 195-204, ISSN 0009-2797

Fernandes, M.A.S., Santos, M.S., Moreno, A.J.M., Chernova, L., Krauze, A., Duburs, G. & Vicente, J.A.F. (2009) Effects of 5-acetyl(carbamoyl)-6-methylsulfanyl-1,4-dihydropyridine-5-carbonitriles on rat liver mitochondrial function. *Toxicology in vitro: an International*

Journal Published in Association with BIBRA, Vol.23, No.7, (October 2009), pp. 1333–1341, ISSN 0887-2333

Gazotti, P., Malmstron, K. & Crompton, M.A. (1979) Laboratory manual on transport and bioenergetics. In: *Membrane Biochemistry*, E. Carafoli & G. Semenza, (Eds.), 62-69, Springer Verlag, ISBN 3540098445, New York, USA

Gornall, G., Bardawill, C.J. & David, M.M. (1949) Determination of serum proteins by means of the biuret reaction. *The Journal of Biological Chemistry*, Vol.177, No.2, (February 1949), pp. 751-766, ISSN 0021-9258

Halestrap AP, Woodfield KY, Connern CP. (1997) Oxidative stress, thiol reagents, and membrane potential modulate the mitochondrial permeability transition by affecting nucleotide binding to the adenine nucleotide translocase. *The Journal of Biological Chemistry*, Vol.272, No.6, (February 1997), pp. 3346-3354, ISSN 0021-9258

Kamo, N., Muratsugu, M., Hongoh, R. & Kobatake, N. (1979) Membrane potential of mitochondria measured with an electrode sensitive to tetraphenylphosphonium and relationship between proton electrochemical potential and phosphorylation potential in stade state. *The Journal of Membrane Biology*, Vol.49, No.2, (August 1979), pp. 105-121, ISSN 0022-2631

Klimaviciusa, L., Klusa, V., Duburs, G., Kaasik, A., Kalda, A. & Zharkovsky, A. (2007) Distinct effects of atypical 1,4-dihydropyridines on 1-methyl-4-phenylpyridinium-induced toxicity. *Cell Biochemistry and Function*, Vol.25, No.1, (February 2007), pp. 15-21, ISSN 0263-6484

Klusa, V. (1995) Cerebrocrast (IOS-1.1212). Neuroprotectant, cognition enhancer. *Drugs of the Future*, Vol.20, No.2, pp. 135–138, ISSN 0377-8282

Kones, R. (2010) Parkinson's disease: mitochondrial molecular pathology, inflammation, statins, and therapeutic neuroprotective nutrition. *Nutrition in Clinical Practice: Official Publication of the American Society for Parenteral and Enteral Nutrition*, Vol.25, No.4, (August 2010), pp. 371-389, ISSN 0884-5336

Kornhuber, J., Bormann, J., Hübers, M., Rusche, K. & Riederer, P. (1991) Effects of the 1-amino-adamantanes at the MK-801-binding site of the NMDA-receptor-gated ion channel: a human postmortem brain study. *European Journal of Pharmacology*, Vol.206, No.4, (April 1991), pp. 297-300, ISSN 0014-2999

McStay, G.P., Clarke, S.J., Halestrap, A.P. (2002) Role of critical thiol groups on the matrix surface of the adenine nucleotide translocase in the mechanism of the mitochondrial permeability transition pore. *The Biochemical Journal*, Vol.367, No.2, (October 2002), pp. 541-548, ISSN 0264-6021

McAllister, J., Ghosh, S., Berry, D., Park, M., Sadeghi, S., Wang, K.X., Parker, W.D. & Swerdlow, R. H. (2007) Effects of memantine on mitochondrial function. *Biochemical Pharmacology*, Vol.75, No.4, (February 2007), pp. 956-964, ISSN 0006-2952

Petrozzi, L., Ricci, G., Giglioli, N.J., Siciliano, G. & Mancuso, M. (2007) Mitochondria and Neurodegeneration. *Bioscience Reports*, Vol.27, No.1-3, (June, 2007), pp. 87–104, ISSN 0144-8463

Pupure, J., Isajevs, S., Gordjushina, V., Taivans, I., Rumaks, J., Svirskis, S., Kratovska A., Dzirkale, Z., Pilipenko, J., Duburs, G. & Klusa, V. (2008) Distinct influence of atypical 1,4-dihydropyridine compounds in azidothymidine-induced neuro- and cardiotoxicity in micee ex vivo. *Basic & Clinical Pharmacology & Toxicology*, Vol.103, No.5, (November 2008), pp. 620-631, ISSN 1742-7835

Ritz, B., Rhodes, S.L., Qian, L., Schernhammer, E., Olsen, J.H. & Friis, S. (2010) L-type calcium channel blockers and Parkinson disease in Denmark. *Annals of Neurology,* Vol.67, No.5, (May 2010), pp. 600-606, ISSN 0364-5134

Rojas, J. C., Saavedra, J. A. & Gonzalez-Lima, F. (2008) Neuroprotective effects of memantine in a mouse model of retinal degeneration induced by rotenone. *Brain Research,* Vol.1215, (June 2008), pp. 208-217, ISSN 0006-8993

Sassa, H., Takaish, Y. & Terada, H. (1990) The triterpenecelastrol as a very potent inhibitor of lipid peroxidation in mitochondria. *Biochemical and Biophysical Research Communications,* Vol.172, No2, (October 1990), pp. 890-897, ISSN 0006-291X

Schapira, A.H. (1999) Mitochondrial involvement in Parkinson's disease, Huntington's disease, hereditary spastic paraplegia and Friedreich's ataxia. *Biochimica et Biophysica Acta,* Vol.1410, No.2, (February 1999), pp. 159-170, ISSN 0006-3002

Schapira, A.H. (2009) Neurobiology and treatment of Parkinson's disease. *Trends in Pharmacological Sciences,* Vol.30, No.1, (January 2009), pp. 41-47, ISSN 0165-6147

Seeman, P., Caruso, C. & Lasaga, M. (2008) Memantine agonist action at dopamine D2High receptors. *Synapse (New York, N.Y.),* Vol.62, No.2, (February 2008), pp. 149-153, ISSN 0887-4476

Sengupta, T., Vinayagam, J., Nagashayana, N., Gowda, B., Jaisankar, P. & Mohanakumar, K.P. (2011) Antiparkinsonian effects of aqueous methanolic extract of Hyoscyamus niger seeds result from its monoamine oxidase inhibitory and hydroxyl radical scavenging potency. *Neurochemical Research,* Vol.36, No.1, (January 2010), pp. 177-186, ISSN 0364-3190

Skolimowski, J., Kochman, A. & Metodiewa, D. (2003) Synthesis and antioxidant activity evaluation of novel antiparkinsonian agents, aminoadamantane derivatives of nitroxyl free radical. *Bioorganic & Medicinal Chemistry,* Vol.11, No.16, (August 2003), pp. 3529-3539, ISSN 0968-0896

Spilker, B.A., Dhasmana, K.M., Davies, J.E. & Claassen, V. (1973) Differentiation between effects of nicotine and DMPP by amantadine. *Archives Internationales de Pharmacodynamie et de Thérapie,* Vol.203, No.2, (June 1973), pp. 221-231, ISSN 0301-4533

Sulzer, D. & Schmitz, Y. (2007) Parkinson's disease: return of an old prime suspect. *Neuron,* Vol.55, No.1, (July 2007), pp. 8-10, ISSN 0896-6273

Tymianski, M., Charlton, M.P., Carlen, P.L. & Tator, C.H. (1993) Source specificity of early calcium neurotoxicity in cultured embryonic spinal neurons. *The Journal of Neuroscience : the Official Journal of the Society for Neuroscience,* Vol.13, No.5, (May 1993), pp. 2085-2104, ISSN 0270-6474

Velena, A., Skujins. A., Svirskis, S., Bisenieks, E., Uldrikis, J., Poikans, J., Duburs, G. & Klusa, V. (1997) Modification of swelling-contraction-aggregation processes in rat muscle mitochondria by the 1,4-dihydropyridines, cerebrocrast and glutapyrone, themselves and in the presence of azidothymidine. *Cell Biochemistry and Function,* Vol.15, No.3, (September 1997), pp. 211-220, ISSN 0263-6484

Vicente, J.A.F., Duburs, G., Klusa, V., Briede, J., Klimaviciusa, L., Zharkovsky, A. & Fernandez, M.A.S. (2006) Cerebrocrast as a neuroprotective, anti-diabetic and mitochondrial bioenergetic effector: A putative mechanism of action, In: *Mitochondrial Pharmacology and Toxicology,* A.J.M. Moreno, P.J. Oliveira & C.M. Palmiera (Eds.), 185-197, Transworld Research Network, ISBN 8178952076, Kerala, India

11

Phosphorescence Oxygen Analyzer as a Measuring Tool for Cellular Bioenergetics

Fatma Al-Jasmi, Ahmed R. Al Suwaidi, Mariam Al-Shamsi,
Farida Marzouqi, Aysha Al Mansouri, Sami Shaban,
Harvey S. Penefsky and Abdul-Kader Souid
United Arab Emirates University, Faculty of Medicine and Health Sciences
Abu Dhabi
United Arab Emirates

1. Introduction

The *"phosphorescence oxygen analyzer"* and its use to monitor O_2 consumption by cells and tissues are discussed in this chapter (Lo et al., 1996; Souid et al., 2003). This analytical tool assesses bioenergetics in cells undergoing apoptosis (e.g., the mitochondrial cell death pathway), in cells exposed to toxins (e.g., loss of viability) and in cells with a genetically altered energy metabolism (e.g., mitochondrial disorders) (Tacka et al., 2004a-b; Tao et al., 2007; Tao et al., 2008a). This method is applicable to suspended (e.g., Jurkat and HL-60 cells) and adherent (TU183 human oral cancer cells) cells and to fresh tissues from humans (e.g., lymphocytes, spermatozoa and tumors) and animals (e.g., liver, spleen, heart, pancreas and kidney) (Badawy et al., 2009a-b; Whyte et al., 2010; Al Shamsi et al., 2010; Al-Salam et al., 2011; Al Samri et al., 2011). The analyzer allows investigating anticancer compounds (single agents or combinations) for dosing, order of administration and exposure (Jones et al., 2009; Tao et al., 2008b; Souid et al., 2006; Goodisman et al., 2006; Tao et al., 2006a-b; Tack et al. 2004b). It can also be used to monitor reactions consuming or producing O_2 (Tao et al., 2008b; Tao et al., 2009).

2. Relevant biological processes

The term *"cellular bioenergetics"* describes the biochemical processes involved in energy metabolism (energy conversion or transformation), while the term *"cellular respiration"* describes delivery of O_2 to the mitochondria, the breakdown of reduced metabolic fuels with passage of electrons to O_2, and the resulting synthesis of ATP. Impaired respiration thus implies any abnormality involving cellular bioenergetics, including glycolysis. The term *"apoptosis"* describes cellular mechanisms responsible for initiating and executing cell death. The initiation step requires a leakage of cytochrome c from the mitochondrial intermembrane to the cytosol. In the cytosol, cytochrome c binds to the apoptotic protease activating factor-1 (Apaf-1), triggering the caspase cascade (a series of cysteine, aspartate-specific proteases). Caspase activation executes mitochondrial dysfunction (Nicholson et al., 1997). This mitochondrial perturbation involves opening the permeability transition pores (accelerating oxidations in the mitochondrial respiratory chain) and collapsing the electrochemical potential $\Delta\psi$ (Ricci et al., 2004). Thus, induction of apoptosis is directly linked to mitochondrial dysfunction (Green and Kroemer, 2004).

3. Expressions of dissolved oxygen

Dissolved O_2 is expressed in mm Hg, mL O_2 per L, mg O_2 per L, or µmol per L (µM). For conversion, a partial pressure of O_2 (PO_2) of 1.0 mm Hg = 0.03 mL O_2 per L; 1.0 mL O_2 per L = 1.4276 mg O_2 per L; and 1.0 mg O_2 per L = 1000/32 µM. In *freshwater* at 760 mm Hg and 20°C, dissolved [O_2] is 9.1 mg/L, or 284 µM. Using a Clark electrode, PO_2 of the reaction mixture phosphate-buffer saline (PBS), 10 mM glucose and 0.5% fat-free bovine serum albumin is 170.5 ± 6.6 mm Hg (n = 4), or 228 ± 9 µM. The 56 mm Hg difference between [O_2] in freshwater and the reaction solution reflects the effect of salinity on dissolved O_2 (Weiss, 1970).

4. Principles and tools of the oxygen measurement

O_2 concentration is determined from the phosphorescence decay rate ($1/\tau$) of the *palladium (II) complex of meso-tetra-(4-sulfonatophenyl)-tetrabenzoporphyrin* (Pd phosphor). This measurement is based on quenching the phosphorescence of Pd phosphor by O_2 (Lo et al., 1996). The probe has an absorption maximum at 625 nm and an emission maximum at 800 nm. Samples are exposed to light flashes (10 per sec) from a pulsed light-emitting diode array with peak output at 625 nm (OTL630A-5-10-66-E, Opto Technology, Inc), Wheeling, IL. Emitted phosphorescent light is detected by a Hamamatsu photomultiplier tube (PMT #928) after first passing it through a wide-band interference filter centered at 800 nm. Amplified phosphorescence is digitized at 1-2 MHz using an analog/digital converter (PCI-DAS 4020/12 I/O Board) with 1 to 20 MHz outputs (Computer Boards, Inc.). Pulses are captured using a developed software program at 0.1 to 4.0 MHz, depending on speed of the computer (Souid, 2003).

The values of $1/\tau$ are linear with dissolved O_2 concentration: $1/\tau = 1/\tau_o + k_q[O_2]$, where $1/\tau$ = the phosphorescence decay rate in the presence of O_2, $1/\tau_o$ = the phosphorescence decay rate in the absence of O_2, and k_q = the second-order O_2 quenching rate constant in sec^{-1} $µM^{-1}$ (Lo et al., 1996).

Cellular respiration is measured at 37°C in 1-mL sealed vials. Mixing is carried out with the aid of parylene-coated stirring bars. The respiratory substrates are the endogenous metabolic fuels supplemented with glucose. In cell suspensions sealed from air, [O_2] decreased linearly with time, indicating the kinetics of cellular mitochondrial O_2 consumption is zero-order. The rate of respiration (k, in µM O_2 min^{-1}) is thus the negative of the slope d [O_2]/dt. Cyanide markedly inhibited respiration, confirming O_2 is consumed mainly by the mitochondrial respiratory chain.

5. Developed software program and instrument description

The software program was developed using Microsoft Visual Basic 6 (VB6) programming language, Microsoft Access Database 2007 (Access) database management system, and Universal Library components developed by the electronic board company, Measurement Computing, for use with Microsoft Visual Basic 6 programming language (http://www.mccdaq.com/daq-software/universal-library.aspx). It allows direct reading from the PCI-DAS 4020/12 I/O Board (http://www.mccdaq.com/pci-data-acquisition/PCI-DAS4020-12.aspx). The software utilizes a relational database that stores experiments, pulses and pulse metadata, including slopes. Pulse identification is performed by detecting 10 phosphorescence intensities above 1.0 volt (by default). Peak identification is performed by the program which detects the highest 10% data points of a pulse and chooses the point in the group that is closest to the pulse's decay curve. Depending on the sample rate, a minimum number of data points per pulse is set and used as a cutoff to remove invalid pulses with too few data points (Shaban, 2010).

Main advantages of the developed program over commercially available packages (e.g., DASYLab™ or TracerDAQ™) are provision of full control and customization of the data acquisition, storage and analysis. The choices of VB6 and Access as programming and storage environments are due to their availability, simplicity, widespread use and VB6 components that read directly from the PCI card made available by Measurement Computing. Table 1 displays identified tasks of the program. Fig. 1 shows a picture of the data acquisition system and the developed software program. Fig. 2 shows a reaction vial.

Experiment identification (title, date, time and sample rate)
Reading directly from the PCI card at the fastest possible rate
Distinguishing pulse data from non-pulse data
Allowing a fuzzy detection of the pulse peak
Calculating the exponential decay rate $(1/\tau)$ and lifetime (τ) of each pulse
Storing each pulse data points, along with the peak, decay and lifetime values
Viewing a representative pulse every 10 sec
Viewing decay rates $(1/\tau)$ in a second graph
Ability to pause and place a marker with a note
Ability to remove erroneous (incomplete) pulses and adjust peak values if necessary
Ability to copy pulse or slope data to clipboard for further analysis
Ability to access a previous experiment, review a pulse with its metadata, markers and associated notes

Table 1. Identified tasks of the customized software program for data acquisition, storage and analysis

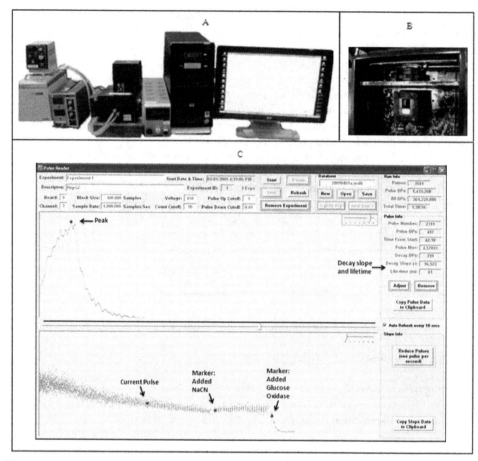

The components (panel A, left to right) are circulating water bath, power supply for the mixer, sample chamber (panel B) attached to PMT, high voltage power supply for PMT, computer with PCI-DAS board, and monitor with developed software running. The PMT is connected to PCI-DAS board on the back of the computer. The developed software program interface is shown in panel C.

Fig. 1. The data acquisition system and developed software program.

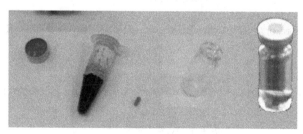

Fig. 2. A sealed reaction vial containing Pd phosphor solution, stirring bar and a mouse liver specimen.

6. Instrument calibration

The instrument was calibrated with β-glucose and glucose oxidase system.

$$\beta\text{-glucose} + O_2 \longrightarrow \text{glucono-}\delta\text{-lactone} + H_2O_2$$

The reaction contained PBS, 3 μM Pd phosphor, 0.5% fat-free albumin, 50 μg/mL glucose oxidase and various concentrations of β-glucose. To achieve a high signal-to-nose ratio throughout the entire range of $[O_2]$, the photomultiplier tube was operated at 450 volts. Representative pulses (with exponential fits) for reactions containing PBS with 0, 125 or 500 μM β-glucose, 50 μg/mL glucose oxidase, 3 μM Pd phosphor and 0.5% fat-free bovine serum albumin are shown in Fig. 3.

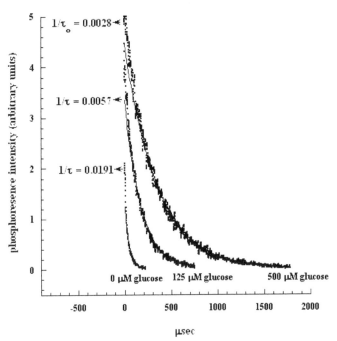

Fig. 3. Representative phosphorescence pulses for reaction mixtures containing 0, 125 or 500 μM glucose. The lines are exponential fits (R^2 >0.924, >0.985 and >0.992, respectively).

The values of $1/\tau$ (mean + SD, n = 1200 over 2 min) as function of [β-glucose] are shown in Fig. 4. The line is linear fit and the value of k_q (101.1 sec^{-1} μM^{-1}) is the negatives of the slope. The value of $1/\tau$ for air-saturated solution (without glucose) was 28,330 sec^{-1} (coefficient of variation, C_v = 12%), for [β-glucose] = 125 μM 5,650 sec^{-1}, and for O_2-depleted solution (with 500 μM β–glucose, $1/\tau_o$) 2,875 sec^{-1} (C_v = 1%). The high values of C_v for the air-saturated solutions were due to the lower phosphorescence intensities with high $[O_2]$ (little light reaching the photomultiplier tube). The corresponding lifetimes (τ) were 52 μsec, 177 μsec and 352 μsec, respectively. Oxygen concentration was calculated using, $1/\tau = 1/\tau_o + k_q[O_2]$.

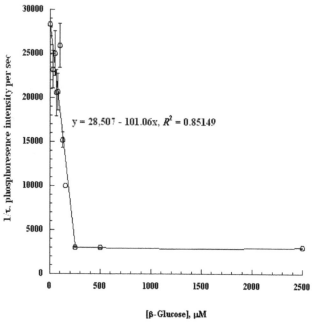

The reaction mixtures contained PBS, 3 μM Pd phosphor, 0.5% fat-free bovine serum albumin, 50 μg/mL glucose oxidase and shown concentrations of β-glucose. The values of $1/\tau$ (mean $\pm$ SD, n = 1200 flashes over 2 min) as a function of [β-glucose] are shown. The lines are linear fits.

Fig. 4. Calibration with β-glucose plus glucose oxidase

7. Aflatoxin B1 impairs human lymphocyte respiration

Aflatoxins (most notably, aflatoxin B1) are highly carcinogenic compounds, commonly found in food contaminated by aspergillus flavus, parasiticus and penicillium species (Williams, et al., 2004; Eaton & Gallagher, 1994). These potent mycotoxins create major health problems, especially where food storage is subjected to heat and humidity. A high rate of dietary exposure is reported in Sahara Africa, China and Taiwan. For example, in eastern China (where liver cancer exceeds 1 per 10,000 population per year), an average human exposure to aflatoxins is estimated to be 2.2 μg/kg/day. For comparison, the exposure in the United States is about 3 orders of magnitude less (Wang et al., 1996).

Biotransformation of aflatoxin B1 is critical for its activation. The parent compound undergoes oxidation by monooxygenases, especially the hepatic cytochrome P450 3A4. The active metabolite, AFB exo-8,9-epoxide, undergoes base-catalyzed rearrangement to a dialdehyde, which rapidly reacts with guanyl N7 in DNA and lysine in proteins (Johnson et al., 1996).

Exposure to aflatoxin B1 has been associated with hepatocellular carcinoma (Montesano et al., 1997), mutagenesis (e.g., in the tumor suppressor gene p53) and immune suppression (Corrier, 1991). Most of the information on immunotoxicity of aflatoxin B1 is derived from animal studies (Stec et al., 2009; Reddy et al., 1987; Reddy et al., 1989; Jiang et al., 2008; reviewed in Williams, et al., 2004]. In healthy humans, exposure to aflatoxin B1 is associated with lower perforin (a cytolytic protein produced by natural killer lymphocytes) expression on CD8+ T-lymphocytes (Jiang et al., 2008). A dose-related decrease in DNA synthesis in lymphocyte cultures (with and without mitogens) is found in mice exposed *in vivo* to aflatoxin B1 (Reddy et al., 1987). A decrease in DNA synthesis is also observed in normal splenic mouse lymphocytes cultured *in vitro* with >10 µM aflatoxin B1; a decrease in RNA synthesis is observed at dosing >25 µM and a decrease in protein synthesis at dosing >100 µM (Reddy et al., 1989).

The phosphorescence oxygen analyzer is used to monitor the effects of aflatoxin B1 on human lymphocyte mitochondrial oxygen consumption. These experiments investigate whether aflatoxin B1 impairs respiration of the lymphoid tissue, an organ that is typically targeted by this potent mycotoxin. Aflatoxin B1 (1.0 mg = 3.2 micromol) was freshly dissolved in 1.0 mL dry methanol and immediately added to cell suspensions with vigorous mixing. Alternatively, aflatoxin B1 powder was directly added to the cell suspension with vigorous mixing. The concentrations were determined by the absorbance at 350 nm (10 µL aflatoxin B1 stock solution or cell-free supernatant in 1.0 mL dry methanol), using an extinction coefficient of 21,500 M^{-1} cm^{-1} (Nesheim et al., 1999); the aflatoxin B1 excitation wavelength is 366 nm and the emission wavelength 455 nm. The reactions were carried out in glass vials and protected from light.

PBMC (0.6 x 10^7 cells/mL) were suspended in 6.0 mL PBS, 10 mM glucose, 3 µM Pd phosphor and 0.5% fat-free bovine serum albumin. The mixture was divided into 2 equal aliquots. Methanol (25 µL per mL, Fig. 5, left panel) or aflatoxin B1 (25 µM, Fig. 5, right panel) was then added and the incubation continued at 37°C (open to air with gentle stirring). At t = 10 and 110 min, 1.0 mL of each mixture was simultaneously placed in the instruments for O_2 measurement. The rate of respiration (k, in µM O_2 min^{-1}) for t = 10 to 78 min for the methanol-treated cells was 2.0 and for the aflatoxin B1-treated cells 2.1. The values of k for t = 110 to 174 were 2.2 and 1.2, respectively (corresponding to 45% inhibition of lymphocyte respiration).

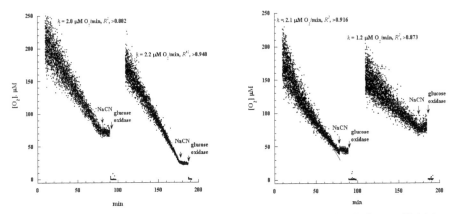

PBMC were incubated at 37°C with 25 µL per mL methanol (left panel) or 25 µM aflatoxin B1 (right panel). Minute zero corresponds to the addition of aflatoxin B1. At t = 10 and t = 110 min, 1.0 mL of each mixture was simultaneously placed in the instruments for O_2 measurement. Rates of respiration (k) were calculated from the best-fit linear curves. Additions of 5.0 mM NaCN and 50 µg/mL glucose oxidase are shown.

Fig. 5. Effect of aflatoxin B1 on human PBMC respiration.

The time-course for aflatoxin B1-induced inhibition of lymphocyte respiration was investigated (Fig. 6). PBMC (1.3 x 10[7] cells/mL) were suspended in 3.0 mL PBS, 10 mM glucose and divided into 2 equal aliquots. Aflatoxin B1 powder was directly added to one aliquot with vigorous mixing (final concentration, ~75 μM). The 2 aliquots were then incubated at 37°C for 60 min (open to air with continuous stirring). At t = 60 min, 5 mg albumin and 2.0 μM Pd phosphor were added to each suspension. The samples were then simultaneously placed in the chambers for O_2 measurement. For the untreated cells, O_2 consumption was linear with time (k = 2.4 μM O_2/min, R^2 > 0.916). For the treated cells, O_2 consumption was exponential with time, R^2 > 0.934. Changes at 5-min intervals for the treated cells showed a sharp decline in the values of k for t = 60 to 90 min, followed by a steady low rate for t = 95 to 175 min. In contrast, the values of k remained relatively stable for t = 60 to 140 min. The mean $\pm$ SD (coefficient of variation) for the values of k for the untreated cells was 2.43 $\pm$ 0.55 (C_v = 23%) and for the treated cells 1.56 $\pm$ 1.80 (C_v = 87%); p-value < 0.02 (Fig. 6, insert).

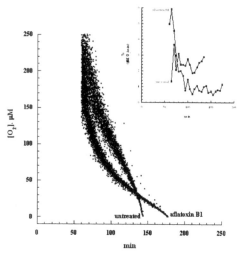

The lines are linear fit for the untreated cells (R^2 > 0.916) and exponential fit for the treated cells (R^2 > 0.936). Insert, changes in the values of k at 5-min intervals.

Fig. 6. Time-course of the effect of aflatoxin B1 on human lymphocyte respiration.

The exponential profile of O_2 consumption in the presence of aflatoxin B1 is similar to dactinomycin (Tao et al., 2006b; Tao et al., 2008a). This pattern of bioenergetic derangements could stem from progressive mitochondrial and metabolic disturbances, ranging from uncoupling oxidative phosphorylation (which accelerates O_2 consumption and rapidly depletes the metabolic fuels) to mitochondrial respiratory chain function collapse. Experimentally, these two phases are clearly distinguishable in our system (Fig. 6, insert).

Caspase activation in lymphocytes treated with aflatoxin B1 was then examined. Many of the caspases (e.g., caspase-3, -2 and -7) target the asp-glu-Val-asp (DEVD) motif and cleave at sites next to the last aspartate residue (Nicholson et al., 1997). Synthetic cell-permeable substrates, such as N-acetyl-DEVD-7-amino-4-trifluoromethyl coumarin (Ac-DEVD-AFC) and N-acetyl-DEVD-7-amino-4-methyl coumarin (Ac-DEVD-AMC) have been used to investigate caspase activities. For example, cleavage of Ac-DEVD-AFC by specific caspases releases the fluorogenic AFC; the latter can be separated on HPLC and detected by fluorescence with a great sensitivity (Tao et al., 2007).

Caspase-3 activity in lymphocytes exposed to aflatoxin B1 is shown in Fig. 7. The purpose of these experiments is to confirm caspases are activated within the time period required for inhibition of respiration. The mixtures (final volume, 0.5 mL) contained 1.5×10^6 cells in PBS, 10 mM glucose and 68 µM Ac-DEVD-AMC (N-acetyl-asp-glu-val-asp-7-amino-4-methyl coumarin, a caspase-3 substrate) with and without 20 µM zVAD-fmk (benzyloxycarbonyl-val-ala-DL-asp-fluoromethylketone, a pan-caspase inhibitor) (Slee et al., 1996). The suspensions were incubated at 37°C for 2 hr without other additions (Fig. 7, left panel) or with the addition of ~100 µM aflatoxin B1 (Fig. 7, right panel). At the end of the incubation period, the cells were disrupted and their supernatants were separated on HPLC and monitored by fluorescence. The results show AMC moieties (the cleavage product of Ac-DEVD-AMC) appear in the cells about 2 hr after the addition of aflatoxin B1. This 2-hr period is the same as that observed for aflatoxin B1-induced inhibition of respiration (see Fig. 5). Thus, the results suggest aflatoxin B1 impairs human lymphocyte mitochondrial function by activating caspases.

The above findings also demonstrate the lymphocyte preparation contain monooxygenases that activate aflatoxin B1. These results are consistent with previous reports (Stec et al., 2009; Rossano et al., 1999; Savel et al., 1970; Wang et al., 1999). In one study, the addition of aflatoxins B1 at concentrations up to 32 µM had a minimum effect on phytohemagglutinin-p-stimulated human lymphocyte proliferation (Meky et al., 2001). However, an earlier study on human lymphocytes by Savel et al. (1970) showed a reduced phytohemagglutinin-p-stimulated lymphocyte proliferation with 16 µM aflatoxin B1. More recently, aflatoxin B1 was shown to inhibit *in vitro* concanavalin A-induced proliferation of pig blood lymphocytes; in 72-hr cultures, the concentration of aflatoxin B1 producing 50% inhibition (IC_{50}) was 60 nM (Stec et al., 2009). In other studies, aflatoxin G1 induced *in vitro* apoptosis in human lymphocytes (Wang et al., 1999; Sun et al., 2002). In summary, the data presented show human lymphocytes exposed *in vitro* to aflatoxin B1 exhibit impairments of cellular respiration, which could result from caspase activation. The results substantiate the potent immunosuppressive activity of aflatoxins in human.

8. Measurement of O_2 consumption in murine tissues

A novel *in vitro* system is developed to measure O_2 consumption by various murine tissues over several hours (Al-Salam et al., 2011; Al Samri et al., 2011; Al Shamsi et al., 2010). Small tissue specimens excised from male Balb/c mice were immediately immersed in ice-cold Krebs-Henseleit buffer (115 mM NaCL, 25 mM $NaHCO_3$, 1.23 mM NaH_2PO_4, 1.2 mM Na_2SO_4, 5.9 mM KCL, 1.25 mM $CaCl_2$, 1.18 mM $MgCl_2$ and 6 mM glucose, pH ~7.4), saturated with 95% O_2:5% CO_2. The samples were incubated at 37°C in the same buffer and continuously gassed with O_2:CO_2 (95:5). Normal tissue histology at hr 5 was confirmed by light and electron microscopy. NaCN inhibited O_2 consumption, confirming the oxidation occurred in the mitochondrial respiratory chain. A representative experiment of pneumatocyte respiration is shown in Fig. 8. The rate of lung tissue respiration incubated *in vitro* for $3.9 < t < 12.4$ hr was 0.24 ± 0.03 µM O_2 min^{-1} mg^{-1} (mean ± SD, n = 28). The corresponding rate for the liver was 0.27 ± 0.13 (n = 11, $t < 4.7$ hr), spleen 0.28 ± 0.07 ($t < 5$ hr, n = 10), kidney 0.34 ± 0.12 ($t < 5$ hr, n = 7) and pancreas 0.35 ± 0.09 ($t < 4$ hr, n = 10), Table 2. This approach provided accurate assessment of tissue bioenergetics *in vitro* over several hours.

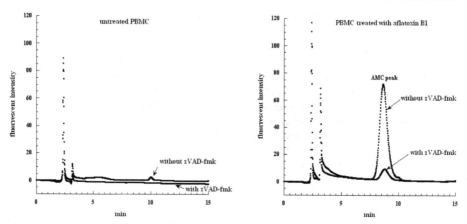

The reactions contained 1.5 x 10^6 cells in PBS plus 10 mM glucose and 68 µM Ac-DEVD-AMC with and without 20 µM zVAD-fmk. The suspensions were incubated at 37°C for 2 hr without other additions (left panel) or with the addition of ~100 µM aflatoxin (right panel). At the end of the incubation period, the cells were disrupted and their supernatants were separated on HPLC and monitored by fluorescence. The retention time for Ac-DEVD-AMC was ~2.4 min and for the released AMC ~8.7 min.

Fig. 7. Caspase activation by aflatoxin B1 in human peripheral blood mononuclear cells parenthesis for PBMC:PBMC

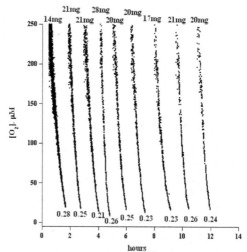

Specimens were excised from the lung of an anesthetized mouse and immediately immersed in ice-cold Krebs-Henseleit buffer saturated with 95% O_2:5% CO_2. The samples were incubated at 37°C in the same buffer with continuous gassing with O_2:5%CO_2. At indicated time periods, specimens were removed from the incubation mixture, weighed and placed in Krebs-Henseleit buffer containing 0.5% albumin and 3 µM Pd phosphor for O_2 measurement. The rate of respiration was set as the negative of the slope of [O_2] vs. time. The weight is shown at the top and the respiration rate (in µM O_2 min^{-1} mg^{-1}) at the bottom of each run (Al Samri et al., 2011).

Fig. 8. Representative experiment of O_2 consumption by lung tissue.

Tissues	Respiration (μM O_2 min^{-1} mg^{-1})
Lung	0.24 ± 0.03
Liver	0.27 ± 0.13
Spleen	0.28 ± 0.07
Kidney	0.34 ± 0.12
Pancreas	0.35 ± 0.09

Values are mean ± SD. For unit conversion, 1.0 mL O_2 = 1.4276 mg or 0.0446125 mmol.

Table 2. O_2 consumption by murine tissues.

9. Biocompatibility of calcined mesoporous silica particles with murine tissue bioenergetics

The *in vitro* system discussed in Section 8 is used to investigate the effects of two forms of calcined mesoporous silica particles (MCM41-cal and SBA15-cal) on cellular respiration of mouse tissues (Al Shamsi et al., 2010; Al-Salam et al., 2011; Tao et al., 2008c). O_2 consumption by lung, liver, kidney, spleen and pancreatic tissues was unaffected by exposure to 200 µg/mL MCM41-cal or SBA15-cal for several hours. A representative experiment of pneumatocyte respiration is shown in Fig. 9.

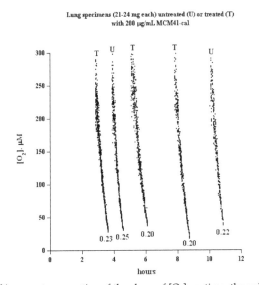

The rate of respiration (k) was set as negative of the slope of [O_2] vs. time; the values of k (in µM O_2 min^{-1} mg^{-1}) are shown at the bottom of the runs. Zero minute corresponds to the addition of the particles. U, untreated; and T, treated.

Fig. 9. Pneumatocyte respiration with and without 200 µg/mL MCM41-cal.

Normal tissue architecture and histology were confirmed by light microscopy. Intracellular accumulation of the particles in the studied tissues was evident by electron microscopy. The results show reasonable *in vitro* biocompatibility of the mesoporous silicas with murine tissue bioenergetics. Therefore, the measurements of respiration can be used to explore biocompatibility and viability of tissues and cells as a result of various treatments.

10. Liver tissue bioenergetics in concanavalin A hepatitis in mice

Concanavalin A (Con A) is a plant lectin from the seeds of Canavalia ensiformis (jack bean). This toxin serves as a polyclonal T-cell mitogen. It produces fulminant hepatitis in mice, a disease that mimics human infection with hepatitis B virus (Tiegs et al., 1992 & 1997). The hepatic injury is typically noted within 3 hr of intravenous injection of > 1.5 mg/kg of Con A and progresses with time (Tiegs et al., 1992). Activation and recruitment of Natural Killer (NK) T-cells and other cells of the innate immune system are early events, which lead to increased secretion of various inflammatory cytokines (e.g., TNF-α, IL-2, IL-10, IL-12 and IFN-γ) (Takeda et al., 2000; Margalit et al., 2005; Chen et al., 2010; Sass et al., 2002). This immune response targets multiple organs including the liver. Its outcome is irreversible hepatotoxicity, which includes inflammatory infiltrates and necrosis (Leist et al., 1996).

The above described *in vitro* system is employed to assess liver tissue respiration in Con A treated C57BL/6 mice. The purpose of the work was to estimate hepatocyte bioenergetics in this well-studied hepatitis model. The mice were injected intravenously with 12 mg/kg Con A or PBS. Specimens (20 to 30 mg each) were cut from the liver of anesthetized (urethane, 100 µL per 10 g body weight, using 25% solution, w/v, in 0.9% NaCl) mice using a sharp scissor (Moria Vannas Wolg Spring, cat. # ST15024-10) (Al Samri et al., 2011). The specimens were immediately immersed in ice-cold Krebs-Henseleit buffer (115 mM NaCl, 25 mM NaHCO$_3$, 1.23 mM NaH$_2$PO$_4$, 1.2 mM Na$_2$SO$_4$, 5.9 mM KCl, 1.25 mM CaCl$_2$, 1.18 mM MgCl$_2$ and 6 mM glucose, pH ~7.2), gassed with 95% O$_2$:5% CO$_2$. Pieces were then weighed and placed in 1-ml Pd phosphor solution (Krebes-Henseleit buffer containing 0.5% albumin and 3 µM Pd phosphor) for O$_2$ measurement. The results are summarized in Table 3.

Strain	Treatment	k_c (μM O$_2$ min^{-1} mg^{-1})	
		mean $\pm$ SD (n)	range
C57Bl/6	PBS	0.26 + 0.04 (5)	0.22 – 0.32
	12 mg/kg Con A	0.18 + 0.03 (5)*	0.13 – 0.20

* *P*-value = 0.005

Table 3. Liver tissue respiration in Con A treated C57BL/6 mice. Mice were injected with Con A or PBS. Liver specimens were collected 12 hr post injection.

A representative experiment following 3-hr treatment is shown in Fig. 10. Liver tissue respiration was measured 3 hr post injection of PBS (Fig. 10, left panel) or 12 mg/kg Con A (Fig. 10, right panel). In untreated mouse, the rates of respiration at $t = 0$ min and $t = 80$ min (post tissue collection) were similar. In Con A-treated mouse, the rate of respiration at $t = 0$ min was high and at $t = 40$ min it was low. Thus, at 3-hr, Con A treatment doubled the rate of liver tissue O$_2$ consumption. However, respiration deteriorated *in vitro* in 40 min.

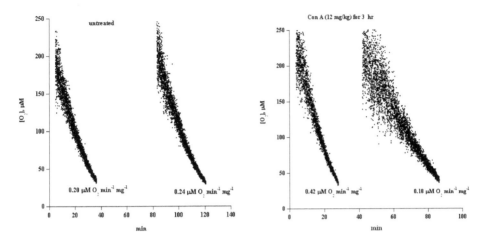

Minutes zero correspond to collecting the liver tissue specimens at 3 hr post injections. Two runs were done for each condition. Rates of respiration (k_c, in μM O_2 min^{-1} mg^{-1}) are shown at the bottom of the runs.

Fig. 10. Representative experiment for liver tissue respiration in C57BL/6 mice 3 hr post injection of PBS (left panel) or 12 mg/kg Con A (right panel)

Thus, Con A treatment produced a concurrent impairment of hepatocyte respiration. The lower rate of respiration at 12 hr post treatment (Table 3) concurred with large areas of necrosis and the enhanced rate of respiration at ~3 hr post treatment (Fig. 10) concurred with inflammatory infiltrates limited to the perivascular space without any notable necrosis. The latter finding suggests a role for inflammatory mediators, such as TNF-α and IL-2 (both known to peak 3 hr post Con A treatment) in modulating hepatocyte energy metabolism (Louis et al., 1997; Gottlieb et al., 2000). The mechanism for the presumed inflammation-induced increase in hepatocyte oxygen consumption could be uncoupling oxidative phosphorylation *vs.* up-regulating the energy metabolism. Nevertheless, for both assumptions, there is a large demand for energy supply to prevent fulminant liver necrosis. In an *in vitro* experiment, liver tissue respiration was measured with and without IL-2 (added directly to the O_2 measuring vial). The rate of respiration without IL-2 was 0.21 μM O_2 min^{-1} mg^{-1} and with IL-2 0.087 μM O_2 min^{-1} mg^{-1} (~60% inhibition). Thus, similar to TNF-α, IL-2 also inhibits cellular respiration *in vitro* (Gottlieb et al., 2000).

11. Spermatozoa respiration

The above *in vitro* system was also used to measure human spermatozoa respiration. O_2 concentrations in solutions containing glucose and human spermatozoa declined linearly with time. Sodium cyanide also inhibited sperm oxygen consumption, confirming the oxidations occurred in the respiratory chain. The rate of respiration (mean $\pm$ SD, n = 10) was 1.0 $\pm$ 0.3 μM O_2 min^{-1} per 10^8 sperm. Immediate decline in the rate of sperm respiration was noted when toxic agents [e.g., 4-hydroperoxycyclophosphamide (4OOH-CP), Δ^9-tetrahydrocannabinol (Δ^9-THC) or Δ^8-tetrahydrocannabinol (Δ^8-THC)] were added to washed sperm or neat semen. The inhibition was concentration-dependent and irreversible (Badawy et al., 2009a-b).

The toxic effect of the cannabinoids was confirmed on isolated mitochondria from beef heart. The effect of Δ^8-THC on respiration of beef heart mitochondria is shown in Fig. 11. The value of k (in μM O_2 min^{-1}) decreased by 64% in the presence of 240 μM Δ^8-THC.

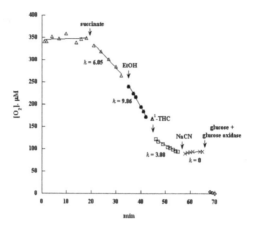

Fig. 11. Δ^8-THC added to isolated mitochondria from beef heart.

12. Phosphorescence O_2 analyzer as a screening tool for disorders of impaired cellular bioenergetics

Disorders of cellular bioenergetics are challenging clinically and biochemically (Chretien and Rustin, 2003; Chretien et al., 1994; Rotig et al., 1990; Rustin et al., 1994). Their manifestations frequently overlap with numerous clinical entities. Furthermore, mutations that limit these processes in humans are incompletely identified (http://www.gen.emory.edu/mitomap.html) (Kogelnik et al., 1997). Therefore, clinicians usually rely on a laborious analysis of skin and muscle biopsies for diagnosis (Chretien and Rustin, 2003; Chretien et al., 1994; Rustin et al., 1994). As suggested by Rustin et al., laboratory evaluation of mitochondrial disorders require testing samples from multiple tissues. The authors also recommended the use of circulating lymphocytes in the initial screening (Rustin et al., 1994). These interrelations justify developing non-invasive simple screening methods that are applicable to various types of samples. Recently, Marriage et al. showed ATP synthesis in permeabilized lymphocytes is an effective screening tool for impaired oxidative phosphorylation (Marriage et al., 2003; Marriage et al., 2004). Decreased ATP synthesis in the lymphocytes was present in the 5 studied mitochondrial disorders (Marriage et al., 2003).

Described herein is the use of the phosphorescence O_2 analyzer to measure lymphocyte respiration in volunteers and a patient. The measurement primarily aimed to show feasibility of using the phosphorescence O_2 analyzer to screen for clinical disorders with impaired cellular bioenergetics. Peripheral blood mononuclear cells (PBMC) were collected from healthy volunteers and patient. The rate of respiration (mean ± SD, in μM O_2 per min per 10^7 cells) for adult volunteers is 2.1 ± 0.8 (n = 18), for children 2.0 ± 0.9 (n = 20), and for newborns (umbilical cord samples) 0.8 ± 0.4 (n = 18, p <0.0001). Representative experiments of the volunteers are shown in Fig. 12.

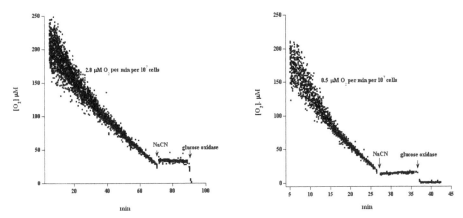

Fig. 12. Left panel: Lymphocytes (1.0 x 10^7 cells/mL) were collected from a 9-year-old girl; the rate of respiration was 2.8 μM O_2 per min per 10^7 cells (R^2 >0.942). Right panel: Lymphocytes (15 x 10^7 cells/mL) were collected from umbilical cord; the rate of respiration was 8.0 μM O_2 per min (R^2 >0.934), or 0.5 μM O_2 per min per 10^7 cells. The additions of 5.0 mM NaCN and 50 μg/mL glucose oxidase are shown.

For an 8-year-old patient with reduced muscle NADH dehydrogenase and pyruvate dehydrogenase activities, the rate was 0.7 ± 0.2 (n = 3) μM O_2 per min per 10^7 cells.

As previously noted in muscle specimens, the rate of lymphocyte mitochondrial oxygen consumption is very similar in adults and children (p =0.801) (Chretien et al., 1994). However, cord blood cells have lower rates of respiration (p <0.001). This finding could be attributed to the high number of nucleated red blood cells in the umbilical cord blood.

Fresh lymphocytes were previously used as a source of tissue for measuring respiratory chain enzymes by polarography (Clark-type O_2 electrode) and spectroscopy (Chretien and Rustin, 2003; Chretien et al., 1994; Rustin et al., 1994). Rotig et al. reported a rate (mean + SD, n=15) of 3.5 ± 0.5 nmol O_2 per min per 10^7 cells (Rotig et al., 1990). Hedeskov and Esmann reported a rate of 2.0 ± 0.07 nmol O_2 per min per 10^7 for cell concentrations >4 x 10^7 per mL and higher rates for less concentrated cells (Hedeskov and Esmann, 1966). Pachman reported rate of 1.0 ± 0.2 nmol O_2 per min per 10^7 equine lymphocytes (Pachman, 1967).

Clinical presentations of entities with impaired cellular bioenergetics vary markedly. Their manifestations may include progressive neuromuscular defects (e.g., psychomotor retardation and hypotonia), heart muscle involvement and encephalopathy. One typical example is Leigh syndrome, which results from an isolated mitochondrial complex I deficiency (Benit et al., 2004). This clinical heterogeneity stems from various mechanisms, including tissue-specific of nuclear-encoded isoforms of the respiratory chain and existence of normal and mutated mtDNA in the same organ (mtDNA heteroplasmy) (Rustin et al., 1994). Therefore, as suggested by Rustin et al., the biochemical analysis should not be limited to skeletal muscle and skin tissues (Rustin et al., 1994). In one study, 42 patients with respiratory chain defects were investigated. The results showed that 50% of the patients had deficiencies in skeletal muscles and lymphocytes, 45% in skeletal muscles only, and 5% in lymphocytes only (Chretien et al., 1994). Patients with Pearson's syndrome on the other hand consistently express defects in the lymphocyte (Rotig et al., 1990).

13. Conclusions

A novel *in vitro* system that allows monitoring of cellular respiration over several hours is described. The method has numerous biological applications, including studying mitochondrial dysfunction during apoptosis or toxic exposure. It also allows screening for metabolic disorders in patients. The procedure is sensitive and reproducible. It is applicable to cells in suspension, adherent cells and various organs, including the heart muscle, liver, spleen, pancreas and kidney.

14. References

Al-Salam, S.; Balhaj, G.; Al-Hammadi, S.; Sudhadevi, M.; Tariq, S.; Biradar, A.V.; Asefa, T. & Souid, A.-K. (2011). *In vitro* study of calcined mesoporous silica nanoparticles in mouse lung. *Toxicology Sciences.* Vol.122, No.1, (Epub 2011 Apr 5), pp.86-99.

Al Samri, M.T.; Al Shamsi, M.; Al-Salam, S.; Marzouqi, F.; Mansouri, A.; Al-Hammadi, S.; Balhaj, G.; Al Dawaar, S.K.M.; Al Hanjeri, R.S.M.S.; Benedict, S.; Sudhadevi, M.; Conca, W.; Penefsky, H.S. & Souid, A.-K. (2011). Measurement of oxygen consumption by murine tissues *in vitro*. *Journal of Pharmacological and Toxicological Methods.* Vol.63, No.2, (Epub 2010 Oct 27), pp.196-204.

Al Shamsi, M.; Al Samri, M.T.; Al-Salam, S.; Conca, W.; Shaban, S.; Benedict, S.; Tariq, S.; Biradar, A.; Penefsky, H.S.; Asefa, T.; Souid, A.-K. (2010). Biocompatibility of calcined mesoporous silica particles with cellular bioenergetics in murine tissues. *Chemical Research in Toxicology.* Vol.23, No.11, (Epub 2010 Oct 20), pp.1796-1805.

Badawy, Z.S. & Souid, A.-K. (2009a). Inhibition of human sperm respiration by 4-hydroperoxycyclophosphamide and protection by mesna and WR-1065. *Fertility and Sterility.* Vol.91, No.1, (2008 Jan 18), pp.173-178.

Badawy, Z.S.; Chohan, K.R.; Whyte, D.A.; Penefsky, H.S.; Brown, O.M. & Souid, A.-K. (2009b). Cannabinoids inhibit the respiration of human sperm. *Fertility and Sterility.* Vol.91, No.6, (2008 Jun 18), pp.2471-2476.

Benit, P.; Slama, A.; Cartault, F.; Giurgea, I.; Chretien, D.; Lebon, S.; Marsac, C.; Munnich, A.; Rotig, A. & Rustin, P. (2004). Mutant NDUFS3 subunit of mitochondrial complex I causes Leigh syndrome. *Journal Medical Genetics.* Vol.41, No.1, pp.14-17.

Chen, F.; Zhu, H-H.; Zhou, L-F.; Li, J.; Zhao, L-Y.; Wu, S-S.; Wang, J.; Liu, W. & Chen, Z. (2010). Genes related to the very early stage of ConA-induced fulminant hepatitis: a gene-chip-based study in a mouse model. *BMC Genomics.* Vol.11, pp.240-250.

Chretien, D.; Rustin, P.; Bourgeron, T.; Rotig, A.; Saudubrary, J.M. & Munnich, A. (1994). Reference charts for respiratory chain activities in human tissues. *Clin. Chim. Acta.* Vol.228, No.1, pp.53-70.

Chretien, D. & Rustin, P. (2003). Mitochondrial oxidative phosphorylation: pitfalls and tips in measuring and interpreting enzyme activities. *J. Inherit. Metab. Dis.* Vol.26, No.(2-3), pp.189-198.

Corrier, DE. (1991). Mycotoxicosis: mechanisms of immunosuppression. *Vet Immunol Immunopathol.* Vol.30, No.1, pp.73-87.

Eaton, D.L. & Gallagher, E.P. (1994). Mechanisms of aflatoxin carcinogenesis. *Annu. Rev. Pharmacol. Toxicol.* Vol.34, pp.135-172.

Goodisman, J.; Hagrman, D.; Tacka, KA. & Souid, A-K. (2006). Analysis of cytotoxicities of platinum compounds. *Cancer Chemotherapy and Pharmacology*. Vol.57, No.2, (2005 Jul 19). pp.257-267.

Gottlieb, E.; Vander Heiden, MG. & Thompson, CB. (2000). Bcl-xL Prevents the Initial Decrease in Mitochondrial Membrane Potential and Subsequent Reactive Oxygen Species Production during Tumor Necrosis Factor Alpha-Induced Apoptosis. *Molecular and Cellular Biology*. Vol.20, No.15. pp.5680–5689.

Green, D.R. & Kroemer, G. (2004). The pathophysiology of mitochondrial cell death. *Science*. Vol.305, No.5684. pp. 626-629.

Hedeskov, C.J. & Esmann, V. (1966). Respiration and glycolysis of normal human lymphocytes. *Blood*. Vol.28, No.2. pp. 163-174.

Jiang, Y.; Jolly, P.E.; Preko, P.; Wang, J-S.; Ellis, W.O.; Phillips, T.D. & Williams, J.H. (2008). Aflatoxin-related immune dysfunction in health and in human immunodeficiency virus disease. *Clin. Dev. Immunol*. (2008:790309).

Johnson, WW.; Harris, T.M. & Guengerich, FP. (1996) Kinetics and mechanism of hydrolysis of aflatoxin B1 exo-8, 9-Epoxide and rearrangement of the dihydrodiol. J Am Chem Soc. Vol.118, No.35, pp.8213-8220.

Jones, E.; Penefsky, HS. & Souid A-K. (2009). Caffeine impairs HL-60 cellular respiration. *Journal of Medical Sciences*. Vol.2, No.2, pp. 61-72.

Kogelnik, A.M.; Lott, M.T.; Brown, M.D.; Navathe, S.B. & Wallace, D.C. (1997). MITOMAP: an update on the status of the human mitochondrial genome database. *Nucleic Acids Res*. Vol.25, No.1, pp. 196-199.

Leist, M. & Wendel, A. (1996). A novel mechanism of murine hepatocyte death inducible by concanavalin A. *J Hepatol*. Vol.25, No.6, pp.948-959.

Lo L-W, Koch CJ, Wilson DF (1996). Calibration of oxygen-dependent quenching of the phosphorescence of Pd-meso-tetra (4-carboxyphenyl) porphine: A phosphor with general application for measuring oxygen concentration in biological systems. *Analytical Biochemistry*. Vol.236, No.1, pp.153-160.

Louis H, Moine O L, Peny M-O, Quertinmont E, Fokan D, Goldman M & Devie`re J. (1997). Production and role of interleukin-10 in concanavalin A–induced hepatitis in mice. *Hepatology*. Vol.25, No.6, pp.1382-1389.

Margalit, M.; Abu Ghazala, S.; Alper, R.; Elinav, E.; Klein, A.; Doviner, V.; Sherman. Y.; Thalenfeld, B.; Engelhardt, D.; Rabbani, E. & Ilan, Y. (2005). Glucocerbroside treatment ameliorates conA hepatitis by inhibition of NKT lymphocytes. *Am J Physiol Gastrointest Liver Physiol*. Vol.289. No.5, (Epub 2005 Jun 23), pp.G917-25.

Marriage, B.J.; Clandinin, M.T.; MacDonald, IM. & Glerum, D.M. (2003). The use of lymphocytes to screen for oxidative phosphorylation disorders. *Analytical Biochemistry*. Vol.313, No.1, pp.137-144.

Marriage, B.J.; Clandinin, M.T.; MacDonald, I.M. & Glenerum D.M. (2004). Cofactor treatment improves ATP synthetic capacity in patients with oxidative phosphorylation disorders. *Molecular Genetics Metabolism*. Vol..81, No.4, pp.263-272.

Meky, F.A.; Hardie, L.J.; Evans, S.W. & Wild, C.P. (2001). Deoxynivalenol-induced immunomodulation of human lymphocyte proliferation and cytokine production. *Food Chem. Toxicol*. Vol.39, No.8, pp.827-836.

Montesano, R.; Hainaut, P. & Wild, C.P. (1997). Hepatocellular carcinoma: from gene to public health. *J. Natl. Cancer Inst*. Vol..89, No.24, pp.1844-1851.

Nesheim, S.; Trucksess, M.W. & Page, S.W. (1999). Collaborative study, molar absorptivities of aflatoxins B1, B2, G1, and G2 in acetonitrile, methanol and toluene-acetonitrile (9+1) (modification of AOAC official method 971.22). *J. AOAC Int.* Vol.82, No.2, pp.251-258.

Nicholson, D.W. & Thornberry, N.A. (1997). Caspases: killer proteases. *Trends Biochem Sci.* Vol.22, No.8, pp.299-306.

Pachman L.M. (1967). The carbohydrate metabolism and respiration of isolated small lymphocytes. In vitro studies of normal and phytohemagglutinin stimulated cells. *Blood.* Vol.30, No.6, pp.691-706.

Reddy, R.V.; Taylor, M.J. & Sharma, R.P. (1987). Studies of immune function of CD-1 mice exposed to aflatoxin B1. *Toxicology.* Vol.43, No.2, pp.123-132.

Reddy, R.V. & Sharma, R.P. (1989). Effects of aflatoxin B1 on murine lymphocyte functions. *Toxicology.* Vol.54, No.1, pp.31-44.

Ricci, J-E.; Munoz-Pinedo, C.; Fitzgerald, P.; Bailly-Maitre, B.; Perkins, GA.; Yadava, N.; Scheffer, IE.; Ellisman, MH. & Green, DR. (2004). Disruption of mitochondrial function during apoptosis is mediated by caspase cleavage of the p75 subunit of complex I of the electron transport chain. *Cell.* Vol.117, No.6, pp.773-786.

Rossano, F.; Ortega de Luna, L.; Buommino, E.; Cusumano, V.; Losi, E. & Catania, M.R. (1999). Secondary metabolites of Aspergillus exert immunobiological effects on human monocytes. *Res. Microbiol.* Vol.150, No.1, pp.13-19.

Rotig, A.; Cormier, V.; Blanche, S.; Bonnefont, J.P.; Ledeist, F.; Romero, N.; Schmitz, J.; Rustin, P.; Fischer, A. & Saudubray, J.M. (1990). Pearson's marrow-pancreas syndrome. A multisystem mitochondrial disorder in infancy. *J. Clin. Invest.* Vol.86, No.5, pp.1601-1608.

Rustin, P.; Chretien, D.; Bourgeron, T.; Gerard, B.; Rotig, A.; Saudubrary, J.M. & Munnich, A. (1994). Biochemical and molecular investigations in respiratory chain deficiencies. *Clin. Chim. Acta.* Vol.228, No.1, pp.35-51.

Sass, G.; Heinlein, S.; Agli, A.; Bang, R.; Schumann, J. & Tiegs, G. (2002). Cytokine expression in three mouse models of experimental hepatitis. *Cytokine.* Vol.19, No.3, pp.115-20.

Savel, H.; Forsyth, B.; Schaeffer, W. & Cardella, T. (1970). Effect of aflatoxin B1 upon phytohemagglutinin-transformed human lymphocytes. *Proc. Soc. Exp. Biol. Med.* Vol.134, No.4, pp.1112-1115.

Shaban, S.; Marzouqi, F.; Al Mansouri, A.; Penefsky, HS. & Souid, A-K. (2010). Oxygen measurements via phosphorescence. *Computer Methods and Programs in Biomedicine.* Vol.100, No.3, pp.265-268.

Slee, E.A.; Zhu, H.; Chow, S.C.; MacFarlane, M.; Nicholson, D.W. & Cohen, G.M. (1996). Benzylooxycarbonyl-Val-Ala-Asp (Ome) fluormethylketone (Z-VAD.FMK) inhibits apoptosis by blocking the processing of CPP32. *Biochemistry Journal.* Vol.315, No. (Pt 1), pp.21-24.

Souid, A-K.; Tacka, KA.; Galvan, KA. & Penefsky, HS. (2003). Immediate effects of anticancer drugs on mitochondrial oxygen consumption. *Biochemical Pharmacology.* Vol.66, No.6, pp.977-987.

Souid, A-K.; Penefsky, H.S..; Sadowitz, P.D. & Toms, B. (2006). Enhanced cellular respiration in cells exposed to doxorubicin. *Molecular Pharmaceutics.* Vol.3, No.3, pp.307-321.

Stec, J., Zmudzki, J., Rachubik, J., & Szczotka, M. (2009). Effects of aflatoxin B1, ochratoxin A, Patulin, citrinin, and zearalenone on the in vitro proliferation of pig blood lymphocytes. *Bull. Vet. Inst. Pulawy*. Vol53, No.1, pp129-134.

Sun, X.M.; Zhang, X.H.; Wang, H.Y.; Cao, W.J.; Yan, X.; Zuo, L.F.; Wang, J.L. & Wang, F.R. (2002). Effects of sterigmatocystin, deoxynivalenol and afaltoxin G1 on apoptosis of human blood lymphocytes in vitro. *Biomed. Environ. Sci*. Vol.15, No.2, pp.145-152.

Tacka, KA.; Dabrowiak, JC.; Goodisman, J.; Penefsky, HS. & Souid, A-K. (2004a). Quantitative studies on cisplatin-induced cell death. *Chemical Research in Toxicology*. Vol.17, No.8, pp.1102-1111.

Tacka, K.A..; Szalda, D.; Souid, A.-K..; Goodisman, J. & Dabrowiak, J.C. (2004b). Experimental and theoretical studies on the pharmacodynamics of cisplatin in Jurkat cells. *Chemical Research in Toxicology*. Vol.17, No.11, pp.1434-1444.

Takeda, K.; Hayakawa, Y.; Van Kaer, L.; Matsuda, H.; Yaqita, H. & Okumura, K. (2000). Critical contribution of liver natural killer T cells to a murine model of hepatitis. *Proc Natl Acad Sci USA*. Vol..97, No.10, pp.5498-5503.

Tao, Z.; Withers, HG.; Penefsky, HS.; Goodisman, J. & Souid, A-K. (2006a). Inhibition of cellular respiration by doxorubicin. *Chemical Research in Toxicology*. Vol.19, No.8, pp.1051-1058.

Tao, Z.; Ahmad, SS.; Penefsky, HS.; Goodisman, J. & Souid, A-K. (2006b). Dactinomycin impairs cellular respiration and reduces accompanying ATP formation. *Molecular Pharmaceutics*. Vol.3, No.6, pp.762-772.

Tao, Z.; Goodisman, J.; Penefsky, HS. & Souid, A-K. (2007). Caspase activation by cytotoxic drugs (the caspase storm). *Molecular Pharmaceutics*. Vol.4, No.4, (Epub 2007 Apr 17), pp.583-595.

Tao, Z.; Jones, E.; Goodisman, J. & Souid, A-K. (2008a). Quantitative measure of cytotoxicity of anticancer drugs and other agents. *Analytical Biochemistry*. Vol.381, No.1, (Epub 2008 Jun 18), pp.43-52.

Tao, Z.; Goodisman, J. & Souid, A-K. (2008b). Oxygen measurement via phosphorescence: reaction of sodium dithionite with dissolved oxygen. *The Journal of Physical Chemistry A*. Vol.112, No.7, (Epub 2008 Jan 30), pp.1511-1518.

Tao, Z.; Morrow, MP.; Asefa, T.; Sharma, KK.; Duncan, C.; Anan, A.; Penefsky, HS.; Goodisman, J. & Souid, A-K. (2008c). Mesoporous silica nanoparticles inhibit cellular respiration. *Nano Letter*. Vol.8, No.5, (Epub 2008 Apr 1), pp.1517-1526.

Tao, Z.; Raffel, RA.; Souid, A-K. & Goodisman, J. (2009). Kinetic studies on enzyme-catalyzed reactions: oxidation of glucose, decomposition of hydrogen peroxide and their combination. *Biophysical Journal*. Vol.96, No.7, pp.2977-2988.

Tiegs, G.; Hentschel, J. & Wendel, A. (1992). A T-cell-dependent experimental liver injury in mice inducible by concanavalin A. *Journal Clinical Investigation*. Vol.90, No.1, pp.196-203.

Tiegs G. (1997). Experimental hepatitis and the role of cytokines. *Acta Gastroenterol Belg*. Vol.60, No.2, pp.176-9.

Universal library for measurements computing devices. http://www.mccdaq.com/daq-software/universal-library.aspx. Access May 2009.

Wang, L-Y.; Hatch, M.; Chen, C-J.; Levin, B.; You, S-L.; Lu, S-N.; Wu, M-H.; Wu, W-P.; Wang, L-W.; Wang, Q.; Huang, G-T.; Yang, P-M.; Lee, H-S. & Santella, R.M. (1996).

Aflatoxin exposure and risk of hepatocellular carcinoma in Taiwan. *International Journal Cancer*. Vol.67, No.5, pp.620-625.

Weiss, R.F. (1970). The solubility of nitrogen, oxygen, and argon in water and seawater. *Deep-Sea Research*. Vol.17, No.4, pp.721-735.

Whyte, DA.; Al-Hammadi, S.; Balhaj, G.; Brown, OM.; Penefsky, HS. & Souid, A-K. (2010). Cannabinoids inhibit cellular respiration of human oral cancer cells. *Pharmacology*. Vol.85, No.6, (Epub 2010 Jun 2), pp.328-335.

Williams, J.H.; Phillips, T.D.; Jolly, P.E.; Stiles, J.K.; Jolly, C.M. & Aggarwal, D. (2004). Human aflatoxicosis in developing countries: a review of toxicology, exposure, potential health consequences, and interventions. *American Journal of Clinical Nutrition*. Vol.80, No.5, pp.1106-1122.

Permissions

The contributors of this book come from diverse backgrounds, making this book a truly international effort. This book will bring forth new frontiers with its revolutionizing research information and detailed analysis of the nascent developments around the world.

We would like to thank Kevin B. Clark, for lending his expertise to make the book truly unique. He has played a crucial role in the development of this book. Without his invaluable contribution this book wouldn't have been possible. He has made vital efforts to compile up to date information on the varied aspects of this subject to make this book a valuable addition to the collection of many professionals and students.

This book was conceptualized with the vision of imparting up-to-date information and advanced data in this field. To ensure the same, a matchless editorial board was set up. Every individual on the board went through rigorous rounds of assessment to prove their worth. After which they invested a large part of their time researching and compiling the most relevant data for our readers. Conferences and sessions were held from time to time between the editorial board and the contributing authors to present the data in the most comprehensible form. The editorial team has worked tirelessly to provide valuable and valid information to help people across the globe.

Every chapter published in this book has been scrutinized by our experts. Their significance has been extensively debated. The topics covered herein carry significant findings which will fuel the growth of the discipline. They may even be implemented as practical applications or may be referred to as a beginning point for another development. Chapters in this book were first published by InTech; hereby published with permission under the Creative Commons Attribution License or equivalent.

The editorial board has been involved in producing this book since its inception. They have spent rigorous hours researching and exploring the diverse topics which have resulted in the successful publishing of this book. They have passed on their knowledge of decades through this book. To expedite this challenging task, the publisher supported the team at every step. A small team of assistant editors was also appointed to further simplify the editing procedure and attain best results for the readers.

Our editorial team has been hand-picked from every corner of the world. Their multi-ethnicity adds dynamic inputs to the discussions which result in innovative outcomes. These outcomes are then further discussed with the researchers and contributors who give their valuable feedback and opinion regarding the same. The feedback is then collaborated with the researches and they are edited in a comprehensive manner to aid the understanding of the subject.

Apart from the editorial board, the designing team has also invested a significant amount of their time in understanding the subject and creating the most relevant covers. They scrutinized every image to scout for the most suitable representation of the subject and create an appropriate cover for the book.

The publishing team has been involved in this book since its early stages. They were actively engaged in every process, be it collecting the data, connecting with the contributors or procuring relevant information. The team has been an ardent support to the editorial, designing and production team. Their endless efforts to recruit the best for this project, has resulted in the accomplishment of this book. They are a veteran in the field of academics and their pool of knowledge is as vast as their experience in printing. Their expertise and guidance has proved useful at every step. Their uncompromising quality standards have made this book an exceptional effort. Their encouragement from time to time has been an inspiration for everyone.

The publisher and the editorial board hope that this book will prove to be a valuable piece of knowledge for researchers, students, practitioners and scholars across the globe.

List of Contributors

Oulès Bénédicte
INSERM U 807, Paris V University, Paris, France

Del Prete Dolores
Istituto Italiano di Tecnologia, Department of Neuroscience and Brain Technologies, Genova, Italy

Chami Mounia
Institute of Cellular and Molecular Pharmacology, Institute of Molecular Neuromedecine, UMR6097 CNRS/UNSA, Valbonne, France

Iseli L. Nantes, Tiago Rodrigues, Mayara K. Kisaki and Vivian W. R. Moraes
Universidade Federal do ABC, Brazil

César H. Yokomizo and Felipe S. Pessoto
Universidade Federal de São Paulo, Brazil

Juliana C. Araújo-Chaves
Universidade de Mogi das Cruzes, Brazil

Alexander G. Trubitsyn
Institute of Biology and Soil Sciences, Far East Division Russian Academy of Sciences, Vladivostok, Russia

Sara Santa-Cruz Calvo, Plácido Navas and Guillermo López-Lluch
Centro Andaluz de Biología del Desarrollo, Universidad Pablo de Olavide-CSI, CIBERER-Instituto Carlos III. Carretera de Utrera Km. 1,Sevilla, Spain

Eldo Campos and Jorge Moraes
Universidade Federal do Rio de Janeiro – Macaé, Brazil
Instituto Nacional de Ciência e Tecnologia - Entomologia Molecular, Brazil

Arnoldo R. Façanha
Instituto Nacional de Ciência e Tecnologia - Entomologia Molecular, Brazil
Universidade Estadual do Norte Fluminense, Brazil

Carlos Logullo
Universidade Estadual do Norte Fluminense, Brazil
Instituto Nacional de Ciência e Tecnologia - Entomologia Molecular, Brazil

Valentin Son'kin and Ritta Tambovtseva
Institute for Developmental Physiology, Russian Academy of Education, Moscow Russian Federation

Rodrigo Zacca and Flávio Antônio de Souza Castro
Universidade Federal do Rio Grande do Sul, Brazil

Diego Mora and Stefania Arioli
University of Milan, Dipartimento di Scienze e Tecnologie Alimentari e Microbiologiche, Milano, Italy

Oliviert Martinez-Cruz and Laura Jimenez-Gutierrez and Adriana Muhlia-Almazan
Molecular Biology Laboratory, Centro de Investigacion en Alimentacion y Desarrollo (CIAD), Hermosillo, Sonora, Mexico

Arturo Sanchez-Paz
Laboratorio de Sanidad Acuicola, Centro de Investigaciones Biologicas del Noroeste (CIBNOR), Hermosillo, Sonora, Mexico

Fernando Garcia-Carreño and Ma. de los Angeles Navarrete del Toro
Biochemistry Laboratory, Centro de Investigaciones Biologicas del Noroeste (CIBNOR), La Paz, Mexico

Linda Klimaviciusa, Nelda Lencberga, Marta Pavasare and Vija Klusa
Department of Pharmacology, Faculty of Medicine, University of Latvia, Riga, Latvia

Maria A. S. Fernandes, Joaquim A. F. Vicente and António J. M. Moreno
IMAR-CMA, Department of Life Sciences, University of Coimbra, Coimbra, Portugal

Maria S. Santos
CNC, Department of Life Sciences, University of Coimbra, Coimbra, Portugal

Catarina R. Oliveira
CNC, Faculty of Medicine, University of Coimbra, Coimbra, Portugal

Imanta Bruvere, Egils Bisenieks and Brigita Vigante
Laboratory of Membrane Active and beta-Diketone Compounds, Latvian Institute of Organic Synthesis, Riga, Latvia

Fatma Al-Jasmi, Ahmed R. Al Suwaidi, Mariam Al-Shamsi, Farida Marzouqi, Aysha Al Mansouri, Sami Shaban, Harvey S. Penefsky and Abdul-Kader Souid
United Arab Emirates University, Faculty of Medicine and Health Sciences, Abu Dhabi, United Arab Emirates

Printed in the USA
CPSIA information can be obtained
at www.ICGtesting.com
JSHW011452221024
72173JS00005B/1042

9 781632 392152